The Ethical Project

The Ethical Project

Philip Kitcher

Harvard University Press

Cambridge, Massachusetts · London, England

2011

Library of Congress Cataloging-in-Publication Data

Kitcher, Philip, 1947-
The ethical project / Philip Kitcher.
p.   cm.
Includes bibliographical references (p.) and index.
ISBN 978-0-674-06144-6 (hardcover : alk. paper)
1. Ethics, Evolutionary.   I. Title.
BJ1311.K53 2011
171'.7—dc22   2010051684

For A.G.P.K.

Contents

Moral conceptions and processes grow naturally
out of the very conditions of human life.
—John Dewey

Introduction

§1. The Shape of Things to Come

Ethics pervades every human society and almost every human life. People deliberate about what they should do on specific occasions, about what is worthwhile, about the kinds of lives they should aspire to lead. In subtle ways, their everyday actions presuppose habits of conduct, roles and institutions current in their societies, endorsed sometimes after serious reflection, often accepted without much thought. With the exception of those afflicted with psychological disruptions that profoundly limit their cognitive capacities or that cut them off from their fellows, we are all embedded in the ethical project.

Yet for ordinary people, and for philosophers too, the status of our ethical judgments and practices is hard to fathom. What exactly do we mean when we praise someone for a correct decision? How could an evaluation like that be grounded? Bertrand Russell famously described mathematics as a subject in which ". . . we never know what we are talking about, nor whether what we are saying is true."[1] The impressive ability of

---

1. "Recent Work on the Principles of Mathematics," *International Monthly*, 4, 1901, 84. I shall not venture into the controversy about how seriously Russell intended his characterization.

mathematicians to reach agreement on conclusions that endure fosters confidence in their power to acquire knowledge, despite the mysteries swirling around the content and grounds of their judgments. By contrast, the persistence of ethical debate reinforces a sense of unease about the status of ethical practice, often leading those engaged in public controversies to shy away from "value judgments," as if any hope of reaching consensus were absurd.

One popular view of the ethical life persists. Many people believe, with Dostoyevsky's Ivan Karamazov, that if ethical precepts were not grounded in God's commands, anything would be permitted. From Plato on, however, the philosophical tradition has frequently—and cogently—questioned the idea of a religious foundation for ethics. Supposing ethics to be grounded in the divine will remains popular because alternatives, including the philosophical alternatives, appear so elusive and unconvincing. Could ethical correctness really consist in representing some independent realm of values? Could ethical judgments really express particular privileged emotions? Could they really be arrived at by fathoming the "moral law within" or by apprehending the deliverances of practical reason?

More than a century ago, Darwin outlined a novel way of thinking about the living world: his fundamental insight was to regard the organisms around us as products of history. We can liberate ourselves from mysteries about many of our current practices by emulating Darwin: think of them, too, as historical products.[2] The aim of this book is to pursue this program in the case of ethics. Ethics emerges as a human phenomenon, permanently unfinished. We, collectively, made it up, and have developed, refined, and distorted it, generation by generation. Ethics should be understood as a *project*—the ethical project—in which we have been engaged for most of our history as a species.

2. Darwin himself made some first efforts to inaugurate this program in the early chapters of *The Descent of Man* (John Murray, 1872). Thomas Kuhn proposed a similar understanding of the natural sciences as historical products (*The Structure of Scientific Revolutions* [Chicago: University of Chicago Press, 1962 and 1970]). I have followed their lead, both for the sciences and for mathematics (in *The Advancement of Science* [New York: Oxford University Press, 1993] and in *The Nature of Mathematical Knowledge* [New York: Oxford University Press, 1983]).

The position to be elaborated—*pragmatic naturalism,* to give it a name—envisages the ethical project as begun by our remote ancestors, in response to the difficulties of their social life. They *invented* ethics. Successive generations have amended the ethical legacy transmitted to them, sometimes, but by no means always, improving it. Doubtless, many traditions have died out, but some have continued into the present, forming the bases of the ways in which people today regulate their conduct. In principle, but not in practice, it would be possible to construct an evolutionary tree, drawing a diagram like the single figure Darwin inserted into the *Origin,* with the important difference that the connecting lines would represent cultural rather than biological descent (so the picture might show fusion, as well as separation of lineages).[3]

As the name suggests, pragmatic naturalism has affinities with both pragmatism and naturalism. In focusing on ethical practice and its history, it attempts to honor John Dewey's call for philosophy to be reconnected with human life.[4] Further, it articulates a Deweyan picture of ethics as growing out of the human social situation; its conception of ethical correctness is guided by William James's approach to truth.[5] The naturalism consists in refusing to introduce mysterious entities— "spooks"—to explain the origin, evolution, and progress of ethical practice. Naturalists intend that no more things be dreamt of in their philosophies than there are in heaven and earth.[6] They start from the inventory of the world allowed by the totality of bodies of well-grounded knowledge (the gamut of scholarly endeavors running from anthropology and art history to zoology), and, aware of the certain incompleteness of the list, allow only such novel entities as can be justified through accepted

---

3. Charles Darwin, *On the Origin of Species* (facsimile of the first edition), edited by Ernst Mayr (Cambridge, MA: Harvard University Press, 1964), between 116 and 117.

4. See *The Quest for Certainty,* vol. 4 in *John Dewey: The Later Works* (Carbondale: Southern Illinois University Press, 1988), 204; and *Reconstruction in Philosophy* (Boston: Beacon Press, 1957), 147.

5. See Dewey and James Tufts, *Ethics,* 2nd ed. (New York: Henry Holt, 1932), 307–9; William James, *Pragmatism* (Cambridge, MA: Harvard University Press, 1978), Lecture VI.

6. The inversion of Hamlet stems from Nelson Goodman: *Fact, Fiction, and Forecast* (Indianapolis, IN: Bobbs-Merrill, 1956).

methods of rigorous inquiry. Appeals to divine will, to a realm of values, to faculties of ethical perception and "pure practical reason," have to go.

Pragmatic naturalism engages with the religious entanglement of ethics more extensively than is usual in secular philosophical discussion—for the pragmatist reason that the entanglement pervades almost all versions of ethical life. Yet, in accordance with its naturalist scruples, it cannot maintain the image favored by those who would ground ethics in the divine will. As we shall see (§27), there are powerful reasons to suppose, even if there were any deity, *ethics* could not be fixed by its (his? her?) tastes. More fundamentally, pragmatic naturalism maintains that, when *religion* is understood as a historically evolving practice, it is overwhelmingly probable that all the conceptions of a transcendent being ever proposed in any of the world's religions are false. For the conceptions introduced in the various religions are massively inconsistent with one another. Each supernaturalist view rests on epistemically similar grounds—typically there was some revelation, long ago, that has been carefully transmitted across the generations to the devout of today—yielding a condition of complete symmetry. Under these circumstances, no believer has any basis for thinking only he and his group are privileged to know the truth about the transcendent realm, while others live in primitive delusions. Further, serious inquiries into the ways in which canonical scriptures are constructed, into the evolution of religions, into the recruitment of converts, into the phenomena of religious experience, demonstrate how radically unreliable are the processes that have yielded the current corpora of belief. Nor can one isolate some core doctrine, shared by all religions, something capable of being viewed as a shared insight. If there are beings of a hitherto unrecognized sort, approximating some idea of the "transcendent," we have every reason to think we have absolutely no clues, or categories, for describing them.[7]

Religious entanglement in ethical practice is no accident. As we shall see (§17), appealing to gods as "guardians of morality" can bring social benefits. Nevertheless, that appeal has distorted the ethical project. Undoing the distortions is not simply a matter of eradicating religion, hack-

---

7. The considerations bluntly advanced here are elaborated at much greater length in the last chapter of my *Living with Darwin* (New York: Oxford University Press, 2007).

ing out the places where false belief has intruded. A secular renewal of the ethical project requires constructive work, positive steps going beyond brusque denial.[8]

Given these clarifications, I can now explain the structure of the following chapters. Part I, Chapters 1–4, elaborates an "analytical history," aimed at providing insight into the evolution of our ethical practice. It provides a basis on which Part II (Chapters 5–7) can explore questions *about* ethics: given this account of the origins and unfolding of ethics, can we make sense of ethical truth or ethical knowledge? The history of Part I and the metaethical account of Part II are then extended, in Part III (Chapters 8–10), into a normative stance, an attempt to suggest how we might best go on from where we are.

It is worth supplementing this bald characterization with a little more detail. A "history of ethical practice" might take many forms, and the one I offer may initially appear strange. Since I suppose our species to have been engaged in the ethical project for tens of thousands of years, it would be hopeless to offer a narrative showing how particular aspects of ethical life have gradually emerged. Until the invention of writing (five thousand years ago), the clues are fragmentary, far scantier than the fossil record, whose poverty provoked Darwin's lament.[9] Primatology, anthropology, and archeology enable us to offer a plausible account of the conditions under which our preethical ancestors lived, but many subsequent steps are beyond our evidential grasp.

The analytical history starts by attempting to understand relevant psychological capacities of the preethical ancestors, and, on that basis, to portray the initial stages of the ethical project. Hominid social life was akin to the contemporary lives of our closest evolutionary relatives: our precursors lived in small groups, mixed by age and sex. For that, they needed a capacity for psychological altruism. Yet the limitations of their altruistic dispositions made living together tense and difficult. The first ethicists overcame some of the problems by agreeing on rules for conduct,

8. See the closing pages of *Living with Darwin*, as well as "Challenges for Secularism," in *The Joy of Secularism*, ed. George Levine (Princeton, NJ: Princeton University Press, 2011).

9. Darwin, *Origin of Species*, chap. 9, esp. 310–11.

rules remedying a few of the recurrent altruism failures that had plagued their group life. Very probably, they began with precepts about sharing scant resources and not initiating violence.

Because the character of early ethical practices is so much simpler than the forms of ethical life visible once written documents are available, it is important to show (Chapter 3), how a series of gradual steps *might* have taken the ethical project from its relatively crude initial phase to the complex articulation of rules and stories found in the first written documents. Thereafter, it is possible to trace, although not with the completeness one might hope for, how *actual* changes in ethical practice have occurred. Chapter 4 considers a few examples from history (rather than prehistory) with the aim of supporting two main theses. First, it is hard to resist the recognition of occasional progress in the evolution of ethics: perhaps ethical progress is rare, but there are transitions (like the repudiation of slavery) in which it seems to occur. Second, even when the records kept by people who participated in apparently progressive ethical change are most extensive, moments of ethical discovery are elusive: there are no analogs of episodes of scientific insight.

The history of Part I offers hypotheses about how the ethical project *actually* began, and how, in recent history, it has *actually* gone. It also addresses concern about the vast difference between the early stages of the project and the rich practices found at the dawn of history by showing how it would have been *possible* for the bare beginnings to evolve, by gradual steps, into the complex systems discernible in the earliest texts. Because the differences in these two modes of explanation need to be clearly appreciated, the next section will address some methodological preliminaries.

How can any history, however carefully focused and articulated, bear on philosophical questions about ethics? One possibility, already illustrated by the example of religion, is that a historical account might undermine current practice. Seeing where our approaches have come from could breed skepticism and disillusionment. In those episodes of ethical change most susceptible to analysis, the participants do not appear to apprehend some previously unrecognized value, or to reason their ways to some novel moral principle. Historical detail, to the extent it can be provided, is inhospitable to philosophical theories about ethical truth

(Chapter 5). Yet the history of Part I also reveals examples of ethical progress. The metaethical perspective of Part II centers on trying to reconcile these points.

A "mere change" view of ethical evolution, in which the history is simply one damned thing after another, conflicts with the pull to characterize some transitions as advances. Chapter 6 resolves the conflict by seeking an account of ethical progress, one that abandons the idea of progress as accumulation of (prior, independent) truth. If this appears a strange idea, we should recall that, in some areas of human practice, progress does not consist in the increase of truth. Technological progress is often a matter of discharging certain functions more efficiently or more fully. Moreover, in line with the history of Part I, the initial ventures in the ethical project are readily conceived as introducing a new—social—technology, aimed at remedying disruptive altruism failures.

Amelioration of altruism failure was the initial function of ethical practice. Yet the obvious differences between the pioneering ventures and the complex codes present at the dawn of recorded history show clearly that other functions have emerged. That is the way with technology in general. People begin with a problem and achieve partial successes in solving it. The successes generate new problems to be solved. Chapter 6 attempts to anchor the concept of ethical progress in the discharging of functions, originating with the problem of remedying failures of altruism, and understanding later functions as generated from the solutions previously obtained.

It thereby paves the way for a concept of ethical truth. Ethical truths are those acquired in progressive transitions and retained through an indefinite sequence of progressive transitions. Pragmatic naturalism proposes that some ethical statements—typically, vague generalizations, commending honesty and disavowing violence, for example—are true. They owe their truth to the role they play in ethical progress: "truth happens to an idea."[10]

To declare that our ancestors *invented* ethics is to deny that they *discovered* it or that it was *revealed* to them. Pragmatic naturalism rejects the idea of a special moment (long ago on Mount Sinai, perhaps) when

10. James, *Pragmatism*, 97.

people received authoritative information about how they should live, and also abandons surrogate philosophical theories about external constraints discovered by special faculties. Yet to declare that ethics is a human invention is not to imply it was fashioned *arbitrarily*. The ethical project began in response to central human desires and needs, arising from our special type of social existence.

There is an obvious concern, one probably already exercising any reader suspicious about misadventures in naturalism. Why should the ethical project, even at its most "progressive," have force on those people who appear late in its evolution? Critics have often charged that naturalism commits a fallacy, and those criticisms need to be addressed. Chapter 7 considers a number of versions of the accusation, attempting to show that pragmatic naturalism has resources equal to those of any nonnaturalistic rival. Yet one important version persists. Because the ethical project generates new functions, not necessarily in harmony with one another, it appears to leave open the possibility of different ways of continuing. To settle worries about radical disagreement, to finish articulating the metaethical perspective, a normative stance is needed. That is the work of Part III.

Parts I and II portray the ethical project as an enterprise in which people work out how to live together. It began without presupposing any sources of truth ethical deliberators sought to fathom—whether those sources lay in a divine will or in any of the philosophical substitutes. Our ancestors needed to fashion their shared life by conversing on equal terms. Pragmatic naturalism denies ethical expertise. The role philosophy plays in ethics can be one only of midwifery: to suggest a direction for renewed conversation and some rules for mutual exchange. Chapters 8 and 9 offer a package of proposals, an egalitarian conception of the good at which we should now aim, together with a method of deliberating under conditions of mutual engagement.

Convincing proposals come with some form of support. Pragmatic naturalism's proposals are motivated by conceiving the current human situation as analogous to that initially prompting the ethical project. As it was in the beginning, so too now—for the conflicts to which our ancestors' lives were subject are mirrored in contemporary hostilities across the human population. According to this vision, the original function of ethics—to remedy altruism failures—remains primary. Challenging the enduring importance of this function fails to achieve an important form of coherence, one pragmatic naturalism attains. So (§56) the deepest challenge—the most important accusation against naturalistic ethics—is turned back.

Finally, Chapter 10 attempts further philosophical midwifery, by suggesting some specific places at which current ethical practices might be amended. It offers ideas for continuing the essential conversation.

Readers familiar with contemporary philosophical discussions about ethics will recognize the large differences between standard approaches and the material I plan to cover. Although my questions connect with those posed constantly in the history of ethics, they are often framed orthogonally to the preferred philosophical formulations of recent decades (at least in the Anglo-Saxon world). At many points, my treatment has been influenced by the writings of my peers (and betters), but any serious attempt to expose connections (especially one attending to the nuances of intricate positions) would require expanding my discussions by hundreds of pages. Pragmatic naturalism aims to steer ethical practice and ethical theory in new directions, and I apologize to those who would like to see it anchored in the "existing literature." I must also ask for patience, if a response to an important worry or objection is postponed. Not everything can be done at once, and the needed resources have to be assembled before they can be applied.

§2. Methodological Preliminaries

Pragmatic naturalism differs from previous attempts to link ethics to our evolutionary past. It does *not* propose to identify ethical properties in evolutionary terms, say, by equating what is good with what is adaptive.[11] Nor does it suppose ethical practice is already present, at least in embryo, in our evolutionary cousins or our hominid progenitors.[12] The

---

11. This idea, defended by some sociobiologists, is criticized in the final chapter of my *Vaulting Ambition: Sociobiology and the Quest for Human Nature* (Cambridge, MA: The MIT Press, 1985).

12. For a forthright attempt to link human ethical practice to the altruistic tendencies of other primate species, see Frans de Waal, *Primates and Philosophers* (Princeton, NJ: Princeton University Press, 2006).

ethical project is not simply the unfolding of previously existent altruistic tendencies—it is more than just a population acquiring capacities for "nice behavior." Ethical practice involves conversation, with others and with yourself, juxtaposing desires you recognize as part of you and other desires you would prefer to move you to action. Neither does it posit a special evolutionary advance, in which our ancestors acquired a "moral instinct," conceived along the lines of our innate capacity for language.[13] Views of this type are in danger of confusing ethics with "nice behavior," and, as we shall see (§14), they underplay the influence of the social environments in which ethical practice occurs.[14]

Given these differences, some familiar methodological concerns about naturalistic ethics do not bedevil pragmatic naturalism. One major worry does arise, however. Because it is important, and because appreciating its significance could easily provoke objections to discussions in Part I, this section endeavors to forestall it.

Darwin's success in applying his historical method to the living world rested on the immense body of evidence amassed in the *Origin*—even hostile reviewers praised him for "his facts." A familiar criticism of later attempts to apply evolutionary ideas to human behavior and to human social life charges Darwin's imitators with failing to live up to the standards set by the master. Evolutionary explanations, it is suggested, run the risk of becoming exercises in storytelling—just-so stories without Kipling's wit.[15] A possible hypothesis for the evolution of some trait is proposed, and, without seriously considering alternatives, discriminat-

13. A view well articulated and defended by Marc Hauser: *Moral Minds* (New York: Ecco, 2006).

14. Finally, pragmatic naturalism also rejects the thought that significant advances can be made in understanding ethical issues by undertaking psychological, or neurological, experiments in which subjects are asked to respond to abstract philosophical scenarios. It is unclear what capacities are being fathomed in posing the questions: for questions and concerns that would arise in everyday life are artificially excluded. Moreover, pragmatic naturalism looks for an alternative to current ethical theorizing, rather than for an experimental extension building on available options.

15. The locus classicus for this accusation is Stephen Jay Gould and Richard Lewontin, "The Spandrels of San Marco and the Panglossian Paradigm: A Critique of the Adaptationist Programme," *Proceedings of the Royal Society*, B, 205, 1979, 581–98. In *Vaulting Ambition,* I argued that storytelling vitiates much work in human sociobiology.

ing them in light of the evidence, it is adopted as describing the actual course of past events.

Reconstructing the *actual* history of the ethical project, from its beginnings to the present, is plainly beyond the evidence available—and probably beyond the evidence anyone could ever hope to obtain. On the account sketched in §1, and developed in Part I, many important changes occurred in the Paleolithic and early Neolithic, in a period lacking any written documents. Visions of human life at that time can be based on just a few tantalizing clues: deposits showing increasing group size, tools found at a far remove from the closest source of material, burial sites, figurines, and cave art. These data are too sparse to screen out rival hypotheses about the sequence of events leading from the beginnings of the ethical project to the complex form in which it appears in the first written texts we have.

Yet some hypotheses about the actual history can be defended. Archeological evidence provides grounds for thinking that, until about fifteen thousand years ago, human beings lived in groups of roughly the size of contemporary bands of chimpanzees, and, like the societies of our evolutionary cousins, these groups were mixed by age and sex. Combining this conclusion with anthropological studies of living people who are closest to the circumstances of our ancestors, researchers have provided a picture of the social context of our forebears, the context in which they began the ethical project. We know what the problems faced by chimpanzees are, and how contemporary bands of hunter-gatherers overcome them. The first ventures in ethical practice probably involved group discussions, on terms of rough equality, directed toward issues of sharing and intragroup aggression.

Those initial efforts likely occurred tens of thousands of years ago, after human beings acquired full linguistic abilities. At a conservative estimate, we can set the age of the ethical project as fifty thousand years. I conclude that roughly the first forty thousand years of the project were directed toward the needs of small groups, whose members worked out their social lives on terms of approximate equality. Their original rules were crude and simple. Out of the social life they permitted came a sequence of dramatic changes, generating societies of greater size, and eventually the hierarchical cities of Mesopotamia and Egypt. That part of the story, I shall argue, can be defended as *actual* history.

How did our ethical ancestors move from those simple beginnings to the complex forms of ethical life recognizable in the world of ancient history? No answer can claim to tell the actual story. Here pragmatic naturalism must face the difficulty of discriminating a preferred story from potential rivals. It is, however, committed to supposing some sequence of transitions led—without revelations, without discoveries of the structure of values or the moral law, without any "spooks"—to an endpoint enormously richer and more complex than the original practices. To answer skeptics claiming that "real ethics" requires resources naturalists cannot allow, some narrative needs to be given. It cannot be advertised as a "how actually" explanation; instead it is a "how possibly" explanation.

Explanations come in many varieties, but for the purposes of this book, these two types will suffice. A historical "how actually" explanation aims to tell the truth about a sequence of events: if it is properly supported, rival options have to be eliminated by the evidence. A historical "how possibly" explanation, by contrast, aims only to tell a story, consistent with the evidence and with background constraints: its status is not impugned by pointing out that there are other options (the more, the merrier). A "how possibly" explanation is important because we sometimes wonder whether a chain of occurrences *could* have occurred, or whether the occurrence of the sequence is permitted by a particular theory. Opponents wonder, for example, if the processes countenanced by Darwin and his successors allow for the evolution of the cell. Answering their doubts requires showing how Darwinian processes *might* have produced the cell. It would be marvelous, of course, to be able to say how the history actually went, but, given the temporal remoteness of the events and the limitations of our evidence, modesty is required. In the context of rebutting the skeptical challenge, modesty—settling for "how possibly"—is enough.

Pragmatic naturalism can advance probable hypotheses about the original state in which the ethical project began, and about the character of the evolution of the project during recorded history. With respect to the transformations that occurred between the early phases and the practices of the ancient world, all that can be claimed is that these *could* have happened without supposing processes or causes of kinds pragmatic naturalism rejects. The history of Part I is self-conscious about the distinction. I have attempted to be clear about the kinds of explanations at which particular discussions aim—and hence the standards they are expected to satisfy. In consequence, the metaethical perspective of Part II can sometimes build on claims about *how* a process unfolded (delivered by "how actually" explanations), and sometimes only on theses *that* particular initial states evolved into particular final states (defended against skeptical challenge by "how possibly" explanations). Explicitness may be pedantic—but pedantry is probably better than arousing suspicion that proper standards have been violated.

The aim is to use history—in the ways, and to the extent, we can reconstruct it—to liberate discussions of ethics from the confining pictures that prompt a sense of mystery. Let us now turn to the work of reconstruction.

PART ONE

An Analytical History

CHAPTER 1

The Springs of Sympathy

§3. Psychological Altruism: Basics

At some point in our evolutionary past, before the hominid line split off from the branch that leads to contemporary chimpanzees and bonobos (possibly quite a long time before), our ancestors acquired an ability to live together in small groups mixed in terms of sex and age. That achievement required a capacity for altruism. It also prepared the way for unprecedented forms of cooperation, and ultimately for the enunciation of socially shared norms and the beginnings of ethical practice. Altruism is not the whole story about ethics, but it is an important part of it.[1]

My analytical history of the ethical project thus begins with a hypothesis about the social groups in which the project originated and about the

1. There is a long tradition, stemming from Hume, Adam Smith, and Schopenhauer, that places a capacity for sympathy at the center of ethics. In recent years, that tradition has been renewed by philosophers (Simon Blackburn, *Ruling Passions* [New York: Oxford University Press, 1999]) and by primatologists (Frans de Waal, *Primates and Philosophers* [Princeton, NJ: Princeton University Press, 2007]). Although the approach I shall defend overlaps with some of the themes of this tradition, it does not ascribe sympathy (or altruism) so dominant a role. For explicit comparisons, see my discussion of de Waal, "Ethics and Evolution: How to Get Here from There," in *Primates and Philosophers*.

psychological capacities of the members of those groups. Fossil evidence, together with the remains found at hominid and early human sites, reveals that our ancestors lived in bands akin to those in which chimpanzees and bonobos live today: the members were young and old, male and female; the band size was (roughly) 30–70.[2] This chapter argues that, to live in this way, hominids and human beings had to have a capacity for altruism, one contemporary people almost certainly retain. To understand the historical unfolding of ethics we shall need to recognize the intricacies of the notion, as well as the varieties and limitations of hominid/human altruism. The next sections supply the necessary preliminaries.

It is important to distinguish three types of altruism. An organism A is *biologically altruistic* toward a beneficiary B just in case A acts in ways that decrease its own reproductive success and increase the reproductive success of B. For a century after Darwin, there was a deep puzzle about how biological altruism is possible. During the past fifty years, however, that puzzle has been solved. Biologically altruistic actions directed toward kin can promote the spread of the underlying genes. Moreover, when organisms interact with one another repeatedly, biological altruism exhibited on some occasions can gain dividends from future reciprocation.[3]

2. Different anthropologists use different methods for estimating hominid group size, some favoring direct comparisons with social groups in other species (either evolutionary relatives or primates with a similar ecology), others taking extant hunter-gatherer bands as models or seeking correlations with measurable anatomical features (e.g., skull size) and extrapolating from the results on hominid skulls (viewed as providing clues to the relative increase in neocortex size). See Robin Dunbar, *Grooming, Gossip, and the Origins of Language* (London: Faber, 1996); Steven Mithen *Pre-History of the Mind* (London: Thames and Hudson, 1996); Christoph Boehm, *Hierarchy in the Forest* (Cambridge, MA: Harvard University Press, 1999); Clive Gamble, *The Palaeolithic Societies of Europe* (Oxford, UK: Cambridge University Press, 1999); and Peter MacNeilage, *The Origin of Speech* (Oxford University Press, 2008). Although I am inclined to accept a relatively small value (30–70), my conclusions would not be greatly affected were this increased to, say, 80–140.

3. The original papers are W. D. Hamilton, "Genetical Evolution of Social Behavior," I, II, *Journal of Theoretical Biology* 7 (1964): 1–52; Robert Trivers, "The Evolution of Reciprocal Altruism," *Quarterly Review of Biology* 46 (1971): 35–57; Robert Axelrod and William Hamilton, "The Evolution of Cooperation," *Science* 211 (1981): 1390–96. Lucid and accessible summaries are available in Richard Dawkins, *The Selfish Gene*, 2nd ed. (New York:

Biological altruism requires no perceptive or cognitive abilities. Even plants can have traits that make them biologically altruistic, for their propensities to form roots or to set seeds can limit individual reproductive success and facilitate the reproduction of neighbors. For animals capable of recognizing the wishes of those around them, however, we can develop a useful behavioral analog of the notion of biological altruism.[4] An animal *A* is *behaviorally altruistic* toward a beneficiary *B* just in case *A* acts in ways that detract from its fulfillment of its own current desires and that promote the perceived wishes of *B*.[5] Behavioral altruists do what they take the animals around them to want. They may act in this way not out of any particular concern for those other animals, but because they think that some of their own wishes will ultimately be well served by doing as they do. Behavioral altruism may be practiced by Machiavellian egoists (and, as we shall eventually see—§11—it can also be practiced by individuals who fall into a category intermediate between egoism and psychological altruism).

Neither biological altruism nor behavioral altruism is of much help in understanding the origins of the ethical project. For our purposes, the significant notion is that of *psychological altruism.* Psychological altruism has everything to do with the intentions of the agent and nothing to do with the spread of genes, or even the successful satisfaction of the wishes of others. Assuming for the moment that there have been human beings who are psychological altruists, the vast majority of them have not known much about heredity, and even those who have were rarely concerned with spreading genes. They acted to promote what they took to be the wishes, or the interests, of other people.[6] Sometimes they succeeded. Yet, even when they did not, their serious efforts to do so qualified them as psychological altruists.

Oxford University Press, 1993); and Robert Axelrod, *The Evolution of Cooperation* (New York: Basic Books, 1984). I shall be exploring these important ideas in §8.

4. For discussions about behavioral altruism, I am indebted to Christine Clavien.

5. There are complications that I glide over here and that will be addressed more thoroughly in treating the third type of altruism, the one pertinent to the examination of ethics. After the presentation of that third notion, it will be easier to see how to characterize behavioral altruism more exactly.

6. As the specification of psychological altruism will show, the account begins with wishes. Interests come later (§21).

Many people believe psychological altruism does not exist, even that it is impossible. Often they are moved by a very simple line of reasoning: when a person acts in a way that could be appraised as altruistic, he or she acts intentionally; to act intentionally is to identify an outcome one wants and to attempt to realize that outcome; hence, any potential altruist is trying to get what he or she wants; but to strive for what you want is egoistic; consequently, the potential altruist turns out to be an egoist after all. The key to rebutting this argument is to distinguish different kinds of wants and goals. Some of our desires are directed toward ourselves and our own well-being; other desires may be directed toward the welfare of other people. Desires of the former type are the hallmark of egoism, but those of the latter sort are altruistic. So altruists are intentional agents whose effective desires are other-directed.[7]

I shall develop this approach to psychological altruism further, by giving a more detailed account of the character of other-directed desires, and thereby bringing into the open some of the complexities of the concept of altruism. In focusing on desires, I ignore for the moment the fact that there are other psychological attitudes—hopes, aspirations, and particularly emotions—that can be properly characterized as altruistic. Attention to these other types of states will occupy us in the next section. Because of the connection of desires with intentions and actions, altruistic desires have a certain priority. They are thus the topic of the basic account.

The other-directed desires central to the defense of the possibility of altruism are desires that respond to the altruistic agent's recognition of the impact of his or her actions on the situations of others. To be an altruist is to have a particular kind of relational structure in your psychological life—when you come to see that what you do will affect other people, the wants you have, the emotions you feel, the intentions you form, change from what they would have been in the absence of that recognition. Because you see the consequences for others of what you envisage doing, the psychological attitudes you adopt are different. You are moved by the

7. This line of response surfaces in the eighteenth century in the famous series of sermons given by Joseph Butler at the Rolls Chapel. Many subsequent writers have followed Butler's lead—as shall I.

perceived impact on someone else. If your response leads you to *act* altruistically, that is because your *desires* have been affected.[8]

So far, that is still abstract and vague. I shall motivate the underlying idea with a simple and stylized example and then offer a more precise definition.

Imagine that you are hungry and that you enter a room in which some food is spread out on a table. Suppose further that there is nobody in the vicinity who might also be hungry and want all or part of the food. Under these circumstances, you want to eat the food; indeed, you want all of it. If the circumstances were slightly different, however, if there were another hungry person in the room or believed to be in the neighborhood, your desire would be different: now you would prefer the outcome where you share the food with the other person. Here your desire responds to your perception of the needs and wants of someone else, so that you adjust what you might otherwise have wanted to align your desire with the wants you take the other person to have.

This is a start, but it is not sufficient to make you an altruist. For you might have formed the new want when you see that someone else will be affected by what you do, because you saw profitable future opportunities for accommodating this other person. Maybe you envisage a series of occasions on which you and your fellow will find yourselves hungry in food-containing rooms. You see the advantages of not fighting and of not simply having all the food go to the first person who enters. You resolve to share, then, because a future of cooperation will be better from your point of view. For real altruism, the adjustment of desires must not be produced by this kind of self-interested calculation.

I offer a definition of "*A* acts psychologically altruistically towards *B* in *C*"—where *A* is the agent, *B* is the beneficiary, and *C* is the context (or set of circumstances). The first notion we need is that of two situations differing from each other in the recognizable consequences for others (people or nonhuman animals). Let us say, then, that two contexts *C* and

---

8. You might be affected by another person's predicament, and form an altruistic emotion, but that might not generate a desire that issues in action. The most basic type of altruism that is of ethical concern is a response to someone else that eventually expresses itself in conduct.

C^* are *counterparts*, just in case they differ only in that, in one (C^*, say) the actions available to A have no perceived consequences for B, whereas in the other (C) those actions do have perceived consequences for B. C^* will then be the *solitary* counterpart of C, and C will be the *social* counterpart of C^*. If A forms different desires in C^* from those A forms in C, the set of desires present in C^* will be A's *solitary* desires (relative to the counterparts C and C^*). Given these preliminary specifications:

> A acts psychologically altruistically with respect to B in C just in case

> (1) A acts on the basis of a desire that is different from the desire that would have moved A to action in C^*, the solitary counterpart of C.

> (2) The desire that moves A to action in C is more closely aligned with the wants A attributes to B in C than the desire that would have moved A to action in C^*.

> (3) The desire that moves A to action in C results from A's perception of B's wants in C.

> (4) The desire that moves A to action in C is not caused by A's expectation that the action resulting from it would promote A's solitary desires (with respect to C and C^*).

Condition 1 tells us that A modifies his or her desires from the way they would otherwise have been, when there is an impact—more accurately, when there is a perceived impact[9]—on the wants of B. Condition 2 adds the idea that the desire, and the behavior it directs, is more in harmony with the wants attributed to B than it would have been if B were unaffected by what was done. (It is possible to modify your desires in response to the perceived wishes of another, but to do so in a way that *diverges* from their perceived wants—that is spite.) Condition 3 explains that the increased harmony comes about because of the perception of B's wants; it is not, say, some caprice on A's part that a different desire comes into play here. Finally, condition 4 denies that the modification is to be un-

---

9. I shall consider cases in which agents have mistaken beliefs later. For the time being, I suppose that the parties get things at least roughly right.

derstood in terms of A's attempt to promote some desire that would have been present in situations where there was no thought of helping or hurting B; this distinguishes A from the food sharer who hopes for returns on future occasions when B is in the position of disposing of the goods. Condition 4 requires that genuine psychological altruists be different from Machiavellian calculators who aim to satisfy the wants they would have in solitary situations (I shall sometimes refer to condition 4 as the "anti-Machiavelli" condition).

Given this account of psychological altruism, it is now possible to characterize *behavioral* altruism more carefully. Behavioral altruists are people who look like psychological altruists. That is, they perform the actions people with psychologically altruistic desires would have been led to perform. In ascribing behavioral altruism, however, we do not suppose any particular psychological explanation of the actions. Perhaps they are indeed the products of psychologically altruistic desires, or perhaps the actions are produced by quite different desires having nothing to do with the satisfaction of the beneficiary—a desire for status, or for feeling oneself in accordance with some socially approved pattern of conduct, or even a self-interested calculation. (We shall explore some possibilities of behavioral altruism later; see §§7, 11.)

The stylized food example allows the introduction of an obvious concept, one that will be important in future discussions, and that further articulates the account of altruistic desires. The altruistic modification of solitary desires can be more or less *intense*. I have spoken—somewhat vaguely—of the altruist as aligning his or her wants with those attributed to the beneficiary.[10] That alignment is often a matter of degree, for example, when there is a continuum of possibilities intermediate between complete egoism (retaining one's solitary desires in the social counterpart) and complete subordination of one's solitary wishes to those one perceives the other to have (where one comes to want exactly what one perceives the other as desiring). In sharing food, this is easily

10. For a more precise and formal discussion of many aspects of altruism, see my essays "The Evolution of Human Altruism," *Journal of Philosophy* 90 (1993): 497–516, and "Varieties of Altruism," *Economics and Philosophy* (2010): 121–148. As I shall note at various places, there are several aspects of the account of altruism provided in this chapter that can be treated mathematically, and these articles make a start on that.

expressed in terms of the mode of division: egoists give nothing, self-abnegating altruists give everything, and in between lie a host of intermediate altruists. One obvious style of altruism is *golden-rule altruism*, distinguished by its equal weighing of the solitary desires and those attributed to the beneficiary.

Inspired by the food example, we can undertake a simple way of representing the intensity of psychological altruism, one that will be useful in some (but by no means in all) instances. Suppose that people's desires can be represented by (real) numbers that correspond to how much they value a given outcome. If one result, eating all the food, say, is worth 10 to me, and another, eating half the food, is worth 7, then I prefer eating everything to eating half, but I also prefer an assured outcome in which I receive half to the state of being awarded all or nothing dependent on the flip of a fair coin. (For, in the latter case, my expected return is measured by 5—half of 10 plus half of 0—which is less than 7.)

When you are in the picture, I also take into account the values you attribute to various outcomes. My social desire could be represented as a weighted average of the values represented in my solitary desires and those I take to measure your solitary desires. Thus, the numbers assigned in my social desires would be given by the simple equation:

$$v_{Soc} = w_{Ego}v_{Sol} + w_{Alt}v_{Ben}$$

where v_{Soc} measures my social desires, v_{Sol} my solitary desires, v_{Ben} the measurements of desire I attribute to the beneficiary (you), and w_{Ego} and w_{Alt} the weights given to my solitary desires and my attributions of desire values to you (so that $w_{Ego} + w_{Alt} = 1$). The intensity of my altruism is represented by the size of w_{Alt}—and hence inversely by the size of w_{Ego}; if $w_{Ego} = 1$ ($w_{Alt} = 0$), then I am, at least with respect to you on this occasion, a psychological egoist; if $w_{Alt} = 1$, then I am a self-abnegating altruist; if $w_{Alt} = 0.5$ ($= w_{Ego}$), then I am a golden-rule altruist.

We should not assume that all types of altruistic alignment with the wishes of others can be conceived in this very simple way. Cases of sharing show that a simple approach sometimes works, and the simple expression of social wants as weighted averages will be useful in explaining and illustrating some of the ideas of later sections.

§4. The Varieties of Altruistic Reactions

As already recognized, altruism is not always about the modification of desire, though we are often reasonably suspicious about alleged examples of altruism that do not change desires in ways leading to action: it is not enough simply to "feel another's pain." We can be moved to share the hopes of others, to modify our own long-term intentions and aspirations to accommodate what we see them as striving for, and, most important, we can feel different emotions because of our awareness either of what they feel or of the situations in which they find themselves. For some kinds of psychological states, hopes and long-term intentions, for example, accounts of altruistic versions of these states can be generated straightforwardly in parallel to the treatment of the previous section. Emotions, however, deserve special consideration, both because they are frequently components of the psychological attitudes with which we shall be concerned, and because they involve types of reactions more broadly shared among animals than the psychological states on which I have so far concentrated.

Altruistic emotional responses to others *might* be—and probably often are—mediated by perception and cognition. We see that another person is suffering—or jubilant, for altruistic emotions are not always dark—and our own emotional state changes to align itself more closely with that attributed to the other. Or, in a different mode of altruistic response, we understand the situation in which another person is placed, and our emotional state changes to take on, to some extent, the feeling(s) we would have if placed in that state.[11] When people, or other animals, have dispositions to modify their emotional states in light of their understanding of the feelings or the predicaments of others, we can treat emotional altruism just as §3 analyzed altruistic desire. The emotional altruist feels one thing in the solitary counterpart and feels differently in the social counterpart; the emotion in the social counterpart is more closely aligned with that attributed to the other (or more closely aligned

---

11. The distinction between these two modes of altruistic emotional response—forms of "sympathy"—was already clearly recognized by Adam Smith in *Theory of Moral Sentiments,* Knud Haakonssen, ed. (New York: Cambridge University Press, 2002).

with the emotion the altruist supposes he or she would feel if placed in the other's shoes), and the alignment comes about because of the recognition of the other's feelings (or of the other's situation); finally, it is not caused by any background solitary emotion or solitary desire. Now, whereas in the understanding of altruistic desire this last condition responds to a genuine worry—for we readily think people can form ostensibly other-directed desires on the basis of selfish calculations (I can want to share with you because I think it will be good for me in the long run)—the anti-Machiavelli condition seems odd and gratuitous in the emotional case. It is natural to think, and it may even be true, that self-directed psychological states simply have no power to generate *emotions* toward others, that our emotional life is not under that sort of control. Emotional responses, one may suppose, are caused by processes more direct and automatic than the perceptions and cognitions figuring in my analyses. Consequently, an account of emotional altruism parallel to the analysis of altruistic desire will be at least incomplete, and perhaps even radically misguided.

This is a serious challenge. To meet it, we shall have to consider, if only briefly, the character of emotions. Without taking sides in unresolved controversies, I shall argue that some kinds of emotional response can be understood along the lines just sketched, while others cannot. An account of more basic altruistic emotional reactions, or "affective states," as I shall call them, provides a valuable supplement to the approach to psychological altruism begun in the previous section.

Emotions involve changes in our physiology, and some students of emotion have identified the emotion with the alteration in physiological state. Others propose that there are important distinctions among emotions that cannot be recognized without supposing those who feel the emotions to have particular beliefs, desires, and intentions: specific forms of awareness are required for guilt and shame, for resentment and indignation, and for certain kinds of contentment and anger. A natural way of responding to the findings of neuroscientists, psychologists, and anthropologists is to suppose that many emotions are complex entities, perhaps processes in which particular types of physiological conditions are accompanied by special kinds of cognitive and volitional states. When someone resents the insensitive remarks made by another, he or she undergoes a physiological response connected to judgments about what has been said and desires about what will happen next. The causal details of these connections are matters of speculation, but, even in advance of knowing them, we can reject an approach to emotions that would leave out either the physiological or the cognitive/volitional features.[12]

Yet there may be emotional states, felt by nonhuman animals and by human infants perhaps, for which the cognitive/volitional component is negligible, even entirely absent. With respect to our own species, it has been argued that there are a number of *basic* emotions, found in all human societies and typically giving rise to the same facial expressions.[13] Although a widespread aspect of human psychology or behavior is often taken as evidence of some biological (typically genetic) basis that generates the common feature across all environments, it is worth treading carefully here. For, trivially, there will be some environments in which members of our species will not develop so as to exhibit the typical reaction—neural and psychological development can be disrupted in many different ways. The interesting questions are whether there are subtle properties of potential human social environments capable of prompting something different (so that the widespread finding of the common feature depends on the absence of those subtle properties in the human societies studied), and whether, if so, the potential environments are in some way pathological. These questions are not settled, but, for the sake of the present discussion, I shall allow that human beings who develop in environments, physical and social, that do not involve damaging disruptions of development, will all share dispositions to *basic affective reactions*—that is, they will have capacities for basic affective states, like disgust and anger and fear, and they will exhibit similar facial expressions characteristic of the individual affective states.[14]

12. Here I am much influenced by the thoughtful and ecumenical approach adopted by Jenefer Robinson in the first three chapters of *Deeper than Reason* (New York: Oxford University Press, 2005).

13. See Paul Ekman and Richard Davidson, eds., *Nature of Emotion* (New York: Oxford University Press, 1994).

14. Since the role of social environments is central to my approach to our altruistic tendencies and the character of the ethical project, my position would be strengthened if this concession proves false.

This concession does not entail any of a number of conclusions sometimes drawn from it. First, there is no implication that the affective reactions in people who belong to different societies will be generated by the
same things, events, and states of the world: virtually all people feel disgust, but different groups find very different things disgusting. Second,
to allow basic affective reactions does not retract the earlier judgment that
many emotions have cognitive/volitional components: every emotion
might involve some basic affective state, but a large diversity of emotions
might be distinguished by the cognitions and volitions connected to that
state. Third, and most important for present purposes, we should resist
the thought that "because they are biological," affective reactions are
based on some mechanism more "immediate," more "primitive" than
human cognition. It is easy to muddle together two different senses in
which a capacity may be "biological," one in which its development occurs across all (nonpathological) environments, and one in which the
ways in which it is activated bypass our beliefs and wishes. Conceding
the "biological" status of affective reactions in the first sense does not
commit us to supposing them "biological" in the second.

We can now address the critical issue concerning altruistic emotions.
Even though it is not *required* that there be affective reactions that do not
depend on the causally prior recognition of the feelings or the predicaments of others (on *beliefs* that they are suffering, for example), it is *possible* that there should be mechanisms prompting particular affective responses, mechanisms not mediated by prior cognition. It is well-known
that infants in the same hospital nursery react to the crying of others: an
initial solo can set off an entire chorus.[15] Supposing the unhappiness of
the one spreads to the many because each believes that someone around
is unhappy strains credulity. A more sober account would view the process
as a kind of contagion, effected by a species-typical neural mechanism,
that transfers the misery of one baby to those around. Recent studies of
the activity of so-called mirror neurons (primarily investigated in
macaques) may offer clues about the potential mechanism. Perceptions,
or even sensations, can cause an animal to activate the same neurons as

15. See Martin Hoffman, *Empathy and Moral Development* (New York: Cambridge
University Press, 2000).

those giving rise to the behavior perceived or causing the sensations: A's observation of B's facial expression produces neuronal firings that tense the pertinent muscles and that result in A's imitation of B; perhaps the sound of another baby crying induces a pattern of neural activity that mirrors that in the source of the crying and thus causes the originally contented baby to cry.[16] Mechanisms of this sort require a different approach to altruistic emotions.

Once we have the challenge clearly in view, however, it is not hard to see how to liberate the account of altruism begun in §3 from its dependence on cognition. The task is to provide a definition of "A feels an altruistic emotion in response to B in C." As before, we shall suppose the notions of solitary and social counterparts. The conditions are as follows:

1. A feels an emotion different from the emotion A would have felt in C^*, the solitary counterpart of C.
2. The emotion A feels is more closely aligned with the emotion A attributes to B in C than the emotion A would have felt in C^*; or it is more closely aligned with the emotion A would have felt had A been in B's position in C; or it is more closely aligned with the emotion B actually feels in C.
3. If the emotion A feels is more closely aligned with that attributed to B or if the emotion A feels is more closely aligned with the emotion A would have felt in B's position in C, then the emotion A feels in C results from A's perception of B's situation in C; if no recognition of B's state plays a causal role and if A's emotion aligns itself with that felt by B, then the emotion felt by A is caused by the operation of some automatic neural mechanism, a mechanism triggered by A's observation of B (the activation of mirror neurons might be one such mechanism).
4. The emotion felt by A in C is not caused by A's expectation that feeling this emotion would promote A's solitary desires (with respect to C and C^*).

16. William Damon, *The Moral Child* (New York: Free Press, 1998); and Hoffman, *Empathy and Moral Development*.

This account introduces clauses into the second and third conditions in order to allow the possibility of altruistic emotions produced in ways that bypass cognition. Although the fourth condition is retained, it is highly plausible that Machiavellian manipulation of our emotional lives is beyond our powers, and, if that is indeed so, this requirement is redundant.

The analysis just given preserves a fundamental feature of my original characterization of psychological altruism (§3): altruists have a particular type of relational structure in their psychological lives—when others are around, the altruist's desires, hopes, intentions, and emotions are different from what they would otherwise have been, closer in some way to those of the others, and the difference is produced by some sort of response to those others, not by something enclosed within the self (calculations of future benefit, for example). What the more complex approach to altruistic emotions adds is the possibility that the generation of the response might involve some precognitive mechanism.

It is easy to overinterpret this last point. One might suppose that affective states are always generated by some mechanism that does not involve cognition—but, not only do I see no basis for holding so sweeping a generalization, but it also seems belied by the fact that affective reactions are often founded in complex and explicit understanding (when I see pictures of Jewish refugee children being greeted at English ports by policemen and willing foster parents, I feel a complex mixture of emotions, surely involving affective states, but these states are clearly dependent on my conscious understanding of what the photographs display). The causal relations among affective and cognitive states may be quite various, and, while we await definitive accounts of them, it is well to suspend judgment and to be open to many possibilities.

Nor should we suppose that noncognitive mechanisms are inevitably involved in whatever altruistic responses occur in nonhuman animals. Although questions about the extent of animal abilities to recognize the wishes and thoughts of their conspecifics are much debated, there is no reason to take an advance stand on these issues.[17] I shall later defend the

17. For defenses of opposing views, see de Waal, *Primates and Philosophers;* and Derek C. Penn and Daniel J. Povinelli, "On the Lack of Evidence that Non-human Animals Pos-

thesis that some of our evolutionary cousins have altruistic desires (in the sense of §3; see §7) and that similar capacities were shared by our hominid ancestors.

§5. Some Dimensions of Altruism

One further aspect of psychological altruism needs to be emphasized before we have all the tools required for probing the hominid preethical state. On the account of the last sections, there are many varieties of altruism. Or, to use a suggestive metaphor, altruism is a multidimensional notion. For animals capable of psychological altruism, each individual occupies a particular place in a multidimensional space where brute (non-Machiavellian) egoism is represented by a single plane, and the various forms of altruism range over the entire rest of the space.[18]

An animal's *altruism profile* (where he or she is located in altruism space) is determined by five factors: the *intensity* of the animal's responses to others, the *range* of those to whom the animal is prepared to make an altruistic response, the *scope* of contexts in which the animal is disposed to respond, the animal's *discernment* in appreciating the consequences for others, and the animal's *empathetic skill* in identifying the desires others have or the predicaments in which they find themselves. Non-Machiavellian egoists never respond to anyone else in any context: for the dimensions of intensity, range, and scope they score 0, 0, and 0; their discernment and empathetic skill can be as you please, for these are never called into play.

Altruists are not like that. They modify their desires and emotions to align them with the perceived desires and (perceived or actual) emotions of at least some others in at least some contexts. As §3 already proposed, their responses may be more or less intense. With respect to altruistic desires, an altruist may give more or less weight to the perceived desire of the beneficiary. My treatment of the stylized example in terms of weighted averaging provides a clear paradigm for intensity—the intensity

sess Anything Remotely Resembling a 'Theory of Mind,'" *Philosophical Transactions of the Royal Society B* 362 (2007): 731–44.

18. For more details about this spatial metaphor, see Kitcher, "Varieties of Altruism."

of altruism is represented by how much of the food you are willing to relinquish. If

$$v_{Soc} = w_{Ego} v_{Sol} + w_{Alt} v_{Ben}$$

egoists set w_{Ego} at 1 and w_{Alt} at 0. People for whom $w_{Ego} = 1 - \varepsilon$, where ε is tiny, are altruists in a very modest sense: they will act to advance the wishes of others only when the perceived benefits to others are enormous compared to the forfeits for themselves—they may suffer the scratching of their finger in order to avoid the destruction of the world, but refuse larger sacrifices. People for whom $w_{Alt} = 1$, by contrast, are completely self-abnegating. They abandon their own solitary desires entirely, taking on the wishes they attribute to the beneficiary. In between, we find golden-rule altruists, for whom $w_{Alt} = 1/2$, who treat the perceived wishes of the other exactly as they do their own solitary desires.

Even when averaging is not appropriate for representing altruistic desires, there will often be a comparable notion of the degree to which one has accommodated the perceived wishes of the other. Moreover, with respect to altruistic emotions there is surely a similar concept. Notoriously, we can be relatively unsympathetic, even with those who are dearest to us, when we are preoccupied or distracted. At other times, we enter fully into the feelings of friends and loved ones, even of strangers. It is not obvious how to delineate the notion of intensity in the emotional case as precisely as the food-sharing example allows, but the varying intensity of altruism in emotional responses is uncontroversial. Notice, however, that it should not be confused with the intensity of *emotion*: intensity depends on the degree of *alignment* with the other's feelings (or with the feeling one would have had in the other's situation), not with the force of what one feels.

Most altruists, indeed probably all, lack a fixed intensity of response, applying with respect to all potential beneficiaries and all contexts. There are many people to whom we would rarely make an altruistic response: these people effectively fall outside the range of our altruism. Even with respect to those to whom we are disposed to respond, there are many contexts in which we do not take their perceived wishes or their feelings into consideration (or into our own minds). For many, perhaps, we are prepared to offer limited forms of aid and support; for a few, we are willing to sacrifice everything. Often our altruistic responses to some are colored by indifference to others: parents who make sacrifices to help their children obtain things the children passionately want frequently do not take into account the wishes of other children (or the altruistic desires of the parents of the other children).

Someone's altruism profile typically shows a relatively small number of people to whom the focal individual responds, frequently with significant intensity, across a wide set of contexts. The beneficiaries lie at the center of the range of altruism for the focal individual, and the scope for these beneficiaries is wide. As we consider other potential beneficiaries more distant from the center, the scope narrows (there are fewer contexts in which the more peripheral people elicit an altruistic reaction) and the intensity falls off, until we encounter people to whom the focal individual makes no altruistic response at all. Henceforth, I shall conceive of the range of A's altruism in terms of the metaphor of center and periphery: the center is the select set of potential beneficiaries for whom A's response is relatively intense across a relatively wide scope of contexts; at the periphery, the intensity of the response and the scope of contexts narrow and vanish.

Someone's character as an altruist is not fixed simply by the factors so far considered—intensity, range, and scope—because there are also significant cognitive dimensions to altruism. A may make no response in a particular context through failure to understand the consequences for B; perhaps A does not differentiate the social from the solitary counterpart. Often this is an excusable feature of our fallibility, for the impact on the lives of others may be subtle; we may just not see that following some habitual practice—buying at the most attractive price, or investing in promising stocks—has deleterious consequences for people about whose welfare we care. Evidently, however, acuity with respect to consequences comes in grades, and we admire those who appreciate the intricate ways in which others can be affected, while blaming those who "ought to have seen" the damage they cause.

Similarly, there are degrees to which people are good at gauging the desires of others. Almost everyone is familiar with the well-intentioned person who tries to advance the projects of an intended beneficiary but who is hopelessly misguided about what the beneficiary wants: almost everyone has had a friend or relative who persists in giving presents no longer appropriate for the recipient's age or conditions of life. It would be hard, I think, to declare that people who attribute the wrong desires to their beneficiaries, or who overlook consequences for those whom they intend to benefit, are not acting altruistically when they carry out their variously misguided plans—their intentions are, after all, directed toward doing good for others—but their altruism needs to be differentiated from that of their more acute fellows. Hence I add two cognitive dimensions, one representing A's skill in understanding the nature of a social counterpart to a solitary context, and one assessing A's ability to empathize with B, to ascribe desires B actually possesses.

A simple reaction to the prospect of human egoism is to propose that people living in community with one another—or even all people—should be altruistic; some even take the second commandment of the New Testament to constitute a complete ethical system. Recognizing the dimensions of altruism undermines that thought. There is no *single* way to be an altruist, and, consequently, the commendation of altruism must be given more specific content. What kind of altruist should we urge someone to be? Moreover, is it right to suppose that the best state of the community (or the entire species) is achieved by having each member (each person) manifest the same altruism profile? You might think the questions have straightforward answers. Along the cognitive dimensions, accuracy is always preferable: ideally people should be aware of the potential impact for others and should understand what others want. For issues of intensity, range, and scope, we ought to aim at golden-rule altruism with respect to all people across all contexts.

The demand for accuracy on the cognitive dimensions is more plausible but still not uncontroversial. Debate about the second part of the proposal arises in obvious ways. It might be valuable for people to develop strong ties with some others—the range of human altruism should have a definite center; from Freud's worries about the "thinning out" of our libido in the development of civilization to familiar philosophical examples about parents who wonder whether they should save the drowning child who *is* closer, when their own drowning child is farther out and harder to rescue, a spectrum of troublesome cases arouses suspicion about completely impartial altruism.[19] Moreover, in a world with finite resources, the desires of others often conflict. If A accurately perceives that both B_1 and B_2 want some indivisible good, it should not be automatic that A's desire should be formed by treating B_1 and B_2 symmetrically. (We may, for example, want A to respond to aspects of the history of the situation, including what B_1 and B_2 have previously done.) None of this is to deny that there may be a level at which we want altruism profiles to respond impartially to others, but merely to insist that the impartiality we want cannot be adequately captured as golden-rule altruism toward all people in all contexts.

Further complexities of the notion of psychological altruism will occupy us later. For the present, however, we have enough to begin charting the history of our ethical practices, by understanding how the most basic forms of psychological altruism could have evolved, and how they formed an important part of the social environment in which the ethical project began.

§6. Maternal Concern

Before our human ancestors invented ethics, they had a capacity for psychological altruism. This thesis might be disputed in any of several ways, but the one of immediate concern recapitulates the skepticism about altruism mentioned earlier (§3). Armed with the elements of an account of psychological altruism, the first task is to decide if any such capacity exists, and if it could plausibly be attributed to contemporary human beings, our hominid ancestors, and our evolutionary cousins. Let us begin with the most straightforward case.

Behavior directed toward the survival of young is quite widespread in the animal kingdom, found, for example, among birds as well as mammals. With respect to some types of animals, the hypothesis that this behavior is directed by altruistic desires appears extravagant, for it pre-

19. Sigmund Freud, *Civilization and Its Discontents* (New York: Norton, 1961); Bernard Williams, "Personhood, Character and Morality" in *Moral Luck* (Cambridge, UK: Cambridge University Press, 1981); Peter Railton, "Alienation, Consequentialism, and the Demands of Morality," in *Facts, Values, and Norms* (Cambridge, UK: University Press, 2003).

supposes the propriety of attributing wants and intentions apparently beyond the cognitive capacities of the pertinent organisms. Nevertheless, we might view the animals as driven by altruistic emotions (or primitive versions of them), generated through the operation of automatic neural mechanisms. Among primates, however, particularly those closest to our own species, our evolutionary cousins the great apes, there is considerable evidence for the ability to have desires and to recognize the desires of others.[20] For the sake of concreteness, we can think of psychologically altruistic dispositions to care for the young as emerging in apelike ancestors of *Homo sapiens,* but it is eminently possible that they evolved much further back in our primate (or even mammalian) past.

Even those who share the orthodox primatological views about the cognitive sophistication of our evolutionary cousins may be skeptical of any hypothesis that parental care is sometimes directed by altruistic desires, in the sense I have explicated in §3. They may wonder, for example, whether any dispositions of this kind could evolve under Darwinian natural selection, or whether the apparently altruistic behavior is really the product of some quite different mechanism. Perhaps the animals are really calculating how to achieve future benefits, violating condition 4 of my account, the anti-Machiavelli condition. Many primatologists take the social organization of primate life to reveal "Machiavellian intelligence," and evolutionary psychologists often propose that increased cognitive powers in hominids reflect the need to manipulate others and to avoid being manipulated oneself.[21] Or perhaps the plausible candi-

20. There are many excellent sources for attributing complex cognitive states to nonhuman primates. See, for example, Dorothy Cheney and Robert Seyfarth, *How Monkeys See the World* (Chicago: University of Chicago Press, 1990), esp. chaps. 3 and 8; Jane Goodall, *The Chimpanzees of Gombe* (Cambridge, MA: Harvard University Press, 1986); C. Bachmann and H. Kummer, "Male Assessment of Female Choice in Hamadryas Baboons," *Behavioral Ecology and Sociobiology* 6 (1980): 315–21; R. Byrne and A. Whiten, eds., *Machiavellian Intelligence* (Oxford, UK: Oxford University Press, 1988), particularly the essay by Nicholas Humphrey ("The Social Function of Intellect," 13–21).

21. Many, though not all, of the essays in Byrne and Whiten, *Machiavellian Intelligence* (see n. 20), adopt this perspective. For a more pronounced articulation of the theme that intelligence is a tool for calculating egoists, see James Barkow, Leda Cosmides, and John Tooby, eds., *The Adapted Mind* (New York: Oxford University Press, 1992). In "The Social Function of Intellect," Nicholas Humphrey offers a broader vision (see esp. p. 23).

dates for altruistic responses to the young are affective and immediate. That would allow for altruistic emotions, even emotions that direct behavior, but not necessarily for altruistic desires. To address this latter concern, I shall begin with an example that involves serious cognition and planning.

Primates roaming on the savannah sometimes encounter carcasses that could serve as food. Imagine a female finding a carcass in the absence of her young. Instead of devouring it on the spot, she quickly summons her young. It is difficult to think of behavior of this sort as an action driven by instincts or emotions. Apparently, the mother has to recognize this as food she can share, and to prefer sharing to devouring it entire. Perceiving the possibilities for her young, she forms a different desire from the one she would have formed had they been out of range or fully mature and dispersed. That desire underlies her efforts to summon them to the scene before the food spoils or is taken by another animal. On the face of it, this is an example of altruistic desires in the sense of §3.

One line of concern about attributing altruistic desires is that capacities for such wishes could not have evolved and been maintained under natural selection. In settling this worry, we can use the tools supplied to solve the problem of biological altruism. Suppose that food has decreasing marginal value (in terms of promoting reproductive success), so that, although eating a whole carcass has a higher effect, on fitness it is considerably less than double the effect of eating just half. Assume that the mother has a disposition to golden-rule altruism (or some approximation of it) with respect to her offspring, and that there is just one of her young in the vicinity. Then it is not hard to show that this disposition can be favored by kin selection.[22]

The more difficult challenge asks whether all the conditions for psychological altruism have been met. Perhaps the adjustment of desires to accommodate the perceived needs of young is based upon "Machiavellian" calculations. What form might these supposedly self-directed processes take? Begin with a style of skeptical argument rarely made explicit, but one underlying the conviction that references to psychological altruism are exercises in sentimental self-deception. According to

22. For details, see Kitcher, "Evolution of Human Altruism."

this line of thought, the benefits to offspring, favoring the evolutionary success of altruism, undermine its genuineness. In the described scenario, however, the mother must do something psychologically sophisticated—she has to recognize this as an occasion for seeking out her young—rather than simply exhibit some instinctive reaction. What, then, is the alternative cognitive account that replaces the disposition to adjust preferences with Machiavellian calculation? It strains credulity to suppose mothers recognize the evolutionary advantages of sharing: only a few very select primates *could* calculate the genetic gains and losses (and those who do make their judgments in this way are, to say the least, misguided). So if she calculates it will have to proceed via proxies, through the attempt to attain selfish goals correlated with increases in reproductive success. What could those be?

The most plausible answer is that maternal care proceeds from expectations of future reciprocity—the child is expected to grow into a future ally, maybe eventually a caregiver. Here, the consequences of the present action would be represented in terms we can imagine being within the mother's conceptual repertoire, but we are supposing animal abilities to abstract from present conditions and to envisage a very different future, to overlook the weak juvenile and see a future strong ally. Even if we allow such amazing foresight, problems remain. If dispositions to share with young evolve under natural selection because of inclusive fitness considerations, then the expectations of future aid ought not to be an accurate guide to the kinds of behavior selection would favor—the alleged proxies do not match up well with the variables (the gene frequencies) that are the "ultimate currency" of evolution. From the standpoint of inclusive fitness, mothers should provide some aid when there is very little chance of reciprocity in the future (simply because, even without reciprocation, helping offspring is a good way to spread the genes), and they should provide extra aid to offspring who can be expected to reciprocate. If the hypothetical calculation is to give values that correlate with inclusive fitness, the perceived gains from reciprocity have to be inflated. Why should mothers think their care will be remembered, or, if recalled, it will trigger a disposition to repay? If sharing is based on the expectation of returns, the young seem bad targets. Other, more mature, members of the group would appear to be better prospects for future aid.

The best version of skepticism invokes psychological variables correlating closely with the well-being of the young, and thus presumably with the spread of the pertinent alleles. Determined "Machiavellians" may concede that the scenario I described—in which mothers bring young to share food—involves cognitive abilities, but they may view the calculations that occur as directed toward benefits arising from simpler, more instinctive, reactions. Start, then, with maternal responses to distress. Here, it might be alleged, mothers promote their own ease by preventing wails, facial expressions, and upsetting bodily gestures; or, more positively, mothers find psychological pleasure in observing smiles or hearing happy gurgles. This, it is conceded, is a form of *emotional* altruism. Hence, on occasions where offspring are present, maternal behavior (hugging, caressing, giving food) is directed by the desire either to avoid an unpleasant state ("the pang," generated from emotional responses to unhappy young) or to attain a pleasant one ("the glow," similarly generated from emotional reactions to contented young).[23] When the young are not directly present, however, but available to be brought to the carcass, the mother recognizes the possibility of attaining the glow (by bringing them to the scene) or the dangers of experiencing the pang (if she devours the whole carcass and then encounters hungry offspring). So she calculates that her own selfish desires can better be satisfied by sharing. Because the anti-Machiavelli condition is violated, she does not count as a psychological altruist.

At least two problems confront this skeptical response. The first, and more obvious, is the highly implausible style of cognition it attributes to the mother at the scene of the carcass. She is supposed to be capable of representing to herself not only her absent offspring and their need for food (as on the interpretation of her as a psychological altruist), but also the ways potential actions will bring about glows and pangs—she has to have such thoughts as "If I find the young and share, I shall enjoy the glow" or "If I devour all the food, I shall suffer a pang when I meet the

---

23. In *Unto Others* (Cambridge, MA: Harvard University Press, 1998), Elliott Sober and David Sloan Wilson rightly regard this kind of skeptical response as the most important challenge to the existence of psychological altruism. I think their way of dealing with it is unnecessarily complex, and offer a simpler treatment. Nonetheless, we are in agreement that the challenge can be met.

offspring." Even the most liberal cognitive ethologist is likely to wonder if thoughts like these are within the repertoire of our primate relatives. Moreover, to deliver the appropriate behavior, the envisaged glow or pang has to be sufficiently vivid to override the present desire for the available meat. Only anti-altruist prejudice could inspire the idea that these hypothetical calculations plausibly reconstruct the animals' psychological lives.

The story already presupposes one type of altruistic tendency: mothers feel altruistic emotions. That was allowed in describing the situations from which the skeptical response sets out, the distress felt in the presence of howling infants, pleasure in smiles and gurgles. By the skeptic's own lights, altruistic emotional responses (in the sense of §4) underlie the Machiavellian calculation. Curiously, the skeptical complaint assumes that these emotional responses engender complicated cognitive and volitional states (beliefs and desires about glows and pangs) but do not issue in much simpler desires. The mother's emotional response to her needy young produces no desire to feed them, but a longing for glows or a fear of pangs. Invoking complex Machiavellian calculation and ignoring the far simpler psychological route leading from emotion to simple desire again looks like an egoist prejudice, not a serious rival hypothesis.

These points can be developed further by temporarily leaving our evolutionary past and focusing on apparent altruism among human parents. Imagine a mother whose child has some serious need, a need difficult to satisfy—the child must be rescued; the mother has to engage in an intricate and risky procedure to have any chance of saving the child's life. Enough determined mothers pursue similar causes with unusual energy and persistence, and for them hypotheses about future reciprocation, respect from third parties, or enhanced social status would be jokes in extremely poor taste. The most difficult form of the skeptical hypothesis proposes that these mothers are driven by internal mechanisms—particularly by desires to avoid the pang. We find it natural to suppose that they "couldn't live with themselves" unless they did everything possible for the child (interestingly, in the human case, we tend to recognize the supposed psychological states, the glows and pangs, as intertwined with matters of conscience, a point that will be important later).

Hence, the skeptic proposes, mothers do the impressive things they do because they want to avoid a future of terrible self-reproach and self-torment.

At least two things cast doubt on the skeptical hypothesis. First, the fact that the mother envisages the future of self-reproach testifies to the motivating power of her recognition of the child's wishes (or, in this instance, more likely the child's *interests*—see §21). It is often preposterous to suppose a mother will reproach herself because she is concerned with attitudes in her society—frequently, those around her would praise her for doing far less than she does, constantly reassure her that she has done more than anyone could possibly expect, and so forth. The drive to pursue every possible avenue comes from within, and it could not be abated by any amount of well-intentioned commendation and comfort. If she fails, the mother will suffer, no matter how much she has done and no matter what others say, and the suffering will stem from her deep desire that the child survive and flourish. So, at least, we might initially believe. On the skeptical hypothesis, however, that desire must be denied. Instead, the mother must be viewed as being able to feel altruistic emotions in response to her child. This ability, and the emotions to which it gives rise, does *not* express itself in a desire for the child's well-being. Instead, the ability leads her to fear a particular type of future state, and the fear replaces the denied desire as the driver of her conduct. We have no grounds for accepting this speculative psychology.

A final—fanciful—way to underscore the point: Our world hardly abounds with clever spirits, willing to offer bargains. Yet the mother might have a particular disposition to react to temptations. Imagine that she were visited by a Mephistophelean figure with a straightforward proposal: "I can give you a pill to ensure you will not feel any guilt should things go badly for your child. The pill will wipe away both the pangs of conscience—you will reflect on your efforts and feel you did your best—and any memory of this conversation and the decision to accept the pill. The downside is truly tiny. The probability of your saving your child if you don't take the pill is p; the probability if you do take the pill is $p - \varepsilon$ (where ε is really infinitesimal). Surely the reasonable thing is to accept?" With respect to many actual mothers, we have no doubt about how they would respond—by telling Mephisto to get lost. They view their future psychological comfort as trivial compared with the value of saving the child—any diminution of the probability of success is a loss for which future amnesia cannot compensate. Their assessment of relative value expresses just the desire for the child's well-being the skeptic attempts to deny.

Psychological altruism is real, it is exemplified in maternal concern, and it originally evolved through the most fundamental type of kin selection. Because it is hard to envisage how psychological altruism could take hold without directing maternal care—no other social bond is as pervasive in our evolutionary past, no other recurrent situation is as relevant to reproductive success—it is the most basic and primitive type of altruism.

§7. Broader Forms of Altruism?

How far does psychological altruism extend? Is it merely something mothers (or parents) direct toward their young?

For a first example, we can turn to the inverse of the relationship just examined, to occasions on which offspring help their parents. In her study of the chimpanzees of Gombe, Jane Goodall relates a moving story about the behavior of an adult female, Little Bee, who tended to her partially paralyzed mother, Madam Bee.[24] On several occasions, Little Bee and her mother lagged behind the rest of the troop, often arriving at the nesting site hours later than the others. Mother and daughter took frequent rests, and, when food was needed, Little Bee climbed trees, collecting fruit to share with Madam Bee. Apparently, Little Bee adjusted her preferences to accommodate the perceived needs of her mother, and by doing so she exposed herself to risks she might otherwise have avoided. Reading Goodall's account, it seems clear that the first three of my conditions for psychological altruism are satisfied. The crucial requirement, where skepticism so often arises, is the anti-Machiavelli condition.

Was Little Bee's adjustment of her preferences based on calculating some narrow advantage for herself? It is hard to think what it might be.

24. Goodall, *Chimpanzees of Gombe*, 357, 386.

There was no realistic possibility of her pronounced efforts on behalf of her mother being reciprocated by some future benefits conferred by Madam Bee. Nor could she obtain extra status among the members of her troop, who were in no position to witness her actions—indeed, because her time for interacting with others in the group was so drastically curtailed, her chances for cooperative interactions with them were diminished. If her behavior resulted from calculation, aimed at advancing her own solitary wants, the only possible conclusion is that she *miscalculated*, but the miscalculations would have been so gross as to be quite at odds with her demonstrated social intelligence. Far more plausible is the hypothesis that Little Bee was what she seemed to be—a psychological altruist.

Similarly for a young male chimpanzee, observed by Frans de Waal:[25] Early one morning, de Waal watched two members of the Arnhem chimpanzee colony enter the outdoor enclosure: Krom, a somewhat retarded mature female, and Jakie, a healthy young male. It had rained overnight, and rain had collected in one of the tires hanging from a horizontal pole attached to the climbing frame. Krom wanted to free that tire, but, unfortunately, it was the innermost of five, and her efforts at removing all five tires at once proved futile. After she sat down in a corner of the enclosure, Jakie approached the frame. Intelligently, he removed the tires one at a time, carefully carried the rain-filled tire to Krom, and set it gently before her. She made no gesture of gratitude.

As with the complex pattern of behavior exhibited by Little Bee, it is very hard to suppose Jakie's action stemmed from the operation of some automatic precognitive mechanism. The whimsical hypothesis that, as he saw Krom's efforts, his own mirror neurons fired in ways producing a readiness for tire-pulling behavior, expressed in imitation of her efforts, could only beguile us if we ignored the direction of his actions toward the release of the *innermost* tire and the subsequent *careful* carrying of that tire *to Krom*. To explain what he did, we must credit him with recognizing that Krom wanted the innermost tire—with the water inside it.

25. Frans de Waal, *Good Natured* (Cambridge, MA: Harvard University Press 1996), 83.

Jakie modified his wishes from what they would have been in Krom's absence, and he did so in light of his perception of her desires. He aligned his wants with hers. Are there grounds for skepticism about his altruism? If so, they must stem from concerns that the anti-Machiavelli condition is violated. Perhaps Jakie expected some future reciprocation— but that would be to impute to him a seriously misguided appraisal of Krom's future abilities to reward him (an appraisal quite at odds with his clear social intelligence; Jakie understands Krom's place in the troop). Perhaps he aimed to impress others—but Jakie was surely aware that the only other primate around was the (socially irrelevant) de Waal. Or should we think Jakie not only feels glows and pangs, but has the cognitive powers to perceive the present causes of their future occurrence? Skeptics about altruism are often moved by the thought that an egoistic story is less extravagant than a hypothesis introducing some ability to identify with others. Here, however, skeptical hypotheses about glows and pangs seem the truly extravagant options.

So we can broaden the domain of psychological altruism in the non-human world, at least a little. This is important for understanding the ethical project, because it allows us to attribute altruistic desires to animals *before ethical considerations are on the scene.* A central theme of my approach to altruism is that there are preethical forms of altruism and that these are realized in animals who have not yet acquired ethical practice. Yet caution is necessary. Besides the striking—and clear—cases, there are many instances of primate behavior suggestive of altruism, in which skeptical challenges are far harder to rebut. Observations of chimpanzees and bonobos frequently inspire the interpretation that particular pairs form genuine friendships, that the mutual adjustment of behavior signals an underlying modification of preferences and intentions, prompted by recognition of the other's wants. When the apparently stable alliance breaks down, when a "friend" deserts a seemingly close ally, there are two possible reactions: one can see this as revealing that the parties were calculating all along, using one another to mutual advantage (or apparent mutual advantage); or one can suppose it exposes the previously unnoticed limits of altruism along one of the dimensions (scope) distinguished earlier. Later in this chapter, my preferred explanation of the evolution of psychological altruism will be used to support the hypothesis that, in some of these cases at least, we find genuine altruism.

Recent studies of human behavior often suggest that altruism is far more prevalent in our own species than in our closest evolutionary relatives, attributing the difference to the power of human cultural evolution. Although this conclusion may be correct, if psychological altruism is understood as in §§3–5, it cannot be established as easily as experimenters often believe. Indeed, as I shall suggest later, experimental results taken to support the "pervasive character of human altruism" are not concerned with *psychological* altruism at all, but with *behavioral* altruism; as we shall see (§11), some of the types of behavioral altruism involved are interesting in their own right.

Participants in interactions where there are possibilities for sharing are willing to divide a pool of money with fellows, even though they have the chance to take everything for themselves, and this finding persists across cultures.[26] The behavior counts as psychological altruism only if these subjects are responding to the wants of their perceived beneficiaries and the response is also not the result of an attempt to satisfy solitary preferences. One might worry about both conditions. First, these participants have little knowledge of the wants of their beneficiaries. It is thus hard to view their response as a modification of preferences through perception of another's wants or needs. Second, the skeptical hypothesis that apparently altruistic behavior is driven by desires to achieve glows or avoid pangs has considerable plausibility in these conditions. It is hard to rule out the suggestion that these people share as they do because they want to accord with (or do not want to violate) canons of approved social behavior. They are behavioral altruists whose motivations are not readily characterized as either altruistic or egoistic.

Reflection on the experiments raises the disturbing thought that there is important kinship between the performances of these behavioral altruists and those of their counterparts in earlier studies of willingness to inflict pain and punishment—to administer electric shocks to people

---

26. I ignore here the variety of ways in which opportunities for sharing arise, and, in particular, the important point that subjects will sometimes give some of their assigned money to "punish" participants who fail to share. For a more extensive discussion, see §11.

who are allegedly being "trained" or to function as an effective "prison guard."[27] In both types of psychological experiments, the behavior elicited, whether apparently callous (even "monstrous") or apparently altruistic, may largely express a desire to conform to social expectations.

Perhaps the precise and imaginative experiments on sharing behavior are not really concerned with *psychological* altruism. Demonstrating the conditions for psychological altruism is demanding. One should conceive altruism as covering both the nonhuman examples discussed earlier and the behavior of the experimental subjects, without raising awkward issues about motivations. For some purposes it is surely more appropriate to concentrate on behavioral altruism—if, for example, one wants to scrutinize the hypothesis that economic agents always behave as rational self-interested agents, exploring the possibilities of behavioral altruism is exactly what is needed.

For our purposes, however, there are two reasons to focus on the more demanding notion of psychological altruism. Those who recognize and respond to the wishes of others are different in important ways from people who are moved to help solely by their desire to be well regarded or to have the narcissistic comforts of self-congratulation. The conjecture that similar motivations pervade the studies of sharing behavior and of willingness to torture brings home the point in a dramatic way—even though we might not want to lump the sharers with those who administered "shocks" in the "very dangerous" range, the recognition of an underlying propensity to conform in both situations reminds us that aiming at conformity can blind one to the wants of others with damaging consequences.

More important, if one hopes to understand how ethical practices grew out of human capacities for psychological altruism, the *conception of psychological altruism will have to be prior to that of behavior done in accordance with, or out of regard for, social norms or ethical maxims.* If, as seems likely, the actions of many of the experimental subjects express their wish to exemplify norms of sharing, then their "altruism," if we call it that, will be a product of their immersion in the ethical practice of their

27. For a concise and informative survey of these experiments, see John Sabini and Maury Silver, *Moralities of Everyday Life* (Oxford, UK: Oxford University Press, 1982), ap. 4.

community. Behavioral altruism of this sort cannot be found in the societies in which the ethical project began. We cannot trace the project to prior dispositions to altruism, if we suppose that the prior dispositions are forms of behavioral altruism grounded in acceptance of ethical maxims.

My persistence in advocating a demanding conception of psychological altruism allows for interesting and valuable forms of human action besides the psychologically altruistic ones. Altruism, to repeat, is a complex notion. As we shall discover later, the taxonomy of human action has further complications—it would be wrong to suppose that everything else, besides psychological altruism, is undifferentiatedly and brutishly selfish. In understanding the evolution of human ethical practice, further distinctions and conceptions will be needed (see §11); at that stage it will be possible to provide a more adequate view of the experimental research alluded to here.

For the time being, it suffices to acknowledge some examples of psychological altruism, manifested in other primate species and in our own, besides the fundamental instances of maternal concern. The next step will be to understand how altruistic dispositions might have originated and been maintained under natural selection. We turn to the second part of the task assigned at the beginning of this chapter: to show that dispositions to psychological altruism were necessary for the type of society shared by our hominid ancestors, chimpanzees, and bonobos.

§8. Possibilities of Evolutionary Explanation

The most fundamental forms of psychological altruism, concern for offspring and, more broadly, altruistic tendencies toward close relatives, can readily be understood in terms of kin selection (as already indicated in §6). If an organism tends to adjust its preferences in response to the perceived wants of others (in accordance with the conditions of §3), if there is an allele (or alleles) that underlies that tendency, if the others who benefit from the tendency are relatives, and if the extent of the benefit is sufficiently large with respect to the personal sacrifices (gauged in terms of reproductive success) made by the altruistic animal, the allele(s) and the tendency will spread under natural selection.[28] Kin selection

28. For details, see Kitcher, "Evolution of Human Altruism."

allows for psychological altruism as *one* mechanism for helping behavior toward relatives, but it will equally favor *any* mechanism achieving the same effects. The fact that psychological altruism issuing in aid toward relatives would have been favored by kin selection does not entail that it must therefore exist. In §6, psychological altruism was defended as the best explanation for some types of sharing and helping behavior toward young. *Given* the altruistic tendency, kin selection is the most likely explanation of its presence. (Of course, it would count against the original attribution of psychological altruism to primate mothers if there were no plausible evolutionary explanation.)

Section 7 began with the poignant example of Little Bee and her mother, and here too there is a ready explanation in terms of kin selection. Imagine an original state in which the only form of psychological altruism is directed toward offspring. Suppose a new variant arises, a genetic change causes (in the pertinent environment) a tendency to broaden the range of altruism, allowing for possibilities that other animals, besides the young, will provoke that modification of preferences constitutive of psychological altruism. Animals with the variant are less fussy about those they want to help, but their altruistic responses are always toward close relatives. For concreteness, assume that an animal with the variant has the original tendency to respond, when a parent, to the perceived needs of the young, as well as other tendencies to respond to the perceived needs of parents and siblings. Helping siblings and parents (although not to the same intensity with which aid is channeled toward one's own young) contributes to the spread of the variant allele: for siblings have chances to produce offspring with that allele, and parents likewise have opportunities for generating further young of the new type. Hence the broadening of the psychological altruism originally focused in maternal concern can be favored by kin selection.

The evolutionary scenario just outlined will account for behavior like Little Bee's. A tendency to respond to the perceived wants and needs of one's mother would be favored by kin selection, for, frequently, the helping behavior produced by the altruistic tendency would increase the mother's expected reproductive success and the frequency of the allele(s) underlying the broadening of psychological altruism. Sometimes, however, animals with the tendency may make sacrifices that far outweigh any expected returns—as exemplified by Little Bee's devotion to Madam

Bee. If their helping behavior were based on calculation, it would be grotesquely misguided, belying the animals' manifest intelligence. It is better viewed as a noncalculational, emotional response, of a type that normally increases inclusive fitness, but that, in the case at hand, has negative effects on the spread of the underlying alleles. (Madam Bee's predicament arouses altruistic emotions in Little Bee—and the disposition to be aroused in this way is generally adaptive; the altruistic emotions give rise to particular altruistic desires; on this occasion, acting on those altruistic desires detracts from reproductive success.)

Will the envisaged evolutionary account extend to the example of Jakie and Krom? Perhaps. Here the relationship is far more distant, but the sacrifice made by Jakie is also quite trivial in comparison with Little Bee's months-long dedication. A tendency to (mild) psychological altruism toward any member of the ambient social group might be favored by kin selection, for there is always a (significant?) chance it will direct aid toward relatives, and thus favor the spread of the relevant allele(s). Any hypothesis along these lines would have to be carefully elaborated—for the reproductive costs and benefits are by no means as easy to assess as in the simpler examples involving close relatives—and it also presupposes that evolution of the traits underlying primate social life can be understood prior to accounting for the spread of psychologically altruistic tendencies to group members. Animals *without* the broader tendencies would have to be able to evolve capacities for group life, so that, with the group in place, the stage would be set for kin selection to favor the expansion of psychological altruism across a broader range. In §9 I shall directly question this presupposition and argue that psychological altruism is fundamental to primate social life.

Kin selection is only one of the two mechanisms whose recognition resolved the long-standing puzzle of the evolution of *biological* altruism. The other is the disposition to reciprocate. Tendencies to engage in a pattern of interaction with other organisms, in which each participant gives up something *on one* occasion and reaps a greater gain in some subsequent encounter, can evolve, thus accounting for cooperation among nonrelatives.[29] The initial thought is simple and elegant. If two animals

29. This approach stems from the important work of Robert Trivers, William Hamilton, and Robert Axelrod. The Trivers-Hamilton-Axelrod approach has given rise t

share a propensity for making small sacrifices (measured in terms of reproductive success) to promote greater (reproductive) benefits for the other, if they interact repeatedly, and if the propensity has a genetic basis, then each may reap (reproductive) advantages from the sequence of interactions. Suppose you and I are the animals in question. Today I help you to some significant biological benefit, at much smaller reproductive cost to myself. Tomorrow, you return the favor. Each of us has made a net gain (measured in terms of reproductive prospects). The longer we continue, the larger the benefits we garner. The apparently pedantic introduction of qualifying terms—"biological," "reproductive"—is important because a mode of evolutionary explanation for *biological* altruism does not automatically provide a convincing account of the evolution of *psychological* altruism. With respect to kin selection, the situation is different, for kin selection is neutral in regard to whether psychological altruism underlies the pertinent forms of helping behavior: where one can argue that psychological altruism is the best explanation of that behavior (as with the case of maternal concern; see §6), viewing the tendency as the product of kin selection does nothing to undermine the argument or its conclusion. Reciprocal altruism, by contrast, precisely because of the simplicity of the idea, invites the skeptical complaint that calculational mechanisms are at work, and that the anti-Machiavelli condition is violated. To put the point bluntly, whenever a tendency to a form of behavior can evolve through reciprocal altruism, it looks as though animals with the cognitive sophistication required for psychological altruism would also have the abilities to make a calculation revealing how the behavioral propensity would satisfy their own solitary preferences; hence there would be grounds for skepticism about any alleged psychological altruism. At the very least, when tendencies to behavior are explained by supposing they evolved through reciprocal altruism, skeptics seem to have a forceful objection to the attribution of

---

extensive series of further investigations. See, for example, Alexander Harcourt and Frans de Waal, *Coalitions and Alliances in Humans and Other Animals* (Oxford, UK: Oxford University Press, 1992); Karl Sigmund, *Games of Life* (New York: Oxford University Press, 1993); and Ronald Noë, Jan van Hoff, and Peter Hammerstein, eds., *Economics in Nature* (Cambridge, UK: Cambridge University Press, 2001).

psychological altruism: the animals can identify the long-term advantages of trading favors.

If reciprocal altruism were the fundamental mechanism through which cooperative behavior between unrelated animals evolved, we should have to meet this concern directly, showing that genuine psychological altruism could emerge and be maintained because of the (reproductive) advantages of reciprocation. I shall proceed differently. Patterns of reciprocation have to rest on something more basic, tacitly assumed by accounts of reciprocal altruism. This more basic evolutionary mechanism favors the emergence of tendencies to psychological altruism. Let us start by reviewing how cooperation among unrelated animals is typically explored.

Interactions among animals can be seen as games, in which the players pursue "strategies" (of which they may or may not be conscious). The outcomes of each combination of strategies are represented by the "payoffs" to the players, assignments of numbers representing the values for them of what occurs (for evolutionary studies, these values are the effects on their reproductive success). Evolutionary game theory approaches reciprocation among nonrelatives by considering games involving possibilities of cooperation and also of competition. One particular game has received great attention, the famous prisoner's dilemma (PD).

In PD, each player has two options: to cooperate or to defect. If one cooperates and the other defects, the former obtains the *sucker's payoff*, while the latter enjoys the *traitor's payoff*. If both cooperate, they reap the *reward for mutual cooperation*. If both defect, they both receive the *punishment for mutual defection*. A table shows the outcomes for both players (with returns to the "Row Player" listed first, and returns to the "Column Player" given second).

	C(ooperate)	D(efect)
C	$\langle R, R \rangle$	$\langle S, T \rangle$
D	$\langle T, S \rangle$	$\langle P, P \rangle$

(Here T is the traitor's payoff, R the reward for mutual cooperation, P the punishment for mutual defection, and S the sucker's payoff.) It supposed that $T > R > P > S$, and that $T + S < 2R$.[30] If the game is played just once, defection (D) is a dominant strategy for both players, since $T > R$ and $P > S$. Rational actors in a socioeconomic interaction of this form are expected to wind up with the noncooperative outcome of mutual punishment, rather than achieving the reward for mutual cooperation— which, if they could be assured of it, they would prefer (since $R > P$). By the same token, if animals sometimes engage in interactions with non-relatives, where the payoffs in units of reproductive success meet the conditions of PD, natural selection would apparently favor strategies of defection.

Not, however, if the interactions are repeated. In an iterated prisoner's dilemma (IPD), players can adjust their strategies to the previous performance of those with whom they interact. A strategy for IPD consists in a choice of how to play on the first round, together with a set of preferred responses to the various potential sequences of choices by one's partner/ opponent. Suppose you know the interaction will be repeated but do not know exactly how many times it will occur.[31] Your strategy is specified by saying how you will begin, and how you will act given any potential history of choices by your partner.

Robert Axelrod investigated the success of various strategies empirically, by inviting scholars to submit their preferred proposals for playing IPD, and staging a computer tournament. In each round of the tournament, different strategies were paired (as in a round-robin), and then played a particular version of PD against each other for a large number of iterations.[32] The winner was one of the simplest strategies submitted, tit for tat (TFT), which begins by cooperating, an-

30. The second condition implies that, if the game is repeated, it is cooperatively better for the players both to play C than to adopt a pattern of alternating C and D (so that, on each occasion, one plays sucker and the other plays traitor, with alternation of roles).

31. This last stipulation is added to address the concern that it will always be preferable to defect on the last round, that once that is a matter of common knowledge it will be rational to defect on the penultimate round, and so on. There are complications here that I shall not explore. For present purposes, it is enough to follow the standard treatment.

32. For details, see Axelrod, *Evolution of Cooperation*. Note that the number of iterations is close to two hundred, and that the payoffs in the game—the values of T, R, P, and —are the same in each iteration and in each round.

swers defection with defection, and responds to cooperation with co-operation. In the common parlance, TFT is "nice, provokable, and forgiving."

Mathematical analyses of populations consisting of variant strategies for playing IPD suggested that TFT is *evolutionarily stable;* that is, in a population in which it is prevalent, it resists invasion by alternatives aris-ing at low frequencies.[33] The analyses accounted for the *maintenance* of cooperative behavior under natural selection, once it has become com-mon, but did not explain how such behavior might *originate,* evolving from an initial state in which it was rare. Unless they are strongly dis-posed to interact with one another rather than with the rest of the popu-lation, TFT variants, arising at low frequencies in groups full of non-cooperators, are driven out by natural selection.

Two problems have now emerged with the hypothesis that psycho-logical altruism toward nonrelatives (or psychological altruism more in-tense than would have been favored by kin selection acting alone) might have evolved through reciprocal altruism. First, while reciprocal altruism may help us understand cooperation, its amenability to predictive calcu-lation raises skeptical doubts about psychological altruism as a mecha-nism for the cooperation. Second, it is hard to understand how disposi-tions to cooperate (Machiavellian or altruistic) could have obtained a first firm foothold. There is a third difficulty, too. The IPD scenario imposes very particular conditions: two animals are designated as partners for a long sequence of PDs with exactly the same structure; at the end of this, they are released and assigned to different partners for a repetition of that sequence of interactions. The idea that anything like this happened

33. For the important notion of evolutionary stability (of an *evolutionarily stable strat-egy*), see John Maynard Smith, *Evolution and the Theory of Games* (Cambridge, UK: Cam-bridge University Press, 1982). From the beginning it was apparent that there were indirect ways in which populations of TFTs could be invaded. In such populations, variants that invariably cooperate would be indistinguishable from the TFTs and could thus enter. Once there were sufficiently many of them, the stage would be set for noncooperative strat-egies to invade through exploiting the undifferentiating cooperators. (See my discussion in *Vaulting Ambition,* [Cambridge, MA: The MIT Press, 1985], 100–101.) Further research revealed that combinations of noncooperative strategies can also invade (Robert Boyd an J. P. Lorberbaum, "No Strategy Is Stable in the Repeated Prisoner's Dilemma," *Nature* [1987]: 58–59).

in our primate past is immensely implausible. Surely no giant hand swooped down on the savannah, locking animals into compelled interactions that recapitulated the same form.

To address the difficulties with reciprocal altruism, start with the last. Far more realistic is a different scenario. Suppose our primate ancestors had recurrent opportunities for interacting with a conspecific, and, on these occasions, they could either engage in that interaction or act by themselves. Assume, too, they could sometimes choose partners for interaction, signaling their willingness (or reluctance) to engage in joint activity. This would replace the standard structure of the IPD, the repeated *compulsory* games, with something different—repeated opportunities for *optional* games (as I shall call them). The framework of optional games is both more realistic and resolves some difficulties besetting the orthodox understanding of reciprocal altruism.

An example helps to fix ideas. Our primate ancestors had to remove parasites from their fur. The task was undertaken repeatedly and could be done in either of two ways. One possibility is self-cleaning—although that poses problems because it is hard to reach some parts of the body. Another is to team up with a partner—but that risks exploitation; after the first animal has provided a thorough cleaning, the second may provide something superficial and then go off to more interesting activities. Primates could have signaled to one another their willingness to engage, issuing, accepting, and turning down invitations, so that partners for interaction could be chosen.

With some plausible assumptions about the benefits of hygiene and the costs of spending time, it can be shown that the scenario envisaged has the structure of an optional PD. If two animals interact with each other, the cooperative strategy is for each to provide a thorough grooming for the other; defecting consists in being quick and sloppy. The best of all outcomes is to receive the thorough attention of one's partner and to provide little in return; slightly less good is to obtain a serious cleaning and to return the favor; significantly less good is to receive a superficial grooming and to give back the same; even worse (although not much worse) is to clean one's partner conscientiously but receive a superficial grooming. Not interacting, "opting out," and cleaning oneself, is intermediate between mutual cooperation and mutual defection. Hence, with a somewhat arbitrary assignment of numbers, the structure of the interaction is as follows:

	C	D
C	<9,9>	<0,10>
D	<10,0>	<1,1>

Interact⟋

Opt out ⟶ 5

Mathematical analyses reveal that high levels of cooperation are likely to develop, and to be sustained, in populations whose members have a sufficiently large number of opportunities for playing optional PD with one another. More exactly, a strategy of *discriminating cooperation* (DC) can originate and be maintained under natural selection.[34] Discriminating cooperators are prepared to interact with any animal that has not previously defected on them; if their only opportunities for interaction involve partners who have previously defected on them, they opt out; whenever they interact, they cooperate. Suppose we begin with a population of *antisocial* animals, beings who interact and defect with one another. In this state *asociality* will be favored: the solo strategy (always opt out) does better. In an asocial population (full of solos), however, a lone DC does equally well; there are no opportunities for interacting, and DCs are left partnerless to behave like solos. Once a second DC is present, however, the two of them team up for a happy life of cooperative interactions that bring large advantages over their asocial fellows. So, from antisociality, the population proceeds via asociality to a state of high levels of cooperation. Those high levels will be maintained until the frequency of *non*discriminating cooperators becomes sufficiently high (among DCs, nondiscriminating cooperators are invisible—they are never exploited) to allow antisocial types to enter and take advantage of them. When that happens, the population can relapse to an antisocial state.

34. The results summarized here were originally presented in Kitcher, "Evolution of Human Altruism." I should note that the strategy DC described here is characterized as DA in the earlier paper ("discriminating cooperator" is a more accurate label than "discriminating altruist").

Computer simulations reveal that the history of high levels of cooperation is quite long.[35]

There are further encouraging results about the mechanisms of cooperation. Suppose we abstract from some of the conditions I placed on
psychological altruism in §3, and, in particular, from the Machiavellian
concerns about calculation. Let *quasi altruists* be individuals who meet
conditions 1–3 but not necessarily condition 4: they adjust their preferences to align them more closely with what they take to be the wishes of
others, but they may do so on the basis of considerations of their own
expected narrow benefit. As in the discussions of §§3 and 5, it is possible
to gauge the intensity of the quasi altruist's response, in terms of the
weight assigned to the perceived wishes of the other. Under a regime of
repeated opportunities for playing optional games of various types, selection will favor quasi altruism of a more intense kind, up to golden-rule
quasi altruism because quasi altruists with more intense responses will
participate in a broader class of profitable interactions with others.[36]

Replacing the scenario of compulsory IPD with the framework of optional games helps. Not only does it offer a more realistic scenario for the
evolution of cooperation, but it overcomes the problem of understanding
how cooperation got going. It even points toward some conclusions about
the mechanisms underlying cooperation: selection will favor tendencies
to respond to the wants of others that give the others' preferences as much
weight as one's own. Plainly, however, the shift does not address the most
fundamental difficulty in using reciprocal altruism to explain the evolution of psychological altruism—*for it preserves the simplicity that invites
skepticism.* Animals with the cognitive resources to count as psychological altruists would be able to see the advantages of discriminating cooperation and of being prepared to cooperate across a wide range of types
of interaction. The scenario thus shows how Machiavellian calculators
might have evolved to behave like golden-rule altruists.

To address this problem, to show how full-fledged psychological altruism of kinds going beyond those favored by kin selection might have

35. See John Batali and Philip Kitcher, "Evolution of Altruism in Optional and Compulsory Games," *Journal of Theoretical Biology* 175 (1995): 161–71.

36. This result is derived in Kitcher, "Evolution of Human Altruism." Note that *quasi*
altruists resemble *behavioral* altruists, although some behavioral altruists may not meet
condition 3 of §3.

evolved, requires a more decisive break with the mechanism of reciprocal altruism. Analyses in terms of both compulsory and optional games
can play a role in understanding human social practice. The evolution of
primate sociality, however, is based on a different scenario, one favoring
the emergence of psychological altruism.

For optional games presuppose certain forms of cooperative abilities
that have not yet been explained.

§9. The Coalition Game

Worries about the realism of the scenarios so far envisaged should remain.
The primatological work of the past decades queries some assumptions
hidden behind the mathematical analyses. Assuming our evolutionary
cousins serve as good models of our primate pasts, can we really suppose our ancestors behaved like discriminating cooperators? On the one
hand, chimpanzees and bonobos seem not to cooperate anywhere near
as much as the conception of them as discriminating cooperators suggests. Moreover, they often fail to cooperate with the "right" partners—in
joint hunting, for example, those who help bring down the prey are not
always rewarded, while those who have not taken part end up with pieces
of the spoils, and yet the dispossessed appear willing to return the next
day for a similar expedition.[37] More generally, chimpanzee and bonobo
societies are pervaded by asymmetries the account fails to recognize.
Grooming partnerships embody some of these asymmetries, and a more
focused look at grooming shows it to be a far more complicated phenomenon than the analysis outlined in §8 pretended. If considerations of
hygiene alone were pertinent, it would be impossible to understand the
enormous amounts of time chimpanzees and bonobos devote to grooming one another. During some periods in the recorded histories of
primate troops, particularly when social tensions are running high, the
animals devote three to six hours per day to plucking and smoothing one
another's fur.[38]

---

37. See, for example, Goodall, *Chimpanzees of Gombe*, 288–89.

38. See Frans de Waal, *Chimpanzee Politics* (Baltimore: Johns Hopkins University
Press, 1984), and *Peacemaking Among Primates* (Cambridge, MA: Harvard University
Press, 1989).

These features of primate societies point to the more fundamental presupposition of the explanations in terms of reciprocal altruism: *these are animals who can endure one another's presence, who can occupy the same region together at the same time.* In the original IPD scenario, that is simply achieved by *force majeure;* the organisms are locked together in their long sequence of interactions. Although the shift to optional games increased the realism, it took for granted the existence of a pool of potential partners. Animals were supposed to encounter others quite frequently and to be able to signal their willingness to interact. For that, a minimal form of sociality must already be in place—the animals must be sufficiently tolerant of one another's presence to form the pool. Reciprocal altruism presupposes an ability to treat others as potential partners and not as dangerous rivals.

That ability should be the first and fundamental target of evolutionary explanation. The processes that gave rise to it generated a capacity for psychological altruism of a more extensive type than those understood in §8 in terms of kin selection.

Begin with some well-established conclusions about social life among the apes. Within this relatively small group, the extent to which social relations, tolerance, and cooperation extend beyond the family varies greatly. Gibbons divide into small family groups (mother, father, and young) that are typically hostile to outsiders. Male orangutans are mostly solitary, ready to defend their territories against incursions from other males; they interact only perfunctorily with the females whose home ranges lie within those territories; the extent of female-female association is a matter of controversy (with older orthodoxy supposing that females travel with one or two offspring, and newer observations pointing to intermittent joining of pairs of females). Groups of gorillas typically contain several adult females but have only one adult male; to a first approximation, gorilla social life involves some cooperation among unrelated females and only aggressive interactions among adult males.[39] For larger social units, with cooperation among unrelated adults of both

---

39. A valuable source for discussions of social life among the apes is Barbara Smuts; Dorothy Cheney, Robert Seyfarth, Richard Wrangham, and Thomas Struhsaker eds., *Primate Societies* (Chicago: University of Chicago Press, 1987).

sexes, we must turn to our evolutionary cousins, chimpanzees and bonobos.

Chimpanzees live in bisexual groups (varying in size from about 20 to approximately 100), within which there are shifting patterns of alliances and dominance relations. Among bonobos, the groups are somewhat larger (roughly 50 to 150), with the same sorts of changing internal structures.[40] A principal difference between the two groups is that the major associations in the wild seem to be among chimpanzee males and among bonobo females, although in both species, there are important social interactions among members of the other sex (and between members of opposite sexes). Study of hominid remains suggests that our ancestors lived in mixed groups and that their size was of the same order as that found in living chimpanzees and bonobos. How did the chimp-bonobo-hominid pattern of sociality evolve?

Any answer to the question must identify the features that distinguish chimps and bonobos from the other great apes. I shall develop an approach originally outlined by Richard Wrangham, who proposed that female behavior is shaped directly by ecological factors, particularly the distribution of the foods consumed by the species; males have to adapt to this distribution, adjusting their behavior to increase the chances of copulating with estrous females.[41] Crucial for our purposes is the conjecture

40. I shall tend to take chimpanzees, rather than bonobos, as the model for our hominid past. This decision rests partly on a sense that many small human societies that live in environmental conditions closer to those of our ancestors appear to share the relative intolerance for neighbors that is so marked in chimpanzee social life, and, more important, on the hypothesis that psychologically altruistic tendencies are more prominent and pervasive in bonobos than in the (common) chimpanzee. Hence I assume that if a compelling story about the evolution of sociality and its roots in psychological altruism can be given for chimpanzees, it would be easier to defend a similar account for bonobos. (Here I am indebted to a valuable conversation with Frans de Waal.)

41. See Richard Wrangham, "On the Evolution of Ape Social Systems," *Social Science Information* 18 (1979): 334–68; "An Ecological Model of Female-Bonded Primate Groups," *Behaviour* 75 (1980): 262–300; "Social Relationships in Comparative Perspective," in *Primate Social Relationships: An Integrated Approach,* ed. Robert Hinde (Oxford: Blackwell, 1983); and "Evolution of Social Structure," in Smuts et al., *Primate Societies,* 282–96. Wrangham bases his analysis on the hypothesis that the principal determinant of female reproductive success will be her access to food and that the principal determinant of male reproductive success will be the ability to copulate as frequently as possible with

that mutually hostile communities of chimpanzees have "evolved from a hypothetical solitary-male system because males could afford to travel in small parties, even though the optimal foraging strategy was to travel alone; they were forced to do so because lone males therefore became vulnerable to attacks by pairs."[42] Abstracting from the emphasis on foraging, one may recognize that, in a world with scarce resources—of whatever kind—competition among vulnerable animals may require their participation in coalitions and alliances. Addressing that problem is prior to realizing possibilities for cooperation: *for understanding cooperative interactions among unrelated animals, PD (whether optional or compulsory) is not fundamental; the framework for the games animals play is set by the problem of forming coalitions and alliances.*[43]

Imagine a population of solitary organisms (the largest units being mothers with dependent young) in an environment in which each must obtain a certain number of resources in order to survive and reproduce. Suppose the resources are scarce, the animals fight over these resources, and the stronger typically win. A five-stage process could have led from the initial situation—no cooperation except for maternal care in early life—to the kind of social structure found in chimpanzees, bonobos, and

---

estrous females. So, for example, on his account, orangutans pursue their relatively solitary lives because females can most efficiently forage for fruit by working alone, and males have physical abilities to defend a territory including the smaller home ranges of several females. I shall make no such specific assumptions. Instead, I abstract from the particularities of Wrangham's discussion, offering a more general model of which his approach would be a special case.

42. Wrangham, "Evolution of Social Structure," 290. Compare Hobbes: ". . . the weakest has strength enough to kill the strongest, either by secret machination, or by confederation with others that are in the same danger with himself" (*Leviathan,* 82). Hobbes, however, would not have thought that this could apply to the brutes, because, without speech ". . . there had been amongst men neither Common-wealth, nor Society, nor Contract, nor Peace, no more than amongst Lyons, Bears, and Wolves" (*Leviathan,* 20). Hobbes underrated the lions and the wolves and knew nothing of the chimpanzees and the bonobos [New York: Oxford University Press (World's Classics) 2008].

43. Some primatologists have recognized the point in the context of their studies of particular societies. See, for example, R. Noë, "Alliance Formation Among Male Baboons: Shopping for Profitable Partners," in *Coalitions and Alliances in Humans and Other Animals,* ed. A. Harcourt and F. de Waal (Oxford, UK: Oxford University Press, 1992), 285–321.

hominids. (Note that what is required here is an account of how a form of social structure we independently know to exist could have emerged and remained stable under natural selection: a "how possibly" explanation.)

1. **Asociality**—animals range alone (at most accompanied by dependent young), finding some resources without contest ("scramble" competition) and competing directly for others ("contest" competition).
2. **First Coalitions**—some animals arise that are disposed to act together in contest and to share the resources obtained (not necessarily equally).
3. **Escalation**—because of the success of the early coalitions, larger coalitions form, sharing the benefits they earn in contests (not necessarily equally, and possibly involving interactions among subcoalitions).
4. **Community Stabilization**—coalition size is ultimately limited by the difficulty of defending all the resources in a range, and the habitat becomes partitioned into ranges defended by stable communities, within which the resources are divided by the formation of subcoalitions.
5. **Cooperation**—by engaging in optional games (some of which may be optional PD) and behaving cooperatively, members of the stable communities increase their fitness.[44]

Without pursuing the technical details, I shall try to show how this process might unfold.

Begin with a more benign version of the initial state, a Rousseauian world that contains more than enough for everyone. As the population expands, competition enters. Eventually, so long as the competition goes on in the assumed way, some animals will not find the resources they need to survive.

---

44. Note that the fitness values that occur in the payoff matrices for the games played by community members, whether optional or compulsory, must reflect the consequences of actions for the underlying alliances to which the animals belong. This recapitulates the point made earlier that the structure of animal interaction cannot be understood in isolation from the demands of the most fundamental game, here seen as the coalition game.

If the animals pursue solitary strategies for gaining resources, as envisaged in stage 1 of the process outlined above, there will be contests for some resources. Assume, for simplicity, that the contests are resolved without actual fighting: the animals simply assess one another's strength, and the weaker one retires (in cases of equal strength, divisible resources are shared equally; indivisible resources are assigned to each animal with probability $1/2$). Throughout the course of their lives, the strength of the animals changes, according to an obvious schedule. Initially, while an animal is under the protection of its mother, it effectively has whatever strength its mother has. Once released from its mother's care, it is at its weakest. Thereafter, strength grows as the animal matures, provided that sufficient resources are obtained; eventually, perhaps, animals that live long enough undergo a slight decline in strength.

Populations faced with these conditions are vulnerable to extinction. For a new generation to arise, the young must survive the critical period after their release from maternal care. During this period, they are the weakest members of the population, and whatever they achieve must be gained by finding resources currently uncontested by others and consuming those resources before a stronger individual arrives to dispossess them. If the competition is sufficiently severe, all resources will be contested, and, after a brief period of maternal care, all the young members of the population die. In a very hostile world, populations stuck at stage 1 are likely to be short-lived. More exactly, the pressure of mortality will cull the population so it is effectively returned to a more benign—Rousseauian—environment.

Suppose, however, variants arise that are disposed to team up with others. Specifically, imagine a variant that is prepared, when weak, to travel around with another animal of similar weakness, to collect resources together and to divide them. (There may be variation in the propensities to tolerate different schedules of division.) If two such variants encounter each other, they form a coalition. Because the members of the coalition have to travel together, the coalition can visit only as many resources as a single individual can. Assume that strength is additive; that is, the strength of a coalition is the sum of the strengths of its members. Each of the variants in a coalitional pair can now increase its access to resources, for the doubled strength will surely provide victory in contests with other weak young animals and may be sufficient to win some encounters with older members of the population. Selection thus favors variants of this type, even if the divisions of the resources acquired are not even.

Plainly, several parameters must be set in developing versions of the scenario I am envisaging, but it is possible to show that, given almost all ways of choosing values for non-Rousseauian worlds, any population at stage 1 will contain at least one pair of organisms who can increase their fitness through coalition formation. That does not mean, of course, that the disposition to team up *must* evolve: there might be no way to generate any such propensity. I shall suggest shortly that more basic capacities for psychological altruism provide a way in which the successful variants might emerge.

Just as stage 1 would favor the emergence of pairwise coalitions, so too the emergence of pairs puts pressure on animals who are working alone. The gains of the animals who team up are obtained by dispossessing those who would otherwise have done better. Any variation that equips them with a disposition to pair with another animal will be favored. As the population becomes full of coalitional pairs competing with one another, the weakest pairs will do better if they are prepared to add single members or merge with other pairs. Selection favors the variants who unite with others at the size required by the actual escalation of coalition formation.

Although the origination and escalation of coalition formation is easy to understand, the termination of the process appears more mysterious. The rationale, however, is a direct consequence of the fact that coalitions have to travel together if they are to exert their joint power. No coalition can visit more resources than a single individual. When the environment is filled with large coalitions, coalition members who receive the smallest shares may have no better option than to resume scrambling for resources the large coalitions are not able to visit. The dynamics of the process leads to a situation in which the habitat is partitioned into territories controlled by sizable coalitions, occasionally with a floating population of individuals who live on the fringes.[45]

---

45. The announced results are not hard to derive analytically. They coincide with the findings of some ingenious computer simulations designed by Dr. Herbert Roseman. See his unpublished Ph.D. dissertation "Altruism, Evolution, and Optional Games," 2008 Columbia University.

This is an evolutionary scenario for the emergence of the social structure found in chimpanzees, bonobos, and hominids, one that will lead to
groups mixed by age and sex, each of which controls a relatively stable
territory that it defends against neighboring rivals. Within these groups,
there will be patterns of alliances bearing on the division of the resources
the group commands. That structure will determine the potential benefits
of various possibilities for cooperation, for there will be gains from
strengthening existing alliances and costs from disrupting them. The ho-
mogeneous pool of partners for optional games, envisaged in the analysis
of §8, is structured by the shapes of previous encounters. Reciprocal altruism and interaction in optional games can be understood only against
the background of the coalitional structure of the group.

So far the conclusions address only animal behavior, with no direct
implications about psychological altruism. To go further, it is necessary
to ask how the variants envisaged, with their disposition to team up with
others, might have been psychologically realized. Answer: this ability
to form coalitions, and ultimately to constitute a stable social group, expresses a further expansion of those fundamental psychologically altruistic tendencies attributed in the case of maternal care.

Mothers have a propensity to modify their wants and preferences from
what they would otherwise have been, to accommodate the perceived
wants of the young. Primates have evolved to broaden this response to
others, so that preferences reflect the perceived wishes of close relatives, a
broadening supported by kin selection and manifest in the behavior of
Little Bee. I propose a further extension: the disposition to adjust wants
and preferences to the perceived preferences of an age-mate, initially triggered in contexts where both animals are weak and vulnerable. This is a
species of psychological altruism, the capacity for early friendship. Pairs
of animals with this broadened altruistic disposition reap the advantages
just outlined. Young animals, no longer under parental protection, need
allies if they are to gain anything in a competitive world. Psychological
altruism of this special type is one way for them to find friends.

Skeptics will suppose there are self-interested routes to the same end.
What would they be? The coalition game is by no means a simple opportunity for reciprocal altruism. It does not present the players with a
compelled or optional iterated prisoner's dilemma, inviting them to cal-

culate a strategy for success. The coalition game is many-personal—and, for the players, the number of participants will typically be unknown. It is not even evident what would count as a "best strategy" for playing it. Whether someone counts as a good ally or not depends on all sorts of delicate facts animals have no way of recognizing. Moreover, working out a good procedure for playing the game challenges the intellectual powers of mathematicians, economists, and philosophers. The best one can do is pick a partner, team up—and hope.

That appears to be just what chimpanzees and bonobos do. Their alliances do not seem to depend on any tallying of costs and benefits. Instead, these animals are prepared to support members of their groups with whom they have a history of interactions, often dating back to periods early in their lives—the strongest alliances descend from that period of juvenile vulnerability.[46] What sorts of calculations might underlie their behavior?

It is natural to believe that the clever head can always substitute for the kindly heart, but that need not always be so. When the problems posed for reasoned selection of the best strategy are sufficiently intractable—as they are in the case of the coalition game—it may not just be that an emotional response to another animal, the transfer of altruistic dispositions to identify with others to a novel sphere, the domain of "early friendship," does no worse than the cunning of the Machiavellian calculator, but that it works *better*. Animals with a disposition to try to work out the costs and benefits suffer from too little information to make good decisions on this basis, and their efforts can easily lead them to abandon an alliance when there are no serious prospects for doing any better. Furthermore, they may hesitate more than their blindly sympathetic counterparts, and indeed be recognizable by others as less reliable and less stable coalition partners.

When weak animals are forced to compete for resources they need, their inability to win contests by themselves confers a selective advantage on a disposition to identify with the interests of conspecifics, particularly with those who are in a similar predicament. That advantage fostered the spread of propensities to psychological altruism antecedently limited,

46. See Goodall, *Chimpanzees of Gombe*, 379–85, 418–24.

first toward young and then toward close relatives. The broadened propensities allowed for the formation of those loose coalitions found in our evolutionary cousins. Far from being anthropomorphic, sentimental, or self-deceiving, the hypothesis advanced here looks like the best explanation of the form of sociality of our hominid past. It also explains why the friendships of youth are so deep and enduring, both in human beings and in other primates, and why newcomers are sometimes accepted into primate social groups when a resident animal has formed social bonds with them in a shared past as juveniles together.[47]

Psychological altruism is the kernel from which ethical practice grows—because it lies at the heart of the type of sociality our hominid ancestors experienced. As we shall discover, however, the plant is far more elaborate than the seed.

47. De Waal relates a striking instance, in which a relatively unprepossessing male (Jimoh) was accepted into a chimpanzee troop because of his prior association with two older females in the group. See de Waal, *Good Natured*, 131–32.

CHAPTER 2

Normative Guidance

§10. The Limits of Altruism

Imagine a population of organisms with altruistic dispositions. For each of these organisms, there is a variety of contexts and a range of other members of the population such that the psychological states of the focal organism—specifically the desires and the emotions—will adjust to reflect that organism's perceptions of the wants, needs, and feelings of the others. These dispositions enable the organisms to function as a population, to live in the same place at the same time and to encounter one another daily without too high an incidence of social friction and violence. But the dispositions are limited: cooperators are sometimes exploited, returns are uneven, and, when there is an opportunity for large selfish benefits, even long-standing allies are sometimes left in the lurch. Defections threaten to tear the social fabric, and, in their wake, much signaling is required; our organisms engage in prolonged bouts of mutual grooming and other forms of physical reassurance.

I shall call these organisms "hominids," although it would be equally apt to dub them "chimpanzees." The limitations of their psychological altruism cause the tensions of their social lives and prevent them from gathering in much larger groups and participating in more complex cooperative projects. A look at their evolved descendants some quarter of a million generations later discloses that the limits have been transcended. Ten thousand years before the present, those descendants have formed settlements that sometimes contain a far larger population; they have learned to interact peacefully with many conspecifics whom they do not encounter on a daily basis; and they have constructed complex systems of cooperation that involve marked differentiation of roles. How has all this been achieved?

One possibility is that they have acquired some new and stronger mechanism for psychological altruism. Conceiving hominid societies as exactly like those of contemporary chimpanzees (or bonobos) is plainly implausible, for the members of the later hominid societies had diverged from their evolutionary cousins five million (or more) years ago. Perhaps as hominid brain size increased, it was necessary for babies to be born at developmentally earlier stages (so their heads would still pass through the birth canal), with the consequence that they were more dependent for a longer period of time. The resulting selection pressure may have favored enhanced altruistic tendencies in the specific context of providing care for helpless young.[1] Without underestimating the importance of steps like these, it is evident that neither hominids nor contemporary human beings have escaped entirely from the difficulties and tensions, the rivalries and conflicts, of chimpanzee social life. If human societies are less vulnerable to breakdown than those of our primate relatives, it is because other modifications have taken place.

What modifications? The hypothesis to be explored is that other changes that have occurred, including in particular the acquisition of language, have made it possible for human beings to reinforce the original limited altruistic tendencies, so members of human societies no longer falter quite as frequently in their cooperation. Because defection is

---

1. Arguments along these lines have been developed by Kristen Hawkes and her colleagues (see, for example, Hawkes, James O'Connell, Nicholas Blurton Jones, Helen Alvarez, and Eric Charnov, "The Grandmother Hypothesis and Human Evolution," in *Evolutionary Anthropology and Human Social Behavior: Twenty Years Later,* ed. L. Cronk, N. Chagnon, and W. Irons [New York: De Gruyter, 1999]) and by Sarah Hrdy, *Mother Nature* (New York: Pantheon, 1999), and *Mothers and Others* (Cambridge, MA: Harvard University Press, 2009).

more often prevented, less time has to be spent in reknitting the social fabric. The cumbersome peacemaking of our original hominids is replaced by a new device, one preempting rupture rather than reacting to it, and in principle capable of operating in a wide variety of contexts.[2] That device is necessary for what we think of as ethical practice. I shall call it a "capacity for normative guidance."

The previous chapter was at pains to defend attributions of psychological altruism and to rebut the skeptical insistence that sees Machiavellian intelligence behind apparently helpful or kindly actions. Its account, however, was entirely consistent with the thesis that the psychological altruism of our hominid ancestors was limited. Recall two of the dimensions of altruism: range and scope. An animal may be disposed to respond altruistically to particular other members of its social group ("close friends") across a relatively broad set of contexts, and to respond to all members of its social group in some contexts (banding together against outsiders, for example), although there are occasions on which it would act selfishly even toward its closest friends and staunchest allies. The limited quality of chimpanzee-hominid altruism, in both range and scope, set the stage for the emergence of normative guidance.

The limits of altruism are most starkly and spectacularly visible when the selfish rewards for deserting erstwhile allies are extremely high—as when a male has the opportunity to achieve dominance in the social group. A study of "chimpanzee politics" in the colony at Arnhem (an environment allowing the animals to retain important features of their life in the wild, but, at the same time, providing opportunities for systematic observation of them) revealed the ways in which three high-status males— Yeroen, Luit, and Nikkie—related to one another and to the high-status females, during times of transitions in power.[3] Each male exhibited social behavior readily interpretable as aimed at retaining dominance, achieving dominance, or, at worst, serving as the principal lieutenant of the dominant male. In the early phases of the struggle, Luit aided the newly

2. It is thus more than a special-purpose mechanism, like the hypothetical emotional disposition that underlies alloparenting.

3. Frans de Waal, *Chimpanzee Politics* (Baltimore: Johns Hopkins University Press, 1984).

adult Nikkie in achieving dominance over the females, while Nikkie's diversionary tactics enabled Luit to dethrone the previously dominant Yeroen. Once he had attained alpha rank, Luit's policy changed. He consolidated his position by siding with the females and with Yeroen against Nikkie. The abruptness and decisiveness of the switch can easily inspire the conclusion that chimpanzee politics is thoroughly Machiavellian. Apparently, friendships among chimpanzees are situation linked.[4] The subsequent twists and turns of the story seem to underscore that judgment. Yeroen deserted Luit to form a coalition with Nikkie, so that Nikkie eventually became dominant with Yeroen as his lieutenant. After a subsequent period of tension between the two allies, Luit reemerged at the top of the hierarchy, apparently in a weak coalition with Nikkie. The uneasy situation was ended by a night fight, in which Luit was fatally injured by the other two.[5]

To say there are no stable friendships within chimpanzee communities is too strong, for some alliances endure for years, even for virtually the entire lifetimes of the animals—as §9 insisted, the friends of one's vulnerable youth are often one's lifelong companions.[6] Moreover, before the political instabilities in the Arnhem colony, Yeroen and Luit had been longtime allies. The fascinating (but sad) story of the months of conflict reveals—as do similar examples, less fully documented, among wild chimpanzees—how the presence of a clear opportunity for self-advancement can expose the limits of altruistic dispositions. Observers

4. De Waal offered a sober judgment of the relationships he observed: "Coalitions based on personal affinities should be relatively stable; mutual trust and sympathy do not appear or disappear overnight. . . . If friendship is so flexible that it can be adapted to a situation at will, a better name for it would be opportunism" (*Chimpanzee Politics*, 128). Readers of de Waal's subsequent books (*Peacemaking Among Primates* [Cambridge, MA: Harvard University Press, 1989]; *Good Natured* [Cambridge, MA: Harvard University Press, 1995]; *Primates and Philosophers* [Princeton, NJ: Princeton University Press, 2006]) may be surprised by his early emphasis on hard calculation—for the later work is softer in tone and more inclined to highlight the "good-natured" aspects of primate behavior. The account I offer in the text supplies a perspective from which all of his evaluations can be endorsed.

5. De Waal, *Peacemaking Among Primates*, chap. 2. De Waal makes the important point that Luit's desire to remain with his social group was so strong that it was difficult to remove him, even after he had been severely wounded.

6. See Jane Goodall, *The Chimpanzees of Gombe* (Cambridge, MA: Harvard University Press, 1986), chap. 8.

have seen enough varied contexts in which two animals respond to each other to assign them to each other's range of altruism—until the animals encounter a new type of context, in which an altruistic response would require the forgoing of huge potential gains. The selfish action in that context is a sign not that everything in the past has been opportunism, but just that the altruistic disposition is incompletely pervasive. Even for animals who are central to the *range* of the altruist's altruism, there are circumstances outside the *scope* of that altruism.

The conception of psychological altruism offered in §§3–5 reveals what is occurring. Chimpanzees (and our hominid ancestors) have regular psychological propensities for making an altruistic response to another member of their group, with the intensity dependent on salient features of the circumstances. Even though an animal frequently displays a tendency to accommodate the wishes and needs of a particular band member—a "friend"—there are environments in which the intensity of the altruistic response drops to zero. In those environments, altruism suddenly vanishes. Friendship is "situation linked" because there is no fixed value of the intensity of the altruistic response depending solely on the strength of the relationship.[7] Even in the most committed mutually altruistic relationships, circumstances offering one party the chance of a huge advantage diminish the intensity of the response. When the stakes are high enough, it disappears entirely.

The struggle for dominance presents in high relief contours visible in more mundane settings. Every day in chimpanzee troops, members who are not one another's principal allies act in blithe indifference to their fellows' obvious plans. Attempts to obtain a valued object are blocked or thwarted, requests to share food are turned down, appeals for aid in conflict are ignored. The animals involved are not entirely indifferent to one another, for they would band and bond together in the face of an externally presented threat. Rather, the scope of their mutual altruism is

7. The approach adopted here has some kinship with Walter Mischel's emphasis on the failure of cross-situational consistency in people who have stable personality profiles. See W. Mischel and Y. Shoda, "A Cognitive-Affective System Theory of Personality: Reconceptualizing Situations, Dispositions, Dynamics, and Invariance in Personality Structure," *Psychological Review* 102 (1995): 246–68. I am grateful to George Mandler for the suggestion that I explore Mischel's work.

very limited; only under the most threatening situations is it exercised. For the rest, they operate on the basis of tolerance of one another's presence, though, when one's indifferent course collides too strongly with the plans of the other, conflict may erupt.[8]

The limitations of psychological altruism thus show up both in the breakdown of close ties under special conditions (the Yeroen-Luit-Nikkie saga) and in the everyday frictions of animals whose altruism toward one another is limited in scope. The bounds of altruism are revealed in a third way. Even when an altruistic response is made, and when it directs a helpful action toward another animal, there are sometimes signs of psychological division. Conflict *within* is occasionally visible. Chimpanzees are openly torn between selfish and altruistic courses of action, making it apt to attribute to them *two* desires, both expressed in facets of their behavior. An animal hesitates. Holding a branch rich in leaves, he is poised to strip them off and eat, and, simultaneously, the set of the body acknowledges the presence of an ally; eventually, the arm is extended, thrusting a small bunch of leaves toward the friend, while the rigidity of the gesture and the averted face show the presence of a contrary desire. The configuration of limbs and muscles is genuinely a mixture. The tension of the moment is apparent.[9]

Animals can have stable dispositions to respond with quite different intensities of altruism to environmental cues that can simultaneously be present. Some features of recurring situations trigger an altruistic response at a particular intensity; in response to different features the animal is disposed to react with a different intensity of altruism, or perhaps to react with zero intensity. No conflict appears until the animal encounters circumstances with both sets of features: the begging gesture elicits the disposition to share; the lushness of the leaves excites the tendency to consume. The conflict may be resolved through an action expressing only one of the desires, or there can be a compromise, a minimal sharing, or the muscular tension expressing a psychological struggle.

8. In terms of the discussion of §9, the extremely limited altruism profiles displayed in such cases express membership in different—and often competing—subcoalitions.

9. Because it is so banal, this phenomenon is rarely described in studies of chimpanzees. Even a few hours of observation will provide instances.

Human behavior reveals similar phenomena. People trying to lose weight are tempted by the aromas from the kitchen. They describe themselves both as wanting and as not wanting the food, and the incompatible wishes are expressed in the active salivary glands and the hasty retreat. Although there is a philosophical temptation to tidy up such cases, to discover a single preference capturing what the person "really" wants, there are, as with the chimpanzees, examples challenging the idea of a single consistent disposition. People who struggle to master a new language or to set themselves a regular regime of exercise can, with equal justice, sometimes be seen as either weak in resolve or healthily unwilling to drive themselves. To find a "real self" free from conflict, we should have to decide which of the candidates is Jekyll and which is Hyde.[10]

If the altruistic dispositions of chimpanzees (and hominids) were limited in the three ways I have described (through breakdown of the most intense responses in extraordinary situations, through the everyday frictions of more casual friends, and through internal conflicts), their social lives would be very difficult. They are (and were). Peace and mutual tolerance are typically hard-won. Precisely because of this, observations of chimpanzee societies disclose periods of intense social interaction, lengthy bouts of grooming undertaken to reassure friends who have been disappointed by recent behavior. At times of great tension within a group, chimpanzees can spend up to six hours a day huddled together, vastly longer than any hygienic purpose demands. Even when daily life is relatively smooth, the minor difficulties and irritations stemming from the incompleteness of altruism, specifically the indifference to one another of animals who belong to different subcoalitions, require an expenditure of time and effort in mutual reassurance. Psychologically altruistic dispositions make it possible for these animals to live together, but the limitations of those dispositions subject their social lives to strain. Day after day, the social fabric is torn and has to be mended by hours of peacemaking.

10. I draw the examples considered here, as well as the helpful Jekyll-Hyde metaphor, from Thomas Schelling's valuable discussion in *Choice and Consequence* (Cambridge, MA: Harvard University Press, 1984), particularly chap. 3, "The Intimate Contest for Self-Command."

Once, that was the predicament of our ancestors, too.[11] They over-came it through acquiring a mechanism for the reinforcement and re-shaping of altruistic dispositions, and for the resolution of conflict. The evolution of that mechanism, the capacity for normative guid-ance, was an important step in the transition from hominids to human beings.

§11. Following Orders

An ability to apprehend and obey commands changed the preferences and intentions of some ancestral hominids, leading them to act in greater harmony with their fellows and thus creating a more smoothly coopera-tive society.[12] A capacity for following orders can be expressed in all sorts of actions, many of which have nothing directly to do with making up for the limitations of altruism. Self-command, a familiar human ca-pacity, can address the kinds of problems just discussed.

Those problems, *altruism failures,* are constituted by occasions on which an animal *A,* belonging to the same social group as an animal *B* toward whom *A* is in other contexts inclined to make an altruistic re-sponse, fails to respond altruistically to *B,* either forming no altruistic preference at all or acting on the basis of a selfish desire that overrides whatever altruistic wishes are present. The simplest—and original—form of normative guidance consists in an ability to transform a situation that would otherwise have been an altruism failure, by means of a com-mitment to following a rule: you obey the command to give weight to the wishes of the other. *A* and *B* belong to the same social group, and, for a range of contexts *R, A* forms preferences meeting the conditions on psy-chological altruism (the conditions of §3). Under circumstances *C,* how-ever, *A* does not respond altruistically to *B* but retains the desire present in *C*,* the solitary counterpart of *C* (or, for examples of internal conflict, it is this selfish desire that leads *A* to action). Under normative guidance,

11. If our hominid ancestors lived in societies more akin to those of contemporary bono-bos, then their situation would have been less tense than under a chimpanzee form of soci-ality. The differences, however, are matters of degree, not of kind.

12. Eventually it also modified our ancestors' emotional lives.

A obeys a command that enjoins behavioral altruism: *A* is to act in the way a psychological altruist would; that is, the desire expressed in the action is more closely aligned with *B*'s wishes than the selfish desire would have been.

Just as the discussion of psychological altruism began from a special example (the sharing of food), so here too a particular case is helpful; complications come later. Imagine two members of the same social group, *A* and *B*. They share with each other across a wide variety of circumstances. Faced with an extremely rich and attractive food item, however, *A* is not disposed to form the altruistic preference generated in other sharing situations; the intensity of *A*'s altruistic response vanishes entirely. (In terms of the averaging model, although *A* sometimes sets the value of w_{Alt} at a value greater than 0, under this particular circumstance, *C*, the value of w_{Alt} is 0.) If *A* is now capable of normative guidance, and if the normative guidance takes the very special form of *A*'s commitment to a command that orders food sharing in *C* (perhaps it is the command: "Always share equally with *B*!"), then the preference *A* forms in *C* will take *B*'s wishes into account, by setting $w_{Alt} > 0$ (if the command enjoins equal sharing, $w_{Alt} = \frac{1}{2}$).[13] If the preference formed leads to action, *A* no longer commits an altruism failure but is behaviorally altruistic. The newly formed desire satisfies conditions 1 and 2 of the account of psychological altruism (§3), but not necessarily conditions 3 and 4. *A*, following orders, need not be responding to any perception of *B*'s wants, nor need *A* be free of the taint of Machiavellianism. Normative guidance transforms the animal's psychological life so that *something that looks, from the outside, like an altruistic preference* is formed (or is operative) across a broader range of contexts.

Psychological altruism was characterized in terms of the difference made to one's own wishes by the perceived presence (and needs) of others; now normative guidance is conceived in terms of the difference made to one's action-guiding preferences by the recognition of

---

13. In the discussion of psychological altruism, where *A*'s own perspective is crucial to the formation of the altruistic preference, I saw that preference as incorporating *A*'s perception of *B*'s wants. Here I imagine the command as requiring alignment with *B*'s actual wants. There will be no discrepancy, when *A* has an accurate perception, and, for the time being, I shall assume that mistakes are not made.

commands.[14] The modified preferences, however, need not be fully psychologically altruistic—they just are different from the blatantly selfish wishes that would have prevailed in their absence. The critical idea is the replacement of a desire that fails to incorporate the perceived wants of another individual with an action-guiding desire that gives the other's preferences some weight. That can be achieved even though the desire is not generated by the perception of the wishes of another, and even though it violates the anti-Machiavelli condition. Behavioral altruism (directed by preferences modified so they are closer to the wants of the beneficiary) will sometimes do.

Normative guidance produces surrogates for psychological altruism in animals who can follow orders. The *products* of normative guidance (in its simplest and original form) are desires that issue in behavioral altruism. To understand the *process* of normative guidance, the following of orders that replaces altruism failure by behavioral altruism, it is necessary to probe psychological causes more thoroughly than has yet been done, both with respect to the lives of normatively guided individuals and with respect to psychological altruists. For it is tempting to adopt an oversimplified (and overly neat) picture of the distinction between normative guidance and the mechanisms behind full psychological altruism.

On this oversimple view, psychological altruism is generated by an *emotional* response to the beneficiary, whereas normative guidance involves the operation of a *cognitive* faculty ("reason," perhaps). Psychological altruism is "hot," normative guidance "cold." Both subtheses should be rejected. Start with the varieties of psychological altruism.

Different kinds of psychologically altruistic individuals are possible. Imagine an altruist who reacts in context *C* by modifying his or her wishes from those occurring in the solitary counterpart *C**, because of his or her perception of the wishes of *B*; the new desire may be accompanied by the presence of an emotion toward *B*, and, if present, the emotion may or may not cause the new preference. Even if we use a crude and unana-

---

14. Plainly, one can recognize commands and act in response to them in ways that have nothing to do with psychological altruism. That will concern us later. For the time being, normative guidance is tied directly to the reshaping of altruism.

lyzed concept of emotion to consider the situation, we can distinguish four cases:

a. *A*'s new desires are caused by an emotional response to *B*.
b. *A*'s new desires are not caused by, or accompanied by, any emotional response to *B*.
c. *A*'s new desires are not caused by any emotional response to *B*, but the factors that generate the new desires also produce in *A* an emotional response toward *B*.
d. *A*'s new desires are not caused by any emotional response to *B*; an emotional response to *B* accompanies those desires, but it is independent of the causal process that generates the new desires.

The oversimple view supposes that cases of type a represent the most fundamental (primitive) form of psychological altruism; cases b–d display responses that could emerge only from normative guidance.

Why should one think this? Underlying the view is an apparently plausible line of argument: the adjustment of desire could result only from the operation of an emotion or the outcome of a process of reasoning; prior to the articulation of ethical practice, the only forms of reasoning available to an agent (human or nonhuman) would have to be calculations of selfish advantage; hence, preethical adjustments of desire based on reasoning would fail the anti-Machiavelli condition; by the same token, the only ways in which obeying commands could produce altruism involve the recognition of reasons for modifying desire.

On the account of §3, all four types count as instances of psychological altruism. The argument just outlined denies that the modification of desire constitutive of *psychological* altruism could occur in cases b–d. To assess it, consider the examples that occupied us in the last chapter. Some of them fit easily into the simple view. Prominent instances of psychological altruism among primates express an emotional reaction to the plight of another animal: mothers' immediate responses to the discomfort of the young, or Little Bee's patience with her mother. It is far from evident, however, that the example of Jakie and Krom can be so easily assimilated. Further, as §6 argued, maternal concern is not always a matter of being prompted by emotion. The primate mother who stumbles across a carcass and views it as an occasion for seeking out her young appears to be undergoing more complicated psychological processes, which are not easily captured in a venerable—but crude— philosophical practice of opposing reason to the passions.

On the ecumenical view adopted in §4, emotions are complex processes typically involving both cognitive and affective states. The causal relations among these states can be quite various, and there is no reason to suppose that the cognition cannot be primary. Perhaps a cognition—recognizing that Krom wants the tire and that she has failed to remove it, seeing that this carcass is food for the young—induces a new affective state. Or perhaps that cognition leaves the prior affective condition of the perceiver unaltered—there is no upsurge of emotion at all, but simply the formation of a new desire on the basis of affective dispositions already present. Animals can have propensities for forming new desires that do not depend on their entering into a new affective state. Consequently, versions of b–d can count as psychological altruism.

Not only can cognition cause affective states, or produce new desires without modifying the affective background, but there can also be intricate chains of causation in which perceptions give rise to new beliefs, the new beliefs generate affective states, these affective states, in turn, lead to altered beliefs, the altered beliefs to novel affective states, all this entangled with the formation of desires: indeed, this may be the stuff of much of our more complex emotional life. The simple vocabulary employed in the examples a–d is inadequate to present clearly all the ways in which psychological altruism can arise (even though we do not yet know just what form a fully satisfactory conceptualization of the emotions would take). Moreover, there is no basis for denying at least some of the complex possibilities to nonhuman animals.

This brief for taking the complexities of our emotional life seriously subverts one-half of the simple view. Troubles also beset the other half, the proposal that normative guidance must be a matter of reasoning. Recent work in neuropsychology suggests that the opposition of "cold" reason to ardent passion is highly problematic and that there is evidence for the role of emotion in what have often been viewed as cool deliberations.[15] Beyond this general point, there are grounds for attributing a major directive role to emotions in some instances of normative guidance.

Consider, first, the way in which the psychology of a normatively guided individual can develop. Initially, a human being, a member of one of those small bands in which our ancestors lived, is disinclined to respond to the predicament of one of his fellows. Capable of normative guidance, he obeys a command to make a behaviorally altruistic response, and his reacting in this way generates in him an emotional response to the beneficiary, a primitive feeling of sympathy (as in case c previously). That feeling is reinforced by the beneficiary's reaction to his behavior, and, perhaps after a few further interactions, this person is able to engage in the behaviorally altruistic conduct *either* on the basis of the original process *or* through a full—psychologically altruistic—identification with the other. An emotional change may thus be the direct product of the commitment to following an imperative: as you come to endorse the command to treat your brother in a particular way, your emotions toward the brother are modified, and the new fraternal feeling gives rise to the desire to treat him in ways you would previously have avoided (or resisted). Initially, normative guidance operates to produce behavioral altruism, but it eventually issues in full psychological altruism.

How is that first step taken? Must it be on the basis of reasoning—perhaps through a Machiavellian recognition of the benefits of complying? Not necessarily. Endorsing the command can embody emotions, sometimes emotions directed toward the commander: you may accept it because you are afraid.

The point may provoke an obvious reaction. If the notion of normative guidance is liberal enough to allow for conformity grounded in fear, acquisition of the capacity for normative guidance cannot be the decisive transition to ethical practice. A dilemma seems to loom. If the ability to follow commands, to obey rules and precepts, is the decisive step in acquiring a *genuinely ethical* practice, then this special sort of ability

15. See Antonio Damasio, *Descartes' Error* (New York: Putnam, 1994), and Marc Hauser, *Moral Minds* (New York: Ecco, 2006).

requires an explanation—for it cannot be rooted in emotions of fear or prudential calculation. On the other hand, if processes in which people comply because they fear the consequences of disobedience were available to our human ancestors and initiated the practice of normative guidance, then only a *simulacrum* of ethical practice has been connected with the prior preethical state; the people in question have not yet made the transition to the *real thing*. These individuals, allegedly "subject to normative guidance," have not yet achieved the distinctively "ethical point of view." The broad conception of normative guidance allows for an evolutionary transition from hominids lacking the capacity to humans who enjoy it, but this continuity is purchased at the cost of losing contact with the proposed goal, to wit, the emergence of ethics. To make normative guidance relevant to ethics, one needs a propensity to act in accordance with commands grounded in a different (and purer) form of psychological causation.

There is no such purer form to be had. At least since the eighteenth century, philosophers who have disputed the character of ethical agents have envisaged an "ethical point of view" in which people give themselves commands—commands that are not external but somehow their own, the "moral law within"—and have regarded this point of view as requiring the subordination, if not the elimination, of emotion.[16] Others have regarded the operation of emotion as central to ethical agency. It is often assumed that the major challenge for a naturalistic approach to ethics consists in showing how the achievement of the "ethical point of view" might have evolved from more primitive capacities; inspired by this thought, naturalistically inclined thinkers frequently address the challenge by attempting a reduction of that "point of view" to the feeling of special types of emotions. Their disputes with their opponents rest on a shared mistake.

The acquisition of a capacity for normative guidance—understood, as above, as an ability to follow orders that issues in surrogates for altruism—

---

16. A prime source of this view is, of course, Kant, and the most sophisticated elaborations of it are offered in the Kantian tradition of ethical theory. Yet Kant's opponents, who often protest the denigration of the emotions, share the emphasis on a distinctively ethical point of view. I am proposing that we reject a precondition of their debates.

does not mark the transition to the "ethical point of view." That is not because there is some further move that does the trick awaited by the critics, one that shows how a very special kind of normative guidance (a special way of internalizing the orders, say) constitutes the "ethical point of view," *but because the entire conception of the "ethical point of view" is a psychological myth devised by philosophers.* There are plenty of ways in which human beings can be led to recognize and to conform to commands. While it is undeniable that some kinds of causal processes make ethical *progress* over others (in ways Chapter 6 explores), we should not infer a binary distinction between those processes that constitute genuinely ethical motivations and those that do not.

Most of the people who have ever lived have embedded their ethical practices in a body of religious doctrine, viewing the precepts to be followed as expressions of the will of gods, spirits, or ancestors (or occasionally as capturing the tendencies of impersonal forces). Fear, awe, and reverence have been parts of the emotional backdrop to most of the important decisions and deliberations these people have made, and virtually all those decisions have been subject to felt concerns about the attitudes of transcendent beings. The fact that these people have presupposed massively false beliefs about the universe does not undermine their status as ethical agents. Neither should the fact that what they want, intend, and do are partially caused by emotions of fear and awe. To insist on an "ethical point of view" liberated from such emotions is to reserve that point of view for a very small number of cool secularists. Moreover, it is reasonable to worry that the alleged ethical point of view is itself only available because of the perspectives previously adopted by those no longer counted as full ethical agents. The ability to "revere the moral law" probably depends, in the evolution of culture and in the development of individuals, on prior emotions, simpler feelings of reverence now written off as ethically primitive.

There are many different ways in which people can be led to behavioral altruism through their commitment to obeying a command. They may explicitly represent to themselves the consequences of disobeying, and find those consequences unpleasant or frightening because of future interference with their bodies, behavior, or projects. They may make no such explicit representation, but be moved by fear, or respect for the commander. They may regard the source of the command as a being greater than themselves, one whom it is important to obey. They may actively want to be in harmony with the wishes of some such being. They may regard the source of the command as part of themselves, and fear the psychic disharmony caused by disobeying it. They may have a general idea of the worth of the situations brought about by commands of a general type to which this particular imperative belongs. They may want to be the sort of person who lives in accordance with a general class of commands they have previously endorsed. They may want to live in harmony with others who expect that commands of this sort will be obeyed. They may have a general ideal for themselves that involves obeying commands current in their social group. Or they may conceive of themselves as members of a joint project, in which commands are issued and obeyed. These surely do not exhaust the possibilities, and some of the considerations can be present together, with different degrees of force.

The merits of a liberal articulation of the concept of normative guidance should now be apparent. Our decisions involve a hodgepodge of considerations and feelings, and it is foolish and unnecessary to limit the full range of psychological possibilities, taking some to be importantly free of emotion and others not, some to be constitutive of "the ethical point of view" and others not, some to accord with the anti-Machiavelli condition and others not. Emotions are complex processes typically involving both cognitive and affective states (§4), causation can run from affect to cognition or in the opposite direction, and our actions sometimes result from intricate cycles involving different types of states. The simple view, against which I have been campaigning, formulates the possibilities using language we know to be inadequate (even though we surely still lack a clear and precise vocabulary for categorizing the relevant states and processes).

Psychological altruism occurs when perception of the wishes of another modifies desires to align them more closely with the perceived wishes. Normative guidance comes about when the recognition of a command leads someone to act in accordance with it and (in the conditions studied so far, the context of the beginnings of the ethical project) to replace altruism failure with behavioral altruism. Emotions, desires, and cognitive states can be entangled in *both* cases. The causes of psychological altruism and of normative guidance are probably highly heterogeneous. There are many ways to be a psychological altruist and, equally, many ways to undergo normative guidance. None of these latter modes is especially privileged as definitive of an "ethical point of view."

No doubt there are extreme cases. Someone who forms the wish to help another, simply because he is commanded to do so and because he recognizes that disobedience will bring painful punishment on himself, is no psychological altruist and (at best) at a rudimentary stage of ethical practice. At the other extreme, a person who has a general conception of the wishes of others, who follows a rule because it is taken to promote the desires of someone else, may be viewed as at least an approximation to psychological altruism and as participating in a more advanced form of ethical practice, despite the fact that the wishes, and even the situation, of some of those she aids are unknown to her, and even though she has a standing desire to be the sort of person who contributes to the satisfaction of others' desires. Normative guidance, as explicated here, applies to individuals of both types, generating behavioral altruism in the one instance and something akin to full psychological altruism in the other.

Given the diversity of causal possibilities, why would one want to take a stand on which of them has to be realized in a genuinely ethical agent? The "ethical point of view" emerges as a challenge for naturalism because it opposes the idea of ethical agents as those sympathetic individuals who respond to the needs of others. While superficially attractive, these people suffer a defect that makes them less than fully worthy.[17] Their kindly emotions are unreliable: it is reasonable to fear that the mind of "the lover of humanity" will sometimes be "clouded," and that, under

---

17. The classic source for the reaction is Kant, *Groundwork of the Metaphysics of Morals*, Mary Gregor, trans., (Cambridge, UK: Cambridge University Press, 1998, Akademie pagination 398). This passage is often viewed as expressing an opposition to Hume, but I suspect that Kant actually had Adam Smith in mind. Not only does Smith develop the notion of sympathy much further than Hume did, but his *Theory of Moral Sentiments* (Knud Haakonssen, ed. Cambridge, UK: Cambridge University Press, 2002) [unlike Hume's *Treatise* (Oxford, UK: Oxford university Press, 1978)] is a work Kant is known to have read.

such conditions, fellow feeling will no longer operate and the person will act selfishly. Yet if we take the concern about reliability seriously, "proper" motivation appears impossible. What basis is there for supposing that carefully restraining the passions and engaging in abstract moral reasoning (of any of the sorts philosophers have commended) will prove reliable? Can't our faculties of reasoning sometimes be "clouded," too? Abstract reflection and reasoning are hardly *more* reliable than the emotional responses dismissed as capricious. Many of the most horrific deeds of the twentieth century were carried out in the name of abstract principles.

As we shall appreciate later, reliability is the issue (§21)—the worry about the "clouding of the emotions" expressed an important point. Yet the search for a single type of psychological causation, invariably reliable or at least always more reliable than its rivals, is foolishly utopian. Different ways of inducing people to modify their preferences and actions through obeying orders have different merits and deficiencies. Normative guidance would work better by taking advantage of the ways in which different psychological processes are suited to different situations. Perhaps normative guidance evolved in parallel fashion to familiar types of organic change, where initially crude systems for producing some important outcome are supplemented with further devices: the organism has a variety of ways of generating what is required and is thus buffered against catastrophe.

Normative guidance almost certainly began with crude external orders, followed out of fear; much normative guidance may have been mediated by respect for the supposed commands of transcendent beings, respect tinged with hopes and fears (§17). Out of those hopes and fears have come quite other emotional resources for motivating obedience, feelings of awe and respect, of social solidarity and of contentment in acting jointly with others, of pride in one's conduct and of responsibility to one's fellows. The history of modes of normative guidance embodies certain kinds of progress, and attempts to act through following dictates the agent sets for himself, considers, and endorses have often been progressive with respect to earlier and cruder forms of psychological causation. These differences, however, are matters of kind rather than of degree. Some processes (perhaps processes involving an especially pure form of emotion, perhaps processes that rein in emotion entirely) are valuable additions to our repertoire, but they have no special standing setting them apart from the modes of normative guidance preceding them. Their merits can be recognized without supposing them to constitute an "ethical point of view," which counts as the last word.[18]

The approach defended here allows a more systematic treatment of the behavior of subjects in economic experiments (§7). These people are recruited by researchers, know little, or even nothing, of one another's wants or needs, and are placed in situations in which they can decide what fraction of a monetary reward to share with fellow participants or how much they will give to punish those who do not act cooperatively. One thing is clear. The participants' preferences cannot be adequately represented by supposing them to be concerned with money and nothing but money: they do not belong to the fictitious species *Homo economicus*.[19] So *why* do they share, or give money to punish? Not because they are moved by the plight of people who would otherwise leave empty-handed, for they lack any basis to make judgments about the impact on these strangers. One explanation, consistent with the evidence, is that some form of normative guidance is playing a role. The participants do what they do, sharing with others, because they follow an order, one they have accepted and endorsed or one they view as current in their society.

If they were genuinely moved by a dedication to fairness, a clear-eyed vision of the value of equality in dividing goods, if this and this alone moved them to want to share (or to punish noncooperators), we might

18. See Thomas Nagel, *The Last Word* (New York: Oxford University Press, 1997), chap. 6; also *The View from Nowhere* (New York: Oxford University Press, 1986), chap. 9.

19. This is already to demonstrate something that is very important for economic research, for it entails that models imputing utility functions that are increasing functions of amounts of money, and of this alone, are unlikely to accord with the behavior of actual agents (for whom other things are important). Indeed, for the project of advancing economics, any concerns about the ways in which the subjects come to the wants they express in their actions are entirely irrelevant. What is far less clear is how these ingenious experiments bear on philosophical concerns about altruism and its role in ethical practice. For an illuminating presentation of the experimental work, see Ernst Fehr and Urs Fischbacher, "Human Altruism—Proximate Patterns and Evolutionary Origins," *Analyse & Kritik* 27 (2005): 6–47.

count their preferences as altruistic. Although they know nothing of the needs of those they reward, they have a general view that outcomes in which those people received nothing (or even received less than half) would be, from the perspective of the beneficiaries, unsatisfactory; the sense of fairness endorses the complaint, and so, without any selfish background motive, they want an outcome of equal division. This conjecture might tell the whole truth about some of the experimental subjects, but we are by no means forced to accept it. For the available evidence leaves open alternative modes of normative guidance: perhaps the participants want the "glow" (or to avoid the "pang"); perhaps they want to be the sorts of people who accord with prevalent social norms of sharing; they know their parents, spouses, friends, or children would disapprove of their greedily making off with everything they can; they may want the approval of the experimenter and not want to go down in his or her records (even if only mentally kept) as "one of those stingy people"; without any clear sense of the virtues of equity, they know this is the sort of thing of which people approve, and the sacrifice does not seem too large (they are going to leave the lab with something in their pockets). Elaborated versions of these psychological scenarios raise serious doubts about whether the anti-Machiavelli condition is satisfied. Even more obviously, the modified wants are not responses to *another person;* indeed, in some experiments, the actual beneficiary is invisible; the dialogue is between the agent and the ambient society (perhaps embodied in the experimenter).[20]

Normative guidance can generate full psychological altruism in situations that would otherwise be altruistic failures. Initially, it almost always generates *behavioral* altruism. Human motivation is sufficiently complex that, in many circumstances including those of the economic experiments, we just cannot tell (at least not without a lot of work—and maybe some luck besides) how exactly to classify people who act to benefit others.[21]

20. Subjects whose primary motivation is to impress (or to avoid disappointing) the experimenter are easily linked to the experimental subjects who were prepared to inflict pain on others.

21. This conclusion motivates the attitudes of the researchers who carry out these experiments, who suppose the important concept is that of behavioral altruism. My account

§12. Punishment

To treat normative guidance in this way has an obvious presupposition. Behind the disposition to follow orders, whether delivered externally or from internalized commands, must stand practices of punishment. Unless there were sanctions for disobedience, fear could hardly be central to the initial capacity for normative guidance. Conversely, when punishment is present in a group, it can make possible the evolution of elaborate forms of cooperative behavior (and much else besides).[22]

Can this presupposition be defended? The *actual* beginnings of the ethical project have been seen as a transition from a state of limited psychological altruism to one in which commands are followed out of fear. The plausibility of that view would be undermined unless there were an explanation of the *possibility* of punishment.[23]

Begin with chimpanzee societies in which a crude precursor of punishment is already present. Conflicts within these groups are often settled through the interventions of a dominant animal.[24] Here rank or physical strength (or both as concomitants of each other) prevail, and a dispute is settled—not always, of course, through the infliction of pain or discomfort on the animal whose initial defection gave rise to the conflict. Allies who might have intervened to protect some of those who receive the rough discipline of the dominant animal anticipate the costs to themselves and hold back.

---

of the ethical project also recognizes the important role of dispositions to psychological altruism. Different concepts are needed in different forms of inquiry and there need be no quarrel about which notion of altruism is the "right one."

22. Here I rely on a brilliant essay by Robert Boyd and Peter Richerson, "Punishment Allows the Evolution of Cooperation (or Anything Else) in Sizable Groups" (originally published in *Ethology and Sociobiology* 13 [1992]: 171–95; reprinted as Chapter 9 of Boyd and Richerson, *The Origin and Evolution of Cultures* [New York: Oxford University Press, 2005]).

23. Here, it is important to recall the methodological points of §2. A hypothesis about the actual origins of the ethical project is supported by evidence about the prior hominid state, and recognition of familiar human capacities to address its social difficulties. That hypothesis must be defended by showing that its presuppositions are compatible with the constraints acknowledged by pragmatic naturalism.

24. Goodall, *Chimpanzees of Gombe*, 321–22; and de Waal, *Peacemaking Among Primates*.

Punishment need not always take so dramatic a form and can be present simply when animals recognize opportunities for cooperation with one another. Once the basic dispositions to altruism toward nonrelatives that underlie chimpanzee-hominid society are present, optional games (§8) are available. There is a pool of potential partners who can be recruited for joint ventures. Because of tendencies to bond with close friends and allies, some kinds of defections in the ventures will be tolerated—animals will not behave with the rigor of discriminating cooperators, refusing invitations to joint activity, when the potential partners are targets of psychological altruism and longtime allies. Nevertheless, as the ties are weaker and the history of interaction more limited, it is to be expected that a strategy like discriminating cooperation will be favored. The altruistic dispositions emerging from the coalition game incline animals to give weight to benefits received by their allies, and thus to increase the value attributed to outcomes in which the ally gains and the focal individual loses; consequently, animals will be less rigorous in dismissing their close friends as potential partners for interaction; as the relationship becomes more distant, however, the deviation from the basic structure of the optional game (for example, optional PD) is much smaller, and the strategy favored will more closely approximate discriminating cooperation, refusing further interaction on the basis of a single defection.

That itself is a form of punishment. To deprive an animal of opportunities for cooperative interaction is to force it sometimes to pursue suboptimal ways of meeting its needs. So long as there are occasions for joint activity with others, allies who remain willing to enter partnerships with the animal in question, the impact need not be severe. If the allies are often unavailable, however, or if the refusal to interact spreads more broadly, life may become quite difficult. Ostracism can be a serious punishment.[25]

The practices just mentioned turn on the responses of individuals toward actions by others, actions they do not like. Those individuals can

---

25. Social confinement and exclusion are used as forms of punishment in small human societies. For a vivid depiction of the effects, see Jean Briggs, *Never in Anger* (Cambridge, MA: Harvard University Press, 1970).

effectively cause pain for the perpetrators, either through their strength (or through force that is unchallenged because of considerations of rank) or through refusal to interact (a response even the weak can usually manage). Social participation in these events is minimal: in the one instance, bystanders behave as mere spectators because of the physical power (or the rank) of the punisher; in the other, their attitudes or actions cannot completely undermine the punisher's success—they may continue to cooperate with the animal whom the punisher has blackballed, but they typically cannot compel the punisher to do so.[26] More sophisticated systems of punishment emerge, as animals form social expectations about the circumstances of punishment.

For an action to be a kind, even a crude kind, of punishment, rather than simply another contribution to the melee, it is important that bystanders not be drawn in. Thus, a first step in the direction of punishment requires that other members of the group, even allies of the threatened animal, should not intervene. There is a regularity—friends of the animal(s) targeted in punishment let it proceed. The next stage couples the mere regularity with an expectation, shared across the population, that others will not interfere in such contexts. The expectation suppresses resistance on the part of the target; the animal picked out expects others not to intervene and merely suffers what happens. A further refinement would be the existence of a regularity concerning the animals who carry out the aggression: perhaps they are animals who bear a particular relation to the context; perhaps they play a particular social role. Finally, there arises an expectation about the identities of the animals who initiate aggression. At this last stage, we have reached the systems of punishment found in contemporary human societies (and in societies for which we have historical records).

The actual evolution of punishment may have diverged from the sequence of steps just envisaged; nor is it necessary to specify a point

---

26. In principle, just as there could be escalation of violence when some animals physically punish others, so too there could be escalation of noncooperation when a discriminating cooperator crosses another individual off the list of potential partners. In the former case, obvious strength or recognition of rank stops the arms race; in the latter, the refusal of *A* to play optional games with *B* is, I suspect, often not recognized and, when it is, does not inspire *B*'s allies to forgo potentially valuable opportunities for cooperation with *A*.

at which "real" punishment is present; nor has it been explained why any hominid lineage went through these stages. Firm views on the last issue ought to be grounded in precise models of the advantages of moving from one stage to the next, and constructing such models would require far more information than we can probably hope to acquire about the causes of reproductive success in the ancestral environment(s).[27] The challenge is not to understand the *actual* evolution of punishment, but to respond to concern that no such evolution is *possible*. Decomposing punishment into conditions that can be sequentially achieved suffices to demonstrate the possibility of gradual evolution. Crucially, to buttress the account of normative guidance, the emergence of punishment does not require the prior achievement of ethical practice.

The early stages of the envisaged sequence could have originated without language: as noted, chimpanzees sometimes resolve conflict by a crude form of punishment, and the possibility of optional games gives rise to another. By contrast, the later steps would be facilitated by prior acquisition of linguistic skills. The emergence of more sophisticated forms of punishment is probably intertwined with the evolution of language—and both are probably entangled with the acquisition of normative guidance.

Suppose a type of altruism failure, keeping food items for oneself, say, regularly elicits aggressive retaliation from others. Chimpanzees and hominids could recognize the regularity, thus allowing for variants who recognize the potential threats to them if they fail to share, and whose fear generates compliance. With the advent of language, descendants of these variants can formulate the command for themselves and for others. Mothers train their young by commanding them to share, and, because of the command, the young stay out of trouble and avoid risks of injury. The repeated commands leave an echo on later occasions, and the

27. It is not hard to construct models allowing for the possibility of adaptive advantages in initiating and refining systems of punishment. Those models serve the function of protecting the hypothesis of a gradual evolution of schemes of punishment against the charge that they are idle fantasies, incompatible with Darwinian evolutionary theory. Yet, without far greater knowledge of the ancestral environments, and hence of the values of pertinent parameters, it would be unjustified to propose that any model of this sort picks out the actual course of the evolution of punishment. Modesty is appropriate here.

original disposition to share is reinforced by the memory of maternal instruction.

Through explicit command and fear of punishment, even the primitive punishment of the earliest stages, normative guidance can obtain a purchase. Animals with a capacity for recognizing and following orders have advantages over their fellows who lack that ability.[28] Once the capacity is present, it can operate to yield the socially coordinated behavior required by the more advanced forms of punishment. Animals—now surely human beings—can formulate descriptions of regularities about the consequences of alternative forms of behavior on the part of bystanders. Bystanders who intervene are seen to encounter the same sorts of trouble as the first-order offenders who perpetrate the failures of altruism that invite punishment. Group members formulate, for themselves, their kin, and their friends, orders to stand back and let the discipline proceed. When these rules become prevalent, each can recognize others as complying, yielding a social expectation that bystanders will do no more than watch. Perpetrators, aware of the expectation, see the futility of resistance, commanding for themselves a strategy of docile submission less dangerous than trying to fight back. So normative guidance, once present, can figure in transitions to more refined forms of punishment. As punishment is refined, further regularities become salient, providing scope for additional occasions of normative guidance.

Recognizing the painful consequences of particular—and tempting—courses of action, our ancestors, prompted by fear of the outcomes, ordered themselves (and their offspring) to hold back. The next step will be to consider how the grip of this capacity for self-command and self-control might be intensified.

---

28. Once again, whether the capacity will be advantageous turns on the details of the situation. If punishment carries even a small probability of serious damage, and if the order-following variant is just slightly more likely to avoid the altruism failure, then the expected gains in terms of staying intact and healthy can outweigh the loss of food that results from sharing. Once again, we cannot know whether this scenario is plausible; this is a "how possibly" explanation.

§13. Conscience

Two prominent Shakespearean figures present a view of conscience. Richard III offers a conjecture about the origins of internal checks on our conduct:

> Conscience is but a word that cowards use,
> Devised at first to keep the strong in awe.

Hamlet, while using similar words, worries about the effects of conscience on behavior, once the tendency for self-regulation is already present:

> Thus conscience doth make cowards of us all,
> And thus the native hue of resolution
> Is sicklied o'er with the pale cast of thought.

Together, the passages suggest an obvious picture: strong people with self-interested intentions are held in check by an internalized mode of normative guidance that substitutes fear for their "native resolution." That picture has sometimes moved thinkers to lament the crippling effects of internalization.[29] Whether or not they are right, pragmatic naturalism needs an explanation of how internalized commands became *possible*.[30]

The first forms of normative guidance, considered in §11, focused on the capacity to follow explicit orders. Human beings (rather than hominids, since they have acquired language) learned the local rules in childhood and later remembered the commands passed on to them. As they grew in strength, however, the memory of older commands might prove

29. Nietzsche's complaint is most evident in the first two essays of On *the Genealogy of Morality* (Cambridge, UK: Cambridge University Press, 1994); similar themes are sounded by Freud, in many later works, but especially in *Civilization and Its Discontents* (New York: Norton, 1989), as well as by William James in his writings on the "strenuousness" of the moral life (James "The Moral Philosopher and the Moral Life" in William James *Writings 1878–1899* (New York: Library of America, 595–617).

30. Once again, the methodological points of §2 are relevant here.

too weak to overlay the "native hue of resolution." They might lapse into the altruism failures from which normative guidance promised liberation.

As more sophisticated systems of punishment are elaborated, however, the ineffectiveness of remembered commands becomes costly both for those who fail to be normatively guided and for other members of their societies. Variant individuals, with a tendency to respond to modes of socialization that reinforced the disposition to self-discipline, would cooperate more thoroughly and encounter less trouble. This extension of normative guidance involves both social innovations and psychological changes in the individuals. On the social side, it requires practices of training the young members of the group so that the prospects of flouting a command become associated with emotions they find unpleasant. On the individual psychological front, it consists in refinements of the emotional lives of these individuals.

The Shakespearean suggestion that fear lies at the root of this process of internalization need not be exclusive: other emotions might be available for recruitment to the cause of normative guidance. Imagine a social group of early humans, able to issue and remember commands, but vulnerable to the flouting of those commands by individuals who think of themselves as strong. An innovation in the training regimes customary among this group, the practice of issuing orders to the young, promotes an enduring fear: perhaps they are lured into violating one of the precepts and then subjected to some extraordinarily harsh and memorable punishment; perhaps this occurs at an especially impressionable age. Thereafter, even as they grow, those trained in this way remain haunted by a sense of dread as they contemplate disobeying certain commands. Conscience does make cowards of them. Yet, similar effects can be achieved in different ways. If the young are induced to identify with some of the orders current in their group, if they see obeying those orders as partly constitutive of belonging to this distinctive social unit, they may feel more complex reactive emotions—pride, perhaps, when they continue to carry out the commands, shame or guilt when they do not. As these reactive feelings attach to outcomes considered in prospect, they may substitute for the raw fear of punishment, promoting the same types of cooperative behavior on a different basis.

We know too little about the intricacies of human emotions to elaborate this scenario in any great detail, but the outline is clear. The simplest modes of internalization trade on the ability of programs of socialization to exploit human fears. More sophisticated methods of training people can foster other emotions, perhaps emotions unavailable in different developmental environments, whose association with potential courses of action reinforces tendencies to behavioral altruism. The result is a society in which cooperation is more broadly achieved and in which costly episodes of punishment are less frequently needed. Further, even at early stages of the ethical project, different groups may have cultivated different emotions, founding their ethical practices in distinctive ways. There may be several ways to build a conscience.

However it is formed, conscience is the internalization of the capacity for following orders. The ably socialized individual does not simply hear the voice of an external commander, or remember the injunctions administered in childhood. The commanding voice seems to come from within, initially and crudely as the expression of fears, later perhaps as the representation of membership in a particular social group. In either mode, it provides a more effective anticipation of the costs of deviating from the approved regularities in conduct than the original tendency to follow and remember external orders. The conscience-ridden human being fits more easily into the social niches, provides less provocation to punishment, and encounters much less trouble.

If, to borrow another phrase from Hamlet, society plays upon the individual as on a pipe, it need not always be the same tune. Successful social inculcation of normative guidance may work through quite different emotional complexes, even though variant group techniques succeed equally in securing cooperative behavior. Although conscience begins in fear, it may later be dominated by shame or guilt, pride or hope, emotions available only in social environments where normative guidance, in some cruder form, has already taken hold.[31]

31. In accordance with my strategy of *outlining* a scenario, I offer no detailed claims about how any of these emotions is to be understood, or whether, as some anthropologists and philosophers influenced by them have suggested, there are cultures in which the emotion of shame is central and others in which the emotion of guilt is central. As noted in the

Nothing follows about the evaluation of internalized normative guidance. Modes of conscience fueled by fear (or other negative emotions) can surely distort and cripple human psychological lives,[32] but whether self-regulation from internalized fears of authorities must always be so baneful in its effects is by no means clear. The consequences from harmonious interactions with others can outweigh sacrifices in expressing selfish desires—indeed, the social involvement may be viewed as a deeper and more significant articulation of what is properly one's own set of wants and aspirations. Much depends, plainly, on the particular orders that the human with a conscience feels compelled to obey, and whether they interfere with yearnings central to a person's life. There are two dimensions to the internalized forms of normative guidance, one characterized by the emotional basis through which compliance is obtained and one depending on the content of the commands. Repressive forms of conscience can be generated along either dimension, if conscience develops in unhealthy ways. Social inculcation that couples all deliberation to fear, shame, and guilt can warp the socialized individuals; equally, massive prohibitions, however backed by emotional responses, can confine someone completely.[33] On the other hand, a person whose conscience expresses itself in a variety of ways, including sometimes through fear, guilt, and shame, can achieve, and recognize herself as achieving, a richer emotional life through the social exchanges conscientious cooperation promotes.[34]

---

text, I do not think these exhaust all the possibilities; nor do I think they exclude one another in the ways often suggested.

32. The point is eloquently expressed by Nietzsche in his critique of the "herd morality" based on *ressentiment*. How to foster forms of conscience that yield the important benefits of internalization without deforming individuals is, of course, a question the ethical project continually has to decide.

33. One way of reading Freud's *Civilization and Its Discontents* is to view him as claiming that any way of achieving the measure of social cooperation required for civilization will have to involve both prohibitions on a massive scale and pervasive negative emotions. His claims rest on very particular ideas about our fundamental desires and drives.

34. This is obviously akin to the Hobbesian perspective on the constructive role of fear that permeates *Leviathan*.

§14. Social Embedding

Members of the human groups envisaged (small societies, akin to the hominid bands preceding them) are socially embedded in two important ways. First, as just supposed, the particular way in which normative guidance is internalized depends upon the training regimes present within the group. Second, the content of the orders given depends on discussions among members of the group. The character of the discussions has varied considerably from group to group, time period to time period, with different degrees of involvement according to age, rank, and sex. Originally, however, an agreed-on code, articulated and endorsed after discussions around the campfire,[35] was transmitted to the young through training regimes that had also been socially elaborated and accepted.

Equality, even a commitment to egalitarianism, was important in the earliest phases of the ethical project. In formulating the code, the voices of all adult members of the band needed to be heard: they participated on equal terms. Moreover, no proposal for regulating conduct could be accepted unless all those in the group were satisfied with it.

Although these theses may appear implausibly strong, they rest on three sources of evidence. Anthropological studies of societies whose ways of life are closest to those of our early human ancestors show the types of equality ascribed.[36] Further, if normative guidance is to resolve the social tensions, discussions must end old conflicts, not generate new ones. Lastly, for a small band, one that must work together and unite against external threats, no adult member is dispensable. These groups are products of the coalition game, and the dynamics of that game create egalitarian pressures.

Equality survives in those contemporary groups whose societies are small and whose relations with neighboring bands are often tense. Our

35. Here my views are close to those of Allan Gibbard, *Wise Choices, Apt Feelings* (Cambridge, MA: Harvard University Press, 1991).

36. See Christoph Boehm, *Hierarchy in the Forest* (Cambridge, MA: Harvard University Press, 1999); Richard Lee, *The !Kung San* (Cambridge, UK: Cambridge University Press, 1979); Raymond Firth *We, The Tikopia* (Boston: Beacon, 1961), Marjorie Shostak, *Nisa* (Cambridge, MA: Harvard University Press, 1981).

ancestors lived like that until roughly ten to fifteen thousand years ago. Consequently, more than three-quarters of the period through which the ethical project has evolved was spent in social circumstances now quite rare. Small societies reasonably fear the interference and predations of neighbors. Social cohesion is vital, and no adult can be marginalized in normative discussion. As the coalition game (§9) already revealed, the hominid bands out of which early human societies grew resulted from the partitioning of the physical environment through coalition building. The stability of the partition depends on the approximate balance among neighboring groups, and, where the groups are small, the contribution of every member is necessary. Discussions that involve all adults, that aim to answer to the needs of all adults, and that blur distinctions of rank and ability were crucial to roughly the first forty thousand years of the ethical project.[37]

Those discussions would have issued in agreed-upon rules for life together—but not merely on that. Ethical codes are multidimensional: besides explicit rules, they involve categories for classifying conduct, stories that describe exemplary actions (both commended and frowned upon), patterns of socialization, and habitual forms of behavior. At the earliest stages, we should think of all these elements as accepted by all members of the group. Around the campfires, they reached agreement on precepts, on stories of model behavior, on ways of training the young, on practices of punishment, on sanctioned habits, perhaps occasionally on changes in the concepts hitherto employed. This form of socially embedded normative guidance set the stage for the evolution of the ethical project.

Ethical codes can pronounce on their own amendment, firmly disallowing any possibilities of change or welcoming revisionary discussion. Perhaps at early stages, there was a common insistence on clear rules, to be followed obediently and never to be modified. The difficulties of

37. My estimates here are speculative. I suppose that the ethical project began with the acquisition of full language, at the latest fifty thousand years ago, and that human societies were small until, at the earliest, fifteen thousand years ago. I conclude that the social egalitarianism observed in contemporary hunter-gatherers, and the kinds of social discussions in which they engage, was central to the ethical project for at least the first thirty-five thousand years.

earlier hominid/human social life were surely sufficiently extensive that initial proposals were incompletely successful, and the social groups that went furthest in resolving their altruism failures almost certainly did so by permitting attempts to adjust what had already been achieved. The codes thus devised and amended are *social* products: they represent a *joint* reaction to the altruism failures previously afflicting the group and they aim to diminish the frequency of similar failures in the future. They presuppose the individual capacity for normative guidance, but how the members are to be guided is a matter for all to decide. The initial function is to reduce the incidence of altruism failures, and codes are fashioned by social apprehension of the ways in which cooperation has broken down.

Does this overemphasize the social character of the ethical project? According to an alternative—"biological"—hypothesis about the origins of ethics, not only did our early human ancestors acquire a disposition to respond to orders—eventually a disposition to command from within—but also the content of the commands given, rather than being fixed through social discussion, embodied shared biases toward particular kinds of rules. Instead of a capacity for normative guidance to be steered in various directions, depending on the ways in which altruism failures are seen as arising (and probably reflecting the actual history of failures of a particular group), the rival conjecture views individuals as evolutionarily biased toward specific modes of self-command.[38]

The biological hypothesis envisions psychological changes. People acquire dispositions to behave in different ways (perhaps sharing more frequently than hitherto), and concomitant capacities to feel particular emotions or to render particular kinds of judgments (negatively directed toward those who do not share). They are furnished with a *moral sense* that redirects some of their own conduct and is expressed in reactions to the actions of others (and sometimes to their own prior behavior). *But the acquisition of this sense would not yet give rise to the ethical project.* Armed with it, members of the group act more frequently in accordance with standards we—we who are participants in the ethical project—

---

38. The type of view considered here is most clearly expressed by Marc Hauser. See his *Moral Minds* (New York: Harper Collins, 2006).

approve, but they, the original agents, do not yet have these standards or yet see a distinction between the behavior they used to exhibit and that which they now perform. From our perspective they may be more just than their predecessors, or kindlier perhaps, but this is not an assessment they can make.

For them to initiate the ethical project they must come to see certain types of behavior as exemplary or particular rules as commanding their obedience. Could they derive any such recognitional ability from their own dispositions and capacities, or from reflection on what they are moved to do? How would they come to see one desire or action-prompting emotion as different in status from others? They feel many kinds of sentiments (although the emotions available to them depend on the social environments in which they live), but how do they ascertain which ones belong to the "party of humanity"?[39] To identify something as a genuine command, they need to distinguish commands from other pressures, and the most evident possibility is to identify a *source*—a *commander.* Given their environment, the only available source consists of their fellow group members. If there were an explicit practice of discussing and formulating rules for the group, they would be able to draw the critical distinctions. Nothing else in their psychology or in the ambient environment can confer that ability on them. The ethical project can only begin, then, when normative guidance is socially embedded.[40]

Even if there are dispositions to behave in ways we think of as ethically progressive—to refrain from violence, to share more, to comfort the suffering, or whatever—these are merely "nice tendencies," ways of *conforming* to regularities (regularities the ethical project, once it gets going, will approve), but they are not abilities to *obey* rules or precepts. To be the beginnings of the ethical project they must be coupled to a capacity to discern and be governed by rules and commands that receive some sort of authority. The ethical project requires normative

39. I borrow the phrase from Hume, *Enquiry Concerning the Principles of Morals* (Indianapolis, IN: Hackett, 1986), 77. It serves as a useful reminder of the fact that those who believe in the existence of particular *moral* sentiments—or *moral* judgments—need to explain how agents are able to identify which ones these are.

40. There are affinities between the line of argument in this paragraph and Wittgenstein's famous private-language argument (*Philosophical Investigations* §§243 ff.).

guidance, and because there are no rival sources of authority to the group (or some subset of it), it demands that normative guidance be socially embedded.

The biological hypothesis needs further refinement if it is to illuminate any aspect of the ethical project. For the novel capacities it posits depend on the social environment.

Consider various forms of the hypothesis. The very strongest would suppose that human beings acquired a tendency to obey particular kinds of rules—or, more properly, to conform to particular kinds of regularities—quite independently of any social backing for those rules. So, for example, with respect to sharing behavior, it might declare that, beyond the limited primate tendencies for sharing, humans acquired a broader disposition compensating for certain kinds of altruism failures. As noted, in this story, normative guidance is not playing any important role; rather, the more extensive human capacities for sharing result from an extra mechanism for psychological altruism. Possibly, our ancestors acquired some such additional mechanism, but no such mechanism could rival the social inculcation of norms in the complex work of enlarging human cooperative tendencies. That is made plain by the prominent part ethical reminders, whether self-given or public exhortations, play in promoting human cooperation—as well as by the controlled experiments on sharing. Effectively, the strong hypothesis must maintain that the large differences between human and nonhuman forms of psychologically altruistic or behaviorally altruistic behavior come about in two distinct ways, some from a strengthened version of the tendencies to altruism already present in other primates and some from human capacities for self-command.

Weaker versions of the hypothesis suppose that evolution under natural selection has equipped people with biases that operate *through* the capacity for normative guidance. Perhaps human beings, placed in any social environment, will develop to feel specific emotions in response to particular types of behavior—positive emotions to sharing (one's own sharing or the sharing actions of others), negative emotions toward failures to share, for example. Social injunctions that direct sharing will thus be more likely to "take" than putative rules prescribing more selfish courses of conduct. At the extreme, it may be supposed that some sets of commands would be impossible for us to follow; they would be analogous to languages we cannot learn.[41]

Experiments in sharing reveal that, in the actual environments in which people grow up, where they acquire from their societies norms prescribing certain types of sharing, laboratory subjects will share with others and will punish those who do not share.[42] Cross-cultural confirmation of the results takes us a little way across the space of potential environments, but it cannot rule out the possibility that common features of contemporary socialization are playing an important causal role. To demonstrate that contrary behavior is impossible for human beings would require showing that *no* environment allows human development to follow a different path. Conclusions of that form are notoriously hard to defend rigorously, because of our massive ignorance of the potential environments.[43] Additionally, we know already that in some environments—unhealthy ones, to be sure—the norms we are supposedly predisposed to follow are violated by human behavior. The ruthlessly self-directed actions of the Ik, the struggles in concentration camps, and the willingness of subjects in psychological experiments to inflict pain on others remind us that, under the right (or, more properly, the wrong) conditions, the supposedly universal effects will not be forthcoming.[44]

---

41. Hauser (*Moral Minds*) uses the analogy, and supposes that there is an ethical counterpart to "universal grammar." For reasons given in the text, I am dubious.

42. The most systematic body of results comes from the work of Fehr and his associates; see the reference in note 19. Hauser lucidly summarizes this.

43. The problem is exactly analogous to one that bedevils many sociobiological and genetic determinist claims—the difficulty of extrapolating a norm of reaction from a small sample of cases. For diagnosis, see Kitcher, *Vaulting Ambition: Sociobiology and the Quest for Human Nature* (Cambridge, MA: The MIT Press, 1985) and "Battling the Undead" in Rama Singh, Costas Krimbas, Diane Paul, and John Beatty (eds) *Thinking About Evolution: Historical, Philosophical and Political Perspectives* (Cambridge, UK: Cambridge University Press, 2001, 396–414).

44. See Colin Turnbull, *The Mountain People* (New York: Simon and Schuster, 1972); Primo Levi, *Survival in Auschwitz* (New York: Touchstone Books, 1996); and John Sabini and Maury Silver, *The Moralities of Everyday Life* (Oxford, UK: Oxford University Press, 1982). Turnbull's ethnography is controversial, but unless all his observations are thoroughly false, there would still be grounds for wondering about the hypothesis that our predispositions make contrary norms impossible for us.

Our tendencies to behavior are most likely quite plastic. Given the hypothetical genomic change that underlies the supposedly broadened altruistic tendencies, there would probably be a range of dispositions to action across the (largely uncharted) space of social environments in which people can live. If the conclusions drawn earlier (§11) about the explanation of the behavior of subjects in experiments on sharing are correct, propensities for conduct are likely to depend on the presence of socially embedded normative guidance and the forms that guidance takes. The weaker version of the biological hypothesis is implausible so long as it insists on a specific type of emotional reaction available across all environments and very particular ways in which that emotional reaction is directed independently of the social milieu.

Far more plausible is the idea that, because of our evolved psychology, not all attempts to inculcate norms will do equally well. Perhaps we do have tendencies for emotional responses to types of actions, so that, in the environments that prevail, following one norm might be uncomfortable for us (in the way experimental subjects feel discomfort as they are following the experimenter's order to inflict "pain"), while following another might be accompanied by feelings of ease. To modify the linguistic analogy, given those social environments so far created, some languages might be more difficult to learn—and some sets of commands similarly hard to follow. Human evolutionary history may have bequeathed to us forms of blindness that make reliable compliance with some prescriptions difficult. Without a proof of impossibility, pragmatism counsels societies to work hard at training their members to follow the precepts they deem most important.

Our early human ancestors, equipped with a capacity for normative guidance, were able to explore various possibilities for social exercise of that capacity. Those explorations proceed along two dimensions, one concerned with the ways in which the young are trained in the ethical code, the other focused on the content of the code. Because we know, as yet, so little about any biases with which our evolutionary past might have equipped us, my account will attend to the more visible, social, features of ethical exploration. To proceed in this way is not to conceive of human beings as infinitely plastic, or (to switch images) as blank slates on which societies can write what they please. The history of the ethical project, from the acquisition of normative guidance to the present, is a history of experiments, carried out by social groups who sometimes may have faced difficulties precisely because they rubbed against the grain of human nature in ways of which neither they nor we are aware.

To recapitulate: hominid societies were confronted with recurrent altruism failures, a predicament limiting their size and level of coopera-tion. Through the acquisition of normative guidance and its social embedding, these failures could be addressed by elaborating ethical codes. The subsequent ethical project is a sequence of ventures in devel-oping such codes, in which—as the next chapter will explain—the domi-nant mechanism is a cultural analogue of natural selection. It is possible that a small portion of the original altruism failures were corrected by an alternative mechanism, some strengthening of the altruistic tendencies already present among primates (although where we have evidence for any such mechanism, the effects are specific to a range of contexts).[45] It is also possible that human psychological evolution equipped human be-ings with biases (as yet uncharted) that interfered with or reinforced spe-cific types of ethical codes. Neither possibility undermines the enterprise of trying to understand the main features of the cultural evolutionary process that the acquisition of normative guidance made possible for us.

45. A prime example is the case of cooperation in child care. See the references in note 1.

CHAPTER 3

Experiments of Living

§15. From There to Here

At the dawn of the ethical project, our ancestors lived in bands small enough so that all adult members could participate in discussions in which each could speak and all could be heard. Around the campfire, in the "cool hour," they sought ways of remedying the altruism failures from which their social lives had suffered. What kinds of problems did they discuss?

Scarcity of resources is a likely candidate. Perhaps times have been hard, and they have often wrangled about the few food items garnered. Suppose today has been a good day; for once each member of the band has had plenty to eat. As they gather together and reflect on their recent squabbles, all of them are able to detach themselves, at least temporarily, from the difficult circumstances, and think in general about possible outcomes when the amount available is too small to give everyone what he or she would like. They imagine possible distributions of that inadequate amount, each considering not only his or her own share but also those of the others, and attempting to recognize the felt consequences for the others. From their reflections and exchanges comes an agreed-on vision of which distributions are preferred and a rule enjoining the dividing of the spoils.[1]

Whether or not they would go as smoothly as just supposed, conversations about sharing are readily imaginable. Equally, the discussants might agree to aim at increasing the food supply, viewing each band member's wish to assuage hunger as something to be supported, or they might all concur in repudiating acts that initiate violence. Socially embedded normative guidance can begin the ethical project, but the precepts it is likely to generate appear simple and crude. How could the project of these pioneers blossom into the ethical richness of contemporary life? How did we get from there to here?

There is no serious chance of answering the second question, of defending some narrative as providing the *actual* evolution of the ethical project. The clues are too scanty. For the fifty thousand (or more) years of the ethical project, we have written records only for the last five thousand. Already, at the dawn of writing, elaborate systems of rules are in place. Evidently, much happened in the Paleolithic and the early Neolithic, leaving only indirect indicators of social change. Knowing the starting point (the small bands of discussants) and the late phases (ethical life today and the historical records of the past few millennia), one can identify *what* changes occurred, even if it would be folly to pretend to know *how* they came about.

Here are some obvious modifications. By five thousand years ago, human beings had assembled in societies vastly larger than the groups in

1. I do not suggest that the rule agreed on need be the first choice of every band member, nor that it take any specific form: perhaps it demands equal division, or equal division among those who have gathered the resources, or division by subunits with special regard to the needs of younger members. The point is both that there is pressure to agree on *something*, and that each of the discussants attempts to accommodate the views of others. This last point is one way in which the approach I favor diverges from that taken by John Rawls (*A Theory of Justice* [Cambridge, MA: Harvard University Press, 1971]). Instead of supposing that the discussants are rational egoists who consider the consequences for themselves under conditions of (partial) ignorance, I take them to be psychological altruists, able to refine that psychological altruism in contemplating a general problem that they face, who deliberate using their knowledge of one another. Further differences lie in the facts that this is no hypothetical contract, and that it is not directed at any "basic structure of society."

which the ethical project began. In those large settlements, the egalitarianism of the early phases had given way to complex hierarchies. Ethical life had become entangled with religion. It had also come to address issues beyond the conceptual horizons of the pioneers: citizens of the polis who inquire into the good life inhabit a different world from those making decisions about how to share scarce resources. New roles and institutions had emerged, generating precepts about property and about marriage. In more subtle ways, an expansion of the notion of altruism, beyond the concepts so far considered, permitted new ethical ideas about human relationships.

There is no doubt *that* these changes occurred. Acknowledging the difficulty of explaining how they *actually* happened leaves pragmatic naturalism with a problem. Skeptics charge that the account of the origins of ethical practice works only by changing the subject—something is shown to emerge, but it is not really *ethics*. Versions of the accusation surfaced in the previous chapter: How did we acquire the "ethical point of view"? How was the commanding voice internalized? How did a system of punishment evolve? The questions gain force by sowing doubt about any possibility of explanation in the terms pragmatic naturalism permits. No available route leads from there to here.

Doubt is settled by telling a story meeting all the constraints. The skeptic denies that something is possible, and an adequate response is to provide a "how possibly" explanation (§2); claiming that this is how things actually happened is not required. In the previous chapter, the challenges were turned back by denying the need for any "ethical point of view" (§11), by offering a scenario for the emergence of systems of punishment, crude and more refined (§12), and by suggesting several possible ways to build a conscience (§13). The goal of the present chapter is to offer something similar for the rich features of the ethical life apparently so far removed from the small groups of the pioneers. Not, then, "how we actually got from there to here" but "how we might have done so."

Fundamental to the "how possibly" explanations to be developed is the increased power of cultural transmission in a species that has acquired language. The first task is to consider this mechanism for change.

§16. Cultural Competition

During the first forty thousand years of the ethical project, our species consisted of a population of small bands, each elaborating a socially embedded mode of normative guidance. Those who framed the first clusters of rules responded to the altruism failures most salient for them, and perhaps there were already intergroup differences here. Or, even if the bands shared a common set of altruism failures—troubles in apportioning scarce resources or controlling violence, for example—the commands accepted in social deliberation varied from group to group. Variation set the stage for a new process: cultural competition.

Differences surely arose both in the content of the rules and in the systems of socialization and enforcement. For simplicity, consider only variation in the rules adopted. Assume the population contains bands with equivalently effective systems of socialization and punishment. One group declares: food acquired is to be equally shared among all; another: food is to be shared only among the participants in foraging efforts; yet another: food is to be divided in accordance with consensus judgments about effort. Each group has the same expected compliance with whatever rules it adopts. These bands engage in "experiments of living."[2] Cultural competition results from the fact that some experiments work more successfully than others.

What does lesser or greater success mean here? One measure recapitulates the fundamental currency of Darwinian evolution, the reproductive success of the members of different bands. So the success of an ethical code is gauged by the extent to which people living in groups adopting that code leave descendants in subsequent generations.[3] That does not imply that there will be an increasing number of subscribers to the code in subsequent generations, for greater success in leaving descendants might be offset by a propensity to desert the code. Imagine two codes, E and F. People in societies adopting E leave, on average,

2. The phrase is John Stuart Mill's: see *On Liberty* [Oxford: Oxford University Press (World's Classics)], 1998 chap. 3.

3. I shall ignore issues about how far one must look into the future to achieve a reliable measure for assessing success. Oversimplifying again, we can suppose that the relative proportions in the first generation of descendants are preserved in subsequent iterations.

three descendants for every two descendants of people living in societies adopting F; if both societies invariably transmit their code to biological descendants, and if there is biological competition in which all individuals prove equally adept, the E societies will grow at the expense of the F societies. But if the F societies invariably transmit their code to biological descendants, while one-sixth of the descendants of people in E societies migrate to an F society, the proportions will remain constant. Thus, codes may have one sort of success (in biological reproduction) without having another (success in commanding adherents).

Cultural competition concerns the latter type of success and is properly measured by the number and size of the groups in which a code is adopted.[4] Separating cultural success, expressed in terms of adherence, whether by individuals or by societies, from reproductive success (the currency of Darwinian evolution) may seem artificial. For, you may suppose, occasions on which the spread of a form of culture is not correlated with any ability to foster reproductive success are likely to be rare and transitory. The fantasy of a striking effect in terms of leaving descendants coupled to, and offset by, a tendency to desert the code is just that—a fantasy. Tendencies of this sort would be opposed by natural selection: variants with a disposition not to switch, but to remain with the biologically more successful culture, would leave more descendants. Hence we should expect a loose correlation between cultures securing many adherents and cultural practices advancing biological reproduction. In a famous slogan, "genes hold culture on a leash."[5]

---

4. Here there is further room for decision, for, if a code commands the obedience of a smaller number of large groups, should it be counted as more or less successful than one that is followed in a larger number of societies whose combined population is less? It is important to appreciate the distinct possibilities, but, for the purposes of the current discussion, no decision about which is the *real* measure of cultural success is required.

5. The slogan was advanced in classic discussions of human sociobiology—see, for example, Charles Lumsden and E. O. Wilson, *Genes, Minds, and Culture* (Cambridge, MA: Harvard University Press, 1982). The details of the argument for thinking the slogan is correct turn out to be much more complex than the simple presentation in the text suggests, and controversial assumptions are required for its derivation [cf. the review by John Maynard Smith and N. Warren, and chap. 10 of my *Vaulting Ambition (Evolution*, 6, 1982, 620–27)]. The most important rebuttal of the idea that cultural success and biological success are likely to be coupled came from articulated accounts of the coevolution of genes and

Detailed attention to mechanisms of cultural transmission and their interaction with Darwinian evolution reveals how the advantages of learning from others allow for processes of imitation, stable under natural selection and sometimes giving rise to biologically maladaptive tendencies.[6] To recognize historical possibilities in the elaboration of ethical practice, no general account is needed—we can manage without any sophisticated theory of gene-culture coevolution. It is, however, important to appreciate the lack of any tie between biological and cultural success. Codes commanding obedience need not be those that further reproductive success. That important point notwithstanding, on occasion some Darwinian consequence of a particular ethical code, for example, the fact that the children of those who subscribe to it tend to survive and flourish, plays a role in the acceptance of that code by other groups.

Cultural competition does not entail that successful codes march en bloc from group to group, in the fashion of colonial conquerors. A collection of rules can spread piecemeal, some of its constituent items being accepted, others rejected. The rule espoused by one band can influence the code accepted by another, even though the latter group does not take over the rule intact: we who apportion the spoils of the hunt according to the perceived contributions of the hunters discover that our neighbors reward all members of the tribe equally, and are inspired to amend our practice in a way that combines aspects of both the extant codes ("Divide the gains equally among those who take part in joint projects!"). New immigrants bring ideas about normative guidance to be aired in discussions, sometimes modifying extant prescriptions, even when they are not taken over wholesale.[7]

---

culture. A seminal analysis was provided by Robert Boyd and Peter Richerson in *Culture and the Evolutionary Process* (Chicago: University of Chicago Press, 1985), and developed further in several essays in their *Origin and Evolution of Cultures* (New York: Oxford University Press, 2005).

6. For a succinct explanation of this important thesis, see Boyd and Richerson, *Origin and Evolution of Cultures*, 8–11; more technical amplifications are provided in chaps. 1 and 2 of the same book.

7. Although the oversimplification I have adopted invites the thought that individual rules are the atomic units transmitted in the history of ethical practice, the possibilities of more subtle influences and modifications reveal that that is inadequate: a search for cultural "atoms" must slice codes more finely. Moreover, there are good reasons for not thinking of

For the first forty thousand years of the ethical project, small bands of human beings regulated their lives by socially embedded ethical codes. Faced with perceived difficulties the extant versions of their codes failed to address, they tried new ideas. Sometimes, they interacted with other bands, in whose practices they saw something to inspire revision of their own rules. Eventually, some groups merged, and aspects of one or both of the antecedent codes endured in the practice of the subsequent society. Some bands simply died out, or dispersed, and their ethical practices withered with them, even though survivors may have brought facets of the previous code into the groups they joined. Sometimes new arrivals, accepted perhaps as mates, brought novel ideas to the campfire discussions, producing a synthesis previously envisaged by neither of the ("parent") groups. Processes of these general types (and probably many more) combined to cause some kinds of rules to be prevalent, others rare.

The most widely shared features of contemporary ethical codes probably emerged in many different ways. If human beings have evolved psychological tendencies to acquire certain kinds of norms, a common rule might reflect these propensities (subject to the qualifications of §14). A rule might be the simplest response to a difficulty faced by all social groups. A rule (or a preliminary version) might originate in a single group and spread to others because it promises to satisfy widely shared desires. Alternatively, groups failing to acquire the rule might suffer some severe disadvantage, so that they had a tendency to die out or to be taken over by outsiders. The features of the ethical codes transmitted to us emerge from these sorts of episodes—and no doubt many more besides.

---

cultural practices generally and ethical codes in particular as collections of discrete atoms that can be shuffled and rearranged in novel combinations. Interactions among such alleged units may be crucial to the nature of the cultural practice, so that there is no stable, practice-independent contribution a cultural "atom" generates. Perhaps only in the context of an entire ethical code does a rule have specific meaning. See, for example, the writings of Dan Sperber, in particular, *Explaining Culture* (Oxford: Blackwell, 1996). I have tried to show the complexities of attempts to build serious theories of culture that mimic biological evolution—theories of the transmission of "memes"—even at the level of kinematics; see my "Infectious Ideas," Chapter 10 of Philip Kitcher *In Mendel's Mirror* (New York: Oxford University Press, 2003).

§17. The Unseen Enforcer

To scotch suspicions about pragmatic naturalism's ability to allow for the emergence of ethics ("real ethics"), it is necessary (and sufficient!) to show how processes of cultural competition could have led from simple early ventures in socially embedded normative guidance to the complex ethical practices of recent millennia. Begin with the entanglement of ethics and religion.

Ideally, discussions in the "cool hour" liberate and expand prior tendencies to psychological altruism. Realistically, however, full engagement with others may only rarely (if ever) be achieved. The deliberations may be conducted by people weary of constant squabbles and yearning for a consensus that will bring peace. They seek shared rules as a matter of convenience, hoping to discipline their fellows who lapse from cooperation—but they are quite ready to break the rules when they think they can get away with it. The discussants engage in a bargain, giving up some limitations on the actions they would like to perform (*genuinely* like to perform) for the sake of the benefits accruing from similar restraints imposed upon others.

It may be a good bargain, in that, with a practice of punishment in place, a significant class of potential altruism failures may be avoided, simply because onlookers can see what is occurring and enforce the agreed-upon rules. Yet when other members of the group are in no position to check whether you are conforming to the rules, you prefer to disobey. If there is an accusing voice from within, it does not sound with any great insistence or volume.

The early history of normative guidance was almost certainly one in which a population of human bands reaped the advantages of public rules, publicly applied in public situations, but in which many—maybe all—individual members were willing not to conform to the rules when they took themselves to be unobserved. Given thoughtful choices by those who introduced and revised the rules, obedience would typically contribute to the average reproductive success of the members of the band—consider, for example, sharing rules that generally ensure food for everyone. Groups would gain in the Darwinian struggle for existence and in cultural competition, through socially embedded normative guidance, even though conformity to the rules was confined to instances in which actors could expect others to monitor their conduct. Advantages in cultural competition might come about in either, or both, of two ways: through others' perception that members of this band could satisfy widely shared wishes, and through the assessment of them as healthier, better fed, or whatever proximate cause contributes to the extra Darwinian fitness. An ability to achieve conformity across a *broader* range of contexts would yield an *extra* edge in cultural competition, while typically also adding to the expected reproductive success of individuals.[8] Techniques for enhancing compliance promote cultural (and probably biological) success.

What techniques? As they reflect together on their ethical practices, the deliberators will recognize noncompliance as often caused by the belief that one is unobserved. They remember rule violators whose confidence that they were not seen—and thus could avoid retribution—turned out to be false. Within some groups, adult members refine programs for socializing the young. Perhaps they inculcate enduring fears of the effects of rule violations, instilling some crude form of conscience (§13) to keep people on track even when there is no obvious observer around. To the extent they are effective in doing this, later generations of the band will tend to comply at higher frequency, with positive Darwinian and cultural effects. But how exactly is this fear to be triggered? Prevalent in human cultures—in the successful surviving experiments—is an appeal to unobservable entities that respond to breaches of ethical codes. Western monotheisms use the device: there is an omniscient deity who observes all, who judges, and who punishes lapses from commandments. Variations on the theme occur in most other religious traditions. The ancestors continue to observe the actions of the descendants and to

8. The plausible assumption that rules directed at remedying altruism failures are likely to promote the average Darwinian fitness of individuals as well as to enhance the cultural competitiveness of the ethical practices realized in social groups enables me to sidestep worries that the forces of cultural and Darwinian competition might tug in different directions. A more general account would explore the ways in which rules antithetical to average Darwinian fitness might be supported because of their efficacy in cultural competition, but my purposes here can be satisfied with far less. Boyd and Richerson, *Origin and Evolution of Cultures*, provide the elements out of which the more general account can be fashioned.

retract their favors if the commands are broken. Spirits are associated with particular places or particular animals and will wreak vengeance on the group if rules are violated. There are hidden forces in nature, with which people must align themselves to be successful, and to deviate from the prescriptions is to endanger or destroy this alignment. Ethnographies testify to the popularity of the idea of unobserved enforcement (typically, but not always, personified): as when informants tell of an "all-father" who "from his residence in the sky watches the actions of men [and] 'is very angry when they do things they ought not to do, as when they eat forbidden food.' "[9]

Once the idea of an unseen enforcer is in place, fear of punishment can be embedded in a complex constellation of emotional responses. Commands promulgated by elders can be identified with the wishes of the gods or spirits (or with the tendencies of the impersonal forces affecting human success). If the gods are local, they may be seen as prescribing particular rules for the group, rules that both express the favor of the deities and constitute the identity of the band. Later phases of a group's ethical practice look back on an episode in which the ancestors obtained the favor of a particular divinity and were also given the divine command(ment)s.[10] Crude fear of punishment is transmuted into more positive emotions—awe, reverence—and the commands are welcomed as a mark of the favor of an extraordinary being. Group members see the rules as constitutive of who they are.

Religious beliefs, beliefs in some kind of "transcendent" reality, tied to the origin and reinforcement of ethical prescriptions, are almost universal across known human societies—at least until recently. Why is this? As noted (§16), there are many possible ways to succeed in cultural competition, and the would-be explainer of the prevalence of any aspect of

9. Edward Westermarck, *The Origin and Development of the Moral Ideas*, 2nd ed. (London: Macmillan, 1926), Vol. 2, 671; the entire chapter (chap. 50) is full of fascinating examples of "Gods as Guardians of Morality."

10. This conception is plainly present in the earliest versions of the Judaic tradition; moreover, as the preambles to the legal codes of the ancient Near East make very clear, very similar ideas appear in societies throughout Mesopotamia and Egypt. I conjecture that these are simply written out versions of oral traditions that thrived and developed over tens of thousands of years.

human culture does well to tread cautiously. In this instance, however, there are grounds for tentatively embracing a historical conjecture: religion permeates human history because groups that did not invent some form of the unseen enforcer were less able to reap the benefits (Darwinian and cultural) of socially embedded normative guidance; with lower levels of cooperation and social harmony, they were losers in cultural competition.

A rival possibility is that Darwinian selection has generated a propensity for conceiving and adopting ideas about transcendent entities who are both the sources of prescriptions and the supervisors of conduct. Yet any thought of a genetic variation inclining individuals to so specific a form of religious belief is utterly implausible: variations, whether point mutations or shufflings of the genome, produce, as their proximate effects, differences in the structures or relative proportions of proteins present in cells, and this kind of change, inserted into some early human environment, could not yield so particular an effect. Nor is cultural diffusion of the idea of the unseen enforcer from some ur-society in which it was first articulated at all likely. Groups lacking this idea, learning of the stories told by outsiders about how beings who especially favored them commanded them to behave according to their particular rules, would hardly be inspired to think that the structure of the account, though none of the details, was applicable to their own case.

The proposed conjecture is far simpler. In a world of apparently unpredictable phenomena and seemingly inexplicable changes, our ancestors responded by invoking unseen entities with extensive powers.[11] Some groups took a further step, attributing to these beings a connection with the social order: impersonal forces would react against those who broke the rules, ancestors or spirits would wreak vengeance on those who failed to conform to the code, deities expressed their wishes in the commands recognized by the band and were able to inspect behavior, even when agents conceived of themselves as "alone." Groups who

11. Dewey favors an anthropological account along these lines, viewing it as the ultimate source not only of religion, but also of philosophy and of science. See the opening chapters of *The Quest for Certainty* vol. 4 of *John Dewey: The Later Works* (Carbondale IL: University of Southern Illinois Press). See also Pascal Boyer, *Religion Explained* (New York: Basic Books, 2001), and Daniel Dennett, *Breaking the Spell* (New York: Penguin, 2006).

took this step gained a powerful mechanism for securing compliance and did better than rivals whose invocation of unseen powers was not connected to the ethical sphere. Religiously entangled ethics is prevalent because the very specific link between unseen powers and ethical conduct bestows significant advantages in cultural competition.

Philosophers have often been unsympathetic to the almost universal historical embedding of ethics within religion. Their arguments, articulated from Plato on, demolish the thesis that religion can provide a particular type of foundation for ethics.[12] They do not, however, touch the thought that religion may be valuable to, even essential for, ethical practice, in virtue of its power to increase compliance. Far from being an irrational idiosyncrasy, divine-command approaches to ethics may reflect a deep fact about cultural competition. Yet, for all the short-term advantages it brings, invoking an unseen enforcer amends the ethical project in potentially dangerous ways. For it threatens the equality that originally reigned in normative deliberations. Those who can convincingly claim to have special access to the will of the transcendent policeman— shamans, priests, and saints—come to have an ethical authority others lack.

Our next task is to examine the breakdown of initial equality more generally, considering divisions by status and role, and the origin of institutions that expand those divisions. How might the ethical project have introduced, tolerated, even favored these differences?

§18. Some Dots to Be Connected

For tens of thousands of years, egalitarian distribution of basic resources was crucial to the ethical project. Vulnerable small groups required the participation of all adults. They surely deployed precursors of the many clever strategies contemporary hunter-gatherers use to promote equality among their members. The !Kung, for example, take steps to ensure that differences in hunting ability are not manifest. They impose serious sanctions for boasting about a kill, cultivate a practice of joking designed to check feelings of pride and arrogance, and have a custom of crediting

---

12. The exact character of the arguments will occupy us in §27.

the kill to the owner of the arrow, which, when combined with a wide-spread practice of arrow sharing, effectively reduces differences in hunting yield.[13] Violation of these conventions is regarded as a way of courting bad luck. Under the circumstances of early human life, groups failing to develop similar strategies would forfeit the advantages normative guidance had brought.[14] The societies visible in the first written records, however, contain fine differentiation of rank and status. What might have produced them?

Archeological evidence of early cities (Jericho, Çatal Hüyük) makes it apparent that, by eight thousand years ago, human beings were able to live in groups far larger than those present at the early stages of the ethical project.[15] When a thousand or more people live within the walls of the same city, strategies of peacemaking through face-to-face reassurance are no longer applicable. There must be a system of agreed-upon rules for forestalling potential conflicts and for dealing with people who are relative strangers. Some extension of the prevailing injunctions to cover transactions with individuals outside the small group of regular associates must have been achieved substantially earlier. By fifteen thousand years ago, at the very latest, bands of human beings were periodically uniting temporarily, for the deposits at some sites testify to a larger association.[16] Moreover, there is indirect evidence for peaceful intergroup associations at earlier stages—and possibly even for the existence of trade between different bands.

13. See Richard Lee, *The !Kung San* (Cambridge, UK: Cambridge University Press, 1979).

14. For general discussion of the importance of egalitarianism, see Cristoph Boehm, *Hierarchy in the Forest* (Cambridge, MA: Harvard University Press, 1999), and for telling examples see Lee,*!Kung San*. The case for a period of egalitarianism in human prehistory, between the hierarchies of apelike hominids and those of the societies for which we have historical records, is succinctly made by B. M. Knauft, "Violence and Sociality in Human Evolution," *Current Anthropology* 32 (1991): 391–428; see, in particular, the famous U-shaped curve.

15. See James Mellaart, *Çatal Hüyük* (New York: McGraw-Hill, 1967).

16. See Clive Gamble, *The Palaeolithic Societies of Europe* (Cambridge, UK: Cambridge University Press, 1999), chap. 8; Paul Mellars, "The Upper Palaeolithic Revolution," in *The Oxford Illustrated Prehistory of Europe*, ed. Barry Cunliffe (Oxford, UK: Oxford University Press, 1994), 42–78.

The earliest hominid technologies were disposable. Two hundred thousand years ago our ancestors made tools as they needed them and left them behind when they moved on. For them, tools posed no important constraint on mobility (people did not need carrying gear), nor did tools figure as a type of property (if someone takes an ax, the maker can easily replace it). But as hominids dispersed, they frequently left the sources of their tools behind them, and, by twenty thousand years ago, bands were foraging in regions a significant distance (a hundred kilometers or more) from the nearest places in which raw materials for their tools were found (the case of tools made from obsidian is particularly striking). Those bands would have needed carrying devices (for understandable reasons, not preserved in the record), and they would also have needed to coordinate their behavior with one another and with other bands so as to make possible either a long-distance trade network or a series of journeys to gather the materials they required.[17] Either instance threatens obvious possibilities of exploitation and aggressive intervention, and the codes of the groups involved would have had to be modified to cope with these dangers. Even if they were not yet practicing trade with one another, their ethical codes would have had to contain rules that forbade harming outsiders, at least under some circumstances. Rules of this sort anticipate the possibilities that flower in the later cities, in Jericho and Çatal Hüyük, Ur, Uruk, and Babylon.

Long before people came together to build pyramids or ziggurats, our ancestors were crafting tools that depended on distant materials, bringing special substances deep into caves to paint animals, and burying their dead with special artifacts. By fifteen thousand years ago, human groups were fashioning statues and leaving them at grave sites, a practice hard to explain without supposing conceptions of transcendent beings whose welfare is a matter of practical concern. Thousands of years earlier, people took time to isolate the pigments needed for decorating the walls

17. The hypothesis of Paleolithic trading networks was originally advanced by Colin Renfrew and his colleagues, based on the discovery of obsidian tools at considerable distances from the nearest source. See C. Renfrew and S. Shennan, eds., *Ranking, Resource, and Exchange* (Cambridge, UK: Cambridge University Press, 1982). The trade-network hypothesis seems superior to the rival idea of long journeys undertaken by members of a band, given the obvious problem of explaining how such journeys might be navigated.

of caves, developing techniques of painting, and producing the extraordinary art of the French and Spanish sites.[18] These activities are unlikely
in any society still struggling to satisfy basic requirements of food and
shelter, improbable also if there is not some incipient division of labor.
By thirty thousand years before the present, the enterprise of framing
rules for life together, the ethical project, must have been quite well
developed.

The early law codes provide the clearest indications of the evolution
of ethical codes that occurred late in prehistory. Ancient Near Eastern
texts include stories embodying ideals of behavior, myths about the
afterlife, and partial codes of laws. The Gilgamesh epic, for example,
provides a picture of what is expected of high-ranking people in the
pyramidal societies of Sumer and Babylon; similarly, the protestation
of innocence in the Egyptian *Book of the Dead*[19] shows us what kinds of
actions were counted as ethical transgressions and thus illuminates the
structure of the ethical code; most obviously, the lists of rules found in
the Mesopotamian codes, from the Lipit-Ishtar code of the early second
millennium, through the code of Hammurabi (a century later) and beyond provide us with a sense of the conduct requiring explicit prohibition and of the relative importance of various social breaches.

The preambles to the law codes constantly emphasize that the lawgiver brings peace and resolution of conflicts; the law is seen as a method
of transcending a social life in which brute force prevails and the strong
oppress the weak. The surviving tablets and stelae do not offer any complete account of the laws in force. They amend a body of existing law,
offering revisions and extensions that address problems arising in the
creation of social order. These "codes" represent a multistage process
of development of social rules extending back to the dawn of writing and
beyond. Their fragmentary character is immediately obvious. Provisions
are made for very specific types of occurrence—whether a "senior" strikes
the daughter of another "senior" and causes a miscarriage, whether an ox

18. See Jean-Marie Chauvet, Eliette Brunel Deschamps, and Christian Hillaire, *Dawn
of Art* (New York: Harry N. Abrams, 1996).

19. *The Egyptian Book of the Dead*, trans. E. A. Wallis Budge (New York: Dover, 1967),
194–98; also in James B. Pritchard, ed., *Ancient Near Eastern Texts* (Princeton, NJ: Princeton University Press, 1969), 34–36.

gores a passerby, whether a woman crushes the testicle(s) of a man who is fighting her husband. The particularity points to new troubles of an increasingly complex society.

The Neolithic pastoralists and farmers of Mesopotamia had already worked out rules for restraining violence, protecting the fruits of their labors, and organizing sexual relations. As they were integrated into larger units in a world dependent on social coordination to supply adequate irrigation, new issues arose—how are measures to be standardized, how does one ensure that land is properly used, how are the public canals and dikes to be maintained? The surviving codes lavish great detail on these questions, as well as addressing the various kinds of violence and sexual relations that emerged from the social friction of large numbers of people occupying a relatively small space. They occur against the background of a general understanding of the ways in which violence is to be contained, sexual relations regulated, and property protected.

Later diffusion of rules from the Babylonian codes reveals the cultural transmission prevalent throughout prehistory (although diffusion would have gone swiftly only once the ethical project had evolved to allow peaceful interactions among bands). The Hebrew Bible takes over parts of the law we find in Sumer and Babylon: Exodus 21:28–29—concerned with control of oxen—recapitulates articles 250–51 of the code of Hammurabi, and Deuteronomy 25:11–12 reaffirms the Mesopotamian prohibition against wifely testicle crushing. Mesopotamian theocracies plainly had complex rules for religious ritual and service to the gods (or their surrogates, the ruler-priests). The code of Lipit-Ishtar already links the law to divine command, and the Egyptian *Book of the Dead* sees the prospects of the afterlife as dependent on present conduct. The idea of the unseen enforcer permeates all these texts.[20]

20. There is even a Babylonian wisdom literature, dating to before 700 B.C.E. (possibly to significantly before this period), in which the attitudes we associate with Christian ethical conceptions are articulated. It reads: "Unto your opponent do no evil / Your evildoer recompense with good; / Unto your enemy let justice [be done]." Another text, of uncertain date but possibly very early, offers the same theme: "Do not return evil to your adversary; / Requite with kindness the one who does evil to you." There is no reason to think that the authors of these texts invented the idea. They, too, like the writers who borrowed the

This fragmentary record provides a small number of dots that must be connected by any adequate account of how the early phases of the ethical project could have given rise to the ethical practices of the present and the historical past. The pieces of evidence constrain a "how possibly" account but are insufficient to yield any confidence that only one narrative will accommodate them (§2). The following sections construct a potential explanation, answerable to the demands of pragmatic naturalism, tracing the emergence of social divisions, trade, the institution of private property, and ultimately of societies in which the most privileged can speculate about the good life.

One important point needs advance consideration. Previous paragraphs have focused on the first written *legal* codes, as if these offered insight into *ethical* practice. Yet, as every beginning philosophy student learns, legal and ethical prescriptions are quite different: there are laws for which compliance is not an ethical matter, as well as ethical maxims not translated into law. To trace the possible evolution of the ethical project, is it legitimate to begin with *social* discussions of regulations for conduct and end with *legal* codes?[21]

For our purposes, boundaries should be blurred. Almost all societies, at almost all times, have socialized new members by inculcating more than *ethical* resources—at least as contemporary philosophy understands the ethical. The young are informed about what is a matter of religious duty, what is a matter of law, what is a matter of politeness and social custom—that is, what *we* see as falling under these categories. The specific conception of the ethical figuring in philosophical discussions grows out of a historical process. Later judgments distinguishing

---

theme of forgiveness from them, probably drew on previous traditions. The ethical codes of prehistory survive in these early texts, accompanied by regulations that deal with novel problems.

21. Many prominent thinkers have been willing to advance views about law quite similar to my proposals about ethics: witness H. J. S. Maine, *Ancient Law* (Tucson: University of Arizona Press, 1986); Benjamin Cardozo, *The Growth of the Law* (New Haven, CT: Yale University Press, 1924); and H. L. A. Hart, *The Concept of Law* (Oxford, UK: Oxford University Press, 1961). Like Dewey, I take the kinship very seriously (as did Cardozo). For encouragement to think about the development of law, I am indebted to Jeremy Waldron, and I am also grateful to Sam Rothschild for valuable conversations.

some obligations as ethical emerge from earlier practices blind to differences among categories of norms.

Divisions of rules into types often makes sense, for rules can conflict and it is sometimes (though not always) handy to supply a general way of deciding what has priority. People are commanded to engage in a particular ritual, but, in the middle of the ceremony, participants hear that the lives of other group members are threatened and immediate attention is required. Does the rule to protect others in the band take priority over the prescription to finish the ritual? Different groups may decide that question differently. Some, perhaps the culturally most successful ones, declare that the command to aid and protect has higher status than the rule requiring the ritual to be carried through to its conclusion. Many societies, contemporary and historical, have divided prescriptions into three (rough) categories. The most fundamental are the commands associated with transcendent beings; these can be used to elaborate, and sometimes override, rules emerging from social discussion, (something like matters of law); both categories take priority over the least important directives, those taken to govern manners and customs. Division into these categories does not settle all issues of priority, for it is possible for two divine commandments to conflict (the rules for worship clash with prescriptions to save others).

These categories, and the ways of deploying them in subordinating some rules to others, are products of the cultural elaboration of normative guidance. There is no inevitability about the outcome. Commitment to a particular hierarchy of types of norms—or, indeed, to *any* such hierarchy—is a matter for potential scrutiny. (There might be an invariant relation among types of commands—type 1 always takes precedence over type 2—or the relative status could be context dependent.) Occasions of conflict among norms provide a spur to the practice of differentiating types of norms, including the very particular practice contemporary philosophy sees as constitutive of ethics.[22]

---

22. It is worth noting that, even within traditions differentiating the ethical from matters of law, religion, and etiquette, some voices speak differently. Pioneering secularists of the past (Hume, Adam Smith) often seem to blur the distinction between morals and manners: eighteenth-century accounts of moral sentiments surprise readers by grouping wit, cheerfulness, and elegance with honesty and generosity. Even thinkers who allow for a

§19. Divisions of Labor

The ethical pioneers lived in the fashion of contemporary hunter-gatherers, in egalitarian societies where almost all adults carry out the same range of daily tasks to ensure individual and social survival, and in which the contributions of all are typically necessary (§§14, 18). Perhaps there was a modest sexual division of labor, centered on special involvement of women with young children and (perhaps) different types of foraging. Deliberations about how to share scarce resources surely acknowledged the basic desires of all members of the band, and *endorsed* those desires, in the sense of preferring everyone's desires to be satisfied provided there is enough to go round. Attitudes of endorsement create pressure to transform conditions of scarcity into a state of greater abundance. Moreover, to the extent that more resources are available to the group, the task of avoiding altruism failure becomes easier. Societies with codes fostering cooperation are more likely to engage in joint projects that garner valuable resources, and the increase in resources yields an enlarged class of occasions on which a socially endorsed outcome is readily seen and relatively attractive, thus promoting more cooperation. As groups develop new strategies for cooperative projects that increase their joint resources, they enter a feed-forward cycle.

One obvious strategy for obtaining more of the things everybody needs is a form of the division of labor. If one of us is better at finding roots and another makes superior arrows, we are likely to acquire more food items, or to acquire the same amount in a shorter time, if the first person concentrates on root finding and the second on arrow making. This is not yet a matter of decomposing tasks into subroutines and

---

separation between religious commands and the requirements of ethics do not always assign priority to the ethical. Kierkegaard is famous (notorious?) for maintaining that the greatness of Abraham, as "the knight of faith," consists in his "suspension of the ethical." To suppose he can be refuted by declaring that it is *constitutive* of ethical maxims to take priority over religious injunctions is no more convincing than specifying that your favorite rule (or strategy) of nondeductive inference must be adopted because it is constitutive of rationality. Better to try to understand why one, rather than the other, might be incorporated in our practices.

assigning people to repeat particular actions.[23] Rather, the spectrum of jobs is partitioned to take advantage of the distribution of skills. Implementing the strategy depends on each individual's doing his or her part, and that requires normative guidance to constrain potential shirkers. Under the aegis of normative guidance, however, the strategy may be to everyone's benefit: the group finds more roots, and its sharper arrows bring down more (or larger) prey; through equal division of the spoils, everyone gains a bigger share.

Sometimes the environment in which the band finds itself is benign, and refined productive strategies are unnecessary. For many, perhaps all, of the Paleolithic bands, times were surely sometimes hard and division of labor correspondingly important. Recognizing the possibility of hard times, some groups could have instituted a practice of storing reserves for future conditions in which even the most efficient distribution of tasks would bring in less than they needed. In a fluctuating environment, division of labor, accompanied by a practice of storing a surplus when life goes relatively easily, promotes both the Darwinian fitness of members of the band and also contributes to the satisfaction of basic desires across a broader period of time.

So far, the bands remain egalitarian: the requisites of life are found or made, and divided among the members according to their needs. The ethical code expands to regulate collective activity, requiring individuals to carry out the tasks assigned to them and to labor to acquire more than is needed to meet current demands. But groups committed to the production of surplus resources prepare the way for a second division of labor. Suppose the demands of a variable environment are sufficiently rigorous that those who do not save do not survive. The result will be a population of small bands all practicing surplus production. Relations among these bands may be wary and suspicious, even aggressively hostile. Nevertheless, if neighboring bands are well matched in size and strength, they will see that little is to be gained by attempts to encroach on others' territory or to take advantage of others' resources. It may also become clear that adjacent groups have analogs of the distribution of

23. As envisaged by Adam Smith, *Wealth of Nations* (New York: Modern Library, 2000), book 1, chap. 1.

skills that made intragroup division of labor profitable. Because of features of the local environment, or because of specially developed techniques, one group may have food, or carrying gear, or tools of kinds or quality the other does not possess. Without endangering their respective abilities to survive periods of environmental challenge, each can recognize a gain in variety of resources—and in overall value of the stock of resources—from giving up some part of its surplus to acquire surplus items from its neighbor. Trade is born.[24]

Once trade begins, there is an impetus to exploit further the initial division of labor. If our group is to exchange with the band across the river, and if we are to retain enough to ensure our survival through possible hard times ahead, we shall have to have more than more than enough. Moreover, the stable pursuit of trade will require a new form of cooperation, peaceful interaction among individuals from different bands whose ancestors viewed one another with suspicion or even hostility. There are new demands on the versions of normative guidance practiced in the neighborhood, additions to the ethical code that regulate behavior with respect to people not previously considered within the framework of commands. This is an important step. With the addition of norms governing interactions with members of other bands, the set of people covered by normative guidance is extended—the circle expands.

This scenario leads from the original small societies, with their rules for remedying the failures of altruism within the group, to later communities, still small, each of which has a stock of collective resources and each of which engages in limited interactions with neighbors. These communities have to extend the division of labor in ways that appear small but are socially and culturally consequential. First, the need to preserve their stock for hard times and to use part of it in barter with neighbors is likely to bring new forms of work. More important, the performance of some of the tasks now carried out requires particular tools or equipment, and if some members of the band spend large periods of time on

24. In telling this story, I diverge from Smith, who appeals to an innate propensity to "truck and barter." Unlike Smith, I have also not assumed that the effect of the division of labor must be greater productivity—some groups may settle for less work and more leisure. See Smith, *Wealth of Nations*, book 1, chap. 2, esp. 16.

the pertinent tasks, while others never do, it will be important to ensure that the users' access to the needed implements is not impeded by the activities of nonusers. These developments introduce an embryonic conception of property.

Even though the basic resources of the group (food, materials for shelter, and so on) are divided, the group as a whole owns the surplus. This is to be used in hard times and to be available for exchange, and it will be important to defend it against those who would take it. The rules allowing for trade must specify that items brought by one trading partner are not simply to be seized by others. Similarly, if the equipment required for performing a particular task is to be available to one group member and not to others, there must be rules allowing the user to keep it and forbidding nonusers to interfere. Those rules will not yet permit the user to transfer the items in question to anyone else he or she chooses—rather, they insist that tools be passed on to the next performer of the task—but they will protect a temporary power to employ the equipment. Communal property and a limited form of individual property have emerged.

§20. Roles, Rules, and Institutions

Divisions of labor assign different tasks to different members of the community and thereby create roles. The band relies on one individual in tracking game, on another in negotiating with neighbors, on yet another in finding or constructing shelter. For these roles to be filled efficiently, their occupants have to be well selected and to behave in ways promoting the ends toward which the role is directed. Groups will make better use of divided labor to the extent they are able to identify reliably the physical and psychological capacities needed for a particular role, and to articulate and enforce rules applying to those who occupy the role.

Even though resources acquired through the division of labor continue to be split equally among group members, the search for an efficient way to apportion roles already brings the beginning of social inequality. Members of the band must attend to individual differences, to the "talents" that some have and others lack, that fit some for particular roles. The simplest ethical prescriptions applying to those roles embody the band's acquired knowledge of how the tasks are best performed, enjoining care, for example, at stages known to be especially crucial. Yet many kinds of performance improve with practice and training, and with respect to these the assignment of roles will go better insofar as the assigners can spot incipient talent and subject its development to rules. Within the socialization of youth, there can arise an appreciation of difference that is valuable to the community, provided rules for self-development are in place.

Becoming good at tracking game or making carrying devices requires the novice to obey the instructions of those who know how and to be diligent in carrying out the exercises prescribed. Some rules governing training will be specific to the task. Others will apply across a wide range of roles. The young are to be obedient and not willful, attentive and not distractible, industrious and not lazy. More general still is a form of prescription combining the idea of differences in propensities for necessary forms of work with the general characteristics required for proficiency in any role: "Develop your talents!" Coordinated group projects thus exert pressures on individual performance, prompting the appreciation of important virtues—industry, courage, prudence, temperance.

Although normative guidance began as a remedy for altruism failures, the ethical codes found in the historical record, from ancient times to the present, contain directives to act in ways without evident impact on the lives of others: people are supposed to be prudent and resolute, even when their imprudence or irresolution would affect only themselves. Where do self-regarding principles come from? A possible answer: divisions of labor introduce the conception of differences in talent, and such differences are potentially valuable in promoting group welfare; the benefit requires talents to be properly developed; once that is understood, ethical codes elaborate to enjoin the development of promise (with derivative rules forbidding laziness). Lurking in the background is still a connection to the original goals, for neglecting one's self-development can be viewed as a kind of altruism failure. But a more personal basis can ensue.

Attention to differences in the propensities of members of a social group probably began long before the articulation of normative guidance. Strategies for playing optional games require recognition of the characteristics of potential partners (§8): at a bare minimum, you have to discriminate those who have been unreliable in the past, and it may also pay to choose individuals with particular physical traits.[25] With developed division of labor, finer scrutiny became pertinent. Scrutiny starts to perturb the egalitarian attitudes of the group—for some of the tasks assigned can be carried out successfully by any of a relatively large subgroup (they require no special talent or training), while others may be difficult to fill well. Roles are more or less demanding, according as they have more or fewer potential occupants.

Suppose a particular role—tracking, say—is both demanding in this sense and requires extensive training and effort in performance. Good potential trackers are rare, their skills need to be honed over a relatively long time, and their tracking activities require searching attentively through a broad area. The group inculcates stringent rules for learning and performing this role. The solitary nature of the task, however, makes enforcement difficult, and the bare idea of unseen enforcement may not prove enough. A new idea is added: significant contribution of successful tracking to the tribal stock is particularly favored by the entities who are the source of the ethical code. Especially pleasing to the gods are those with rare talents, who develop those talents to the full and use them energetically in service of the common good. In societies elaborating the division of labor in this fashion, new human desires readily emerge: people come to want the approbation, even admiration of their fellows; they wish to enjoy the favor of the gods.

So elaborated, the code begins to advance a new conception of the good human life. Its earlier forms identified the common good in terms of basic desires, viewing human lives as going well when those basic desires were met. Early stages of the ethical project introduced rules whose intended effect was to improve the prospect of satisfying more of the basic desires and thus living better. Introducing the unseen enforcer connected the rules with the wishes of a great being with special concern for

25. Consider a joint hunting venture. If two equally reliable partners are available, and one is quicker than the other, that person may be a better bet for bringing home the game. That can easily be offset by other considerations: if the slower individual is a longtime associate, and failure to interact on this occasion would prompt the person then to refuse future invitations, the physical superiority of the rival candidate is better ignored.

the group.[26] Now the development and exercise of rare talent—in service of meeting the basic desires of members of the band—is seen as favored by this being. Well-socialized group members want this sort of approval. For the specially gifted, at least, to live well involves gaining the being's favor.

Once the broader idea of a good human life has been introduced as an incentive for the rarely talented, it can be extended to others. Although it is less important to encourage those who fill less demanding roles, general diligence benefits the band. The myth that divine approval descends on those who fit themselves to their station and discharge its duties with energy is a valuable extension of the idea of unseen enforcement. It detaches the rules enjoining development of talents from their derivative status, as consequences of more general principles about contributing to cooperative projects, and locates them in the direct command of the unseen lawgiver. Perfecting one's talents may contribute to the success of the group, but it is required of each member because it is the divine will.[27]

Cultural competition can favor an evolutionary transition from an initial stage at which ethical codes are directed only at altruism failures to more internally complex societies, with divisions of labor, prescriptions for interactions with members outside the group, specific roles, rules for carrying out those roles, and injunctions for behavior even when the impact on others is not of central concern. At this stage in the ethical project, continued discussion of the prevalent ethical code will sometimes need to consider the *institutions* of the group: prescribed patterns of behavior focused on some domain of the band's life. My "how possibly" story concludes with the emergence and evolution of an institution presupposed by well-known ethical maxims: property.[28]

26. To keep things simple, I offer a version in which unseen enforcement is personified.

27. Once the injunction to develop talent has been detached from the consequences for society, it can be maintained as a freestanding, self-regarding maxim, even when the idea of a divine backing for it is abandoned. The attitude of the citizens of Plato's Republic is, presumably, one of seeing lives as good in terms of the perfection of talent—for this is what the organization of the city aims to do—even though they have read the *Euthyphro*.

28. Both the seventh commandment and the tenth presuppose the institution of private property.

Trade already introduces a notion of communal property, for the partners each have to subscribe to rules forbidding violent appropriation of the other's resources. Division of labor gives rise to a weak conception of private property within the group, in that those assigned to certain tasks are awarded privileged access to whatever equipment those tasks require. The privilege depends on proper use by those who occupy the relevant roles: the maker of carrying equipment or the digger for roots is not free to abandon, or blunt, or misuse the special hand ax or the spade—or even to let the tools sit idly by; the items assigned are to be employed in the pertinent tasks, and, after the labor is over, to be preserved in a socially expected modification of their original form.[29] Nor can the user transfer the equipment to anyone he or she chooses; once the user's career in this role has ended, the new occupants of the role will acquire the privileges he or she now enjoys.

How might a stronger conception of private property, one that allows owners to use and dispose of their belongings as they choose, have emerged? One obvious thought: productive performers of particular tasks, particularly tasks playing a large role in the group's life, might be rewarded by giving them power to dispose of resources previously owned by all; their production might be encouraged by such rewards.[30] So long as one considers contexts of privileged access to tools, any transition of this sort looks mysterious, for it is hard to conceive how a power to dispose of equipment previously owned by all might be any significant reward or motivation to diligent exertion. If we examine stages close to the historical present, the mysteries dissolve.

By ten thousand years before the present (five thousand years before the invention of writing), our ancestors had learned how to domesticate plants and animals. The plants and animals owned by a group

29. Often, tools will be supposed to show a little, but only a little wear; there may be occasions, however, when extensive performance involves considerable damage to a tool. The point is that the extent to which this is to occur is not for the user to decide.

30. This is a standard conception in the history of political economy (developed by Smith and by many others), and the thought that some kind of private property is a necessary incentive for hard work occurs even in a thinker as worried about the notion as John Stuart Mill (see *Principles of Political Economy, Works* [Toronto: University of Toronto Press, 1963], 2:207, 2:225–26, 3:742–55).

constitute resources temporarily in the charge of occupants of specific roles—tenders of gardens, herders. Unlike the equipment assigned in a privileged fashion to particular workers, these resources are attractive potential rewards. Imagine a social innovation. Within a pastoral society, the deliberators resolve that shepherds may hold back a small fraction of the healthy lambs born, disposing of them as they see fit. Assuming the members of the group would like to have this new power,[31] the shepherds would be motivated to work in ways conducive to the reproductive success of the flock, expending extra effort on protection against predators, discovering good pasturage, and nurturing young lambs.[32] The innovation could increase the success of the group, as measured by the growth of the communal and the individual stock; it could generate ever-more successful trade with other bands—as well as producing pronounced inequalities within the pastoral society. If asymmetries in command of resources translate into differences in power, greater weight given to some voices in deliberations about rules, the apportioning of resources between public and private ownership could be increasingly tipped in the direction of the latter, until the communal stock was insignificant in comparison with private ownership. In tandem with this movement, the idea of private control of resources could easily be extended into other spheres—most notably when domesticated animals are exchanged with the products made by members of other groups.

A word of caution is appropriate here. The thought of private ownership as a motivator to group productivity has been so overused in economic discussions that it is essential to be specific. The "how possibly" story does not view private property as an essential outgrowth of human society, supposing that once we had engaged in normative guidance, division of labor, trade, and the domestication of animals it was inevitable. Instead, under particular conditions, against a scheme of socialization

31. This is a nontrivial supposition, for one can imagine groups in which practices of socialization rendered it quite repugnant.

32. My account echoes a famous *Genesis* story, the deal Jacob strikes with his father-in-law, Laban. Similarly, the earlier discussion of saving surplus in good years, in preparation for hard times, recalls the policies Joseph institutes in Egypt. Might we consider the myths of ancient religions as embodying records of transitions in human prehistory retrospectively seen as important forms of social and ethical advance?

producing certain kinds of desires—desires for the control of resources previously communally owned—an institution of private property could succeed in cultural competition.[33]

§21. Altruism Expanded

Previous sections have attempted to show how stepwise evolution of the ethical project might transform the social environment and thereby make new kinds of desires, aspirations, and emotions possible. The final task is to show how these transitions can expand the scope and the character of altruism. Originally, normative guidance was seen as generating *behavioral* altruism in response to altruism failures. As the ethical project evolves, it can generate *psychological* altruism, even in more elaborate forms than hitherto considered.

Altruism failures can be remedied by harnessing a number of emotions: fear, dread of the unseen enforcer, awe and reverence, a positive desire to be in harmony with the deity's plans and wishes, even a sense of identity with the society blessed with divine favor. The same ends can be achieved by inspiring people not simply to simulate altruism but to have altruistic propensities across a wider set of contexts. Well-socialized people then act to help others through a mixture of motives—through taking others' wishes seriously, through sympathetic emotions, through respect for the supposed source of the ethical code, through a sense of identity with a group, through worries about the results of breaking the rules. No special sort of psychological process is likely to be better at producing appropriate behavior across all circumstances; the mind of "the friend of humanity" may cloud over, but, equally, his or her reason may go astray (§11). Reliability is an entirely appropriate measure, for, from the perspective of achieving cultural success, the goal is to arrive at strategies of socialization for eliciting preferred behavior on as many occasions as possible. Pluralism has evident advantages. The group that supplies a variety of psychological dispositions for altruistic response obtains greater relief from altruism failures.

---

33. Much later (in §62) we shall take up this institution with more critical eyes.

Cultural success exerts pressure to develop schemes of socialization extending the scope of psychological altruism. That can result from effective techniques of promoting behavioral altruism. Change behavior patterns so people engage in a larger number of friction-free interactions with one another, and extended psychological altruism may follow. Participating in cooperative ventures with *B* inclines *A* to think of *B*'s wishes as good to satisfy and engenders feelings of warmth toward *B*. Skillful socialization reinforces the effects. Parents acquire an arsenal of techniques to induce rivalrous siblings to get along—and their methods are not all recent inventions.

How far successful projects in socially embedded normative guidance extend genuine psychological altruism, rather than replacing some previous altruism failures with behavioral altruism, is unclear—as unclear as the categorization of contemporary people, in many everyday contexts of helping and in the special circumstances of experiments in sharing, into real psychological altruists and others (§11). Under normative guidance, psychological altruism is also extended in other ways, and the rest of this section is concerned with modifications of the notion that introduce further complications to the account previously offered (§§3–5).

I begin from an obvious point, one that recognizes familiar types of altruism not involving any positive response to the *actual* desires of the beneficiary (or to the desires that would *actually* be attributed to the beneficiary). The altruistic mother does not align her wants with the wishes of the young child who vigorously resists the medicine. Yet the mother is surely responding to *some* sort of attributed wish: it is as though she envisages the future life of her child, recognizing wishes that would arise later, given various sorts of response now. In parallel fashion to the account of §3, we can approach altruism in terms of responses to the perceived *interests* of others.

How should the distinction between interests and mere wishes be drawn? Many thinkers are tempted to identify interests with the wants those others would have if they were clearly (and coolly) to deliberate on the basis of all the facts, but this approach threatens to collapse into triviality. (One of the facts is about what we would want if we knew everything.) Yet there is an insight here: we separate the wish someone expresses from his or her genuine interest by attributing to the person some type of ignorance. A current wish diverges from a real interest when the wish would give way to the preference marked by the interest, were the person relieved of some current misconception or form of ignorance, and when the modified preference would be retained in light of further knowledge.[34]

In responding to young children, accommodating wishes rather than interests is often a defective form of altruism—perhaps not even worthy of the name. Does this require the account of altruism to be rewritten to focus on attributed interests rather than attributed wishes, so paternalism would be preferred across the board? Reflecting on our ordinary notion of psychological altruism, *framed as it is by the ethical project*, you might say this: to be an altruist is to identify with the other person, and that is to take the person seriously as an agent (at least once he or she is mature); hence, even if you think the person's wishes misguided, as unlikely to promote what he or she would want were he or she to know more, those actual wishes are to be respected. Or you might say something different: to be an altruist is to care about the other's good, and that is not what the person actually—and myopically—wants, but rather what he or she would want were he or she better situated to judge; so one should align one's desires with the person's interests, the person's wants as they would be if he or she were aware of crucial facts.

Once the ethical project has introduced the ideas of "identifying with others" and of "the other's good," both thoughts are available: there are paternalistic and nonpaternalistic forms of altruism. Ethical considerations now figure in decisions about psychological altruism. On some occasions, it would be arrogant to substitute one's own judgment about what the intended beneficiary would want, given the benefit of an idealized

---

34. A fully rigorous account would need further qualifications, since it is possible for someone to acquire misleading information that subverts the modified wish. It would do so, of course, by coming with another type of misconception or ignorance, from which yet further knowledge could relieve the person. Perhaps the best way to approach the notion of interests is to start with the idea of a *remedy for ignorance*, conceived as the clearing up of misconceptions or a new piece of knowledge. An interest is a wish one would have, given a remedy for current ignorance, and a wish that would survive any further acquisition of knowledge provided the acquisition was supplemented with an appropriate remedy for ignorance. Whether this disposes of all the difficulties is not obvious. In any event, for our purposes, the simpler approach of the text will do.

perspective. If *A* has evidence that would support the judgment that *B* has thought hard about his or her valuations of outcomes, if *A*'s own reflections on those outcomes are hasty and uncritical, *A* is quite wrong to override *B*'s expressed wants, even though, by chance, *A*'s particular judgment on this occasion would be closer to what *B*, given more information and cool reflection, would actually desire. By the same token, if *A* has excellent evidence that *B* is missing a crucial item of information, if there is no opportunity to present the salient facts to *B*—and thus induce a change in *B*'s desires with which *A*'s own valuations could then be aligned—then responding to *B*'s actual wants would seem to rest either on indifference to *B*'s welfare or on disrespect for *B*'s powers of rational revision.

There is a *preethical* notion of psychological altruism, out of which the ethical project grows, but also *ethically charged* notions of psychological altruism emerging later. The latter can revise previous judgments about "altruistic" responses. Before the introduction of the first agreed-on rules, an agent may have been inclined to respond positively to the desire of a fellow in a particular context: *B**, a member of the band who does not often associate with *A* and *B*, has found a coveted resource and offers to share with *B* on an equal basis; *A* perceives *B* as wanting more than an equal share and acts to help *B* grab the whole.[35] Once a socially embedded system of normative guidance has included the command that those who volunteer to share what they have found should not be interfered with (in the way *B* intends), the status of *A*'s intervention is changed. *A* no longer endorses *B*'s desire: *A* has agreed to a rule that distinguishes *B*'s wish from the desires *A* wishes to be satisfied. The first modification of the concept of psychological altruism consists in moving from a notion that treats *any* desire of the beneficiary as an occasion for positive response, to one that restricts psychological altruism to those desires that accord with the ethical code. Initially, because the original rules are

---

35. Note that the imagined scenario does not involve any direct benefit for *A*; for *A* there are only costs (use of time and energy, risk of harm from the resentful *B**). In terms of the approach of §9, we should take *A* and *B* to belong to a subcoalition that does not include *B**; *B** is a more distant member of the band, in that the first subcoalition that includes *A* and *B** is bigger and more inclusive than the first subcoalition that includes *A* and *B* (and similarly for *B* and *B**).

to remedy altruism failures, desires ruled out by this restriction will be those embodying a failure of altruistic response: you do not count as an altruist if you respond positively to the desire of someone whose wish represents an altruism failure.[36]

Once ethical practice includes self-regarding commands (§20), some desires are viewed as defective—and thus not to be promoted by others—because they prevent a person from developing in a particular way. So arises the vision of mature, well-socialized people, the people who engage in the deliberations about the commands of the code, whose "real" desires are to be identified by the things they would express in such deliberations. The deviant wishes expressed on different occasions can even be conceived as altruism failures, where the agent and the potential beneficiary are one and the same person. To want to behave in this way now is to fail to respond to the wishes of the person you wanted to become. Paternalism enters the picture, as altruists respond not to the actual wishes of those buffeted by fluctuating forces but to the wishes endorsed by people in "the cool hour," wishes in accordance with the ethical code.

Part of altruism consists in advancing the (endorsed) altruistic wishes of others: A can be an altruist in virtue of wanting to advance B's altruistic wish to help B^*. One of those toward whom B has altruistic desires can, of course, be A, and this forms the basis for a distinctive sort of altruism, *higher-order altruism*, as I shall call it (although, if the term had not already been preempted, "reciprocal altruism" might be better).

Sometimes it is altruistic to allow others to express their altruism toward you, even though your own solitary wishes are thereby satisfied. You have a long-term history of psychologically altruistic interaction with someone else. Often you wish to do something together, although each of you has different ideas about what this should be. Were both of

36. Here I am indebted to Jennifer Whiting, whose perceptive comments on an earlier discussion of psychological altruism brought home to me the importance of this kind of restriction. As Whiting noted, the "infection" of what superficially look like altruistic responses can proceed along a chain of indefinite length (A promotes B_1's wish to advance B_2's wish to . . . to promote B_n's wish to do something that would be counteraltruistic toward B_{n+1}), making the provision of sharp conditions on psychological altruism extremely messy.

you to act selfishly, you would perform the individually preferred actions, but you would forfeit what is primary for each of you—to wit, being together. Were both of you to act as psychological altruists, as so far construed, the situation would be even worse: each would do the *less-*preferred action and still not have the benefit of acting together. To escape the bind, one of you has to be a different type of psychological altruist, an altruist who adjusts wishes to align them with the other's altruistic desire.[37]

Although it is evident that the concept of higher-order altruism can be abused, providing cover for people to pursue their selfish wants, the anti-Machiavellian condition (§3) discriminates cases. Egoists simply have their solitary wants, or see the simulation of an altruistic response to others as a good strategy for achieving those wants ("Of course, if you really want to help me by doing that, I don't want to stand in your way"). Psychological altruists reflect on their partners' wishes, factor in their own desires to promote those wishes, and, if they accept the altruism of a partner, do so because they view it as based on a wish more central to the partner's life than any they would express by promoting the partner's nonaltruistic desires.[38]

Often-repeated interactions among people, in which altruistic responses are expressed by both parties, bring with them the possibility of an importantly different form of higher-order altruism, one in which the *processes* through which outcomes are reached become sources of happiness for the participants. The original solitary value ascribed to an outcome is sometimes negligible in comparison to the value that whatever outcome is reached results from a serious process of mutual engagement with the wishes of another person. Adjusting our actions to one another can be more important for us than what those actions actually achieve.

37. For a more developed account, see Kitcher, "Varieties of Altruism."

38. I strongly doubt that the considerations in play are ever very precise—one simply has a feeling that expressing altruism in a particular way really matters to a friend, a spouse, a parent, or a child. Because those judgments are not exact, considerations of turn taking often play a role. You allow your friend to be generous because you were able to express your friendly feelings on the last similar occasion. This is one place in which reciprocation does play a role in genuine altruism—simply because what is reciprocated is a genuinely altruistic response.

Our experiments of living began when a primitive system of rules was used to make up for salient altruism failures. The desires targeted by those first efforts were quite basic. The elaboration of normative guidance generates new desires, eventually desires for interactions revealing the mutual expression of altruistic responses. Successful extension of altruism can produce series of occasions on which people promote one another's wishes, taking pleasure both in the pleasure given and received and in social approval of that pleasure. Institutions like marriage (and perhaps other forms of partnership) give rise to such series, and those who participate come to view efforts at accommodation, even when not completely successful, as valuable. In turn, the recognition of value in mutual response can reshape the institution of marriage or the most important kinds of friendship. From individuals for whom normative guidance is a way of ameliorating social trouble, a sequence of experiments of living can produce people for whom mutual recognition in an enduring relationship is central among their desires.

The expansion of human desires was surely coupled to the refinement of our emotional lives. Through the evolution of the ethical project, even if our affective responses remained unaltered, new types of cognitions and desires became attached to the affective states. Positive affective responses might be triggered, even amplified, by the recognition of the ways of attuning our desires to those of others and theirs to ours. We can dimly apprehend the origins of love.

All this changed our ancestors' conceptions of what it is to live well. To be secure, to be healthy, to eat, and to copulate is no longer enough. My "how possibly" story ends with a vastly enriched notion of the good life. Desires to develop one's talents become central, the active contribution to our community is important to us, and particular relationships are more significant than anything else. By gradual steps, the ethical project could evolve, from the simple beginnings of socially embedded normative guidance to the ethical sensibilities we discern in ancient Greece. Plato is a footnote to the history of ethical practice.

CHAPTER 4

One Thing after Another?

§22. Mere Change?

Pragmatic naturalism aims to understand the character of the ethical project by exposing major features of its evolution. Probing the deeper past is difficult, for clues are fragmentary. The invention of writing, however, enhances the opportunities to investigate the evolution of ethics: the records of the past five thousand years might reveal how contemporary societies have come to their present practices. More specifically, historical investigation promises to address challenging questions, issues of immediate concern.

Is the evolution of ethics a matter of *mere change*? Is it analogous to a Darwinian picture of the history of life, revealing only local adaptations without any overall upward trend? Do ethical codes diffuse and metamorphose through processes having no connection with truth or knowledge or progress? Is it just one damn thing after another?

These worries express in a temporal context concerns much bruited with respect to cultural variation. As anthropologists documented the diversity of cultural practices (often framed only with difficulty by using the concepts of Western ethical thought), and as they argued for understanding these practices on their own terms, rather than dismissing them as primitive, ethical relativism began to be taken seriously.[1] The core relativist idea denies any standard, or measure, independent of the ethical practices of different societies, against which the code of one society can be judged as superior to the code of another. The idea provokes an obvious reaction. Consider groups of people you view as having done horrible things. Familiar examples: the Nazi attempt to purge Europe (and potentially the world) of "vermin" or the killing fields of the Khmer Rouge. Many people feel a powerful urge to protest the behavior and whatever ethical prescriptions are brought forward in its defense, to say there is something *objectively wrong* about what was done, to deny that condemnation only expresses a local perspective, to protest that those condemned cannot, with equal justice, criticize their critics. There must be some external standard to which ethics is answerable.

Exploring ethical variation across time avoids some of the tangles figuring in cross-cultural debates about relativism. Historical study promises examples of societies, not merely distantly related, but actively engaged with rival options for ethical transition. It might show the "mere change view" to be correct or, by disclosing how people make "objective" decisions, provide clues about the constraints ethical deliberators sense. As we shall learn, the task is harder than it might initially seem, but two useful conclusions emerge. First, there are compelling examples of transitions that look progressive—the mere-change view is hard to sustain. Second, to the extent that decisions made by the pioneers who first took progressive steps can be scrutinized, they are not readily viewed as responses to external constraints. These points generate the predicament Part II addresses.

The crucial episodes are those in which a society makes an ethical innovation that appears not simply to articulate ideas already in place, and that also seems to represent ethical progress. Initially, I shall look

---

1. Characterizing what relativism actually claims, and assessing its credentials, turns out to be a complex matter. For perceptive discussions of the cluster of problems here, see Gilbert Harman, *The Nature of Morality* (New York, Oxford University Press, 1977); Michele Moody-Adams, *Fieldwork in Familiar Places* (Cambridge, MA: Harvard University Press, 1997); and Carol Rovane, "Relativism Requires Alternatives, Not Disagreement or Relative Truth" Blackwell Companion to Relativism, ed. S. Hales (Blackwell, 2011); Rovane's views about the claims of relativism are developed further in a forthcoming book.

briefly at three examples challenging the mere-change view, but not providing insight into the processes underlying the apparently progressive transitions: the transformation of the lex talionis in the ancient world, the change from a heroic ethos in ancient Greece to the ideal of the citizen of the polis, the emphasis on compassion introduced by Christianity. For clearer ideas about how the participants made their decisions, we shall need more recent cases.

§23. Three Ancient Examples

In the earliest legal codes, the idea of exact retribution -eye for eye, tooth for tooth, life for life—is construed in an oddly literal (and, by our lights, repugnant) fashion. If someone causes the death of "the daughter of a senior," "that man's daughter" is to be put to death.[2] Although the surviving references more often concern daughters, the law does not appear to rest upon the invisibility of *women* as independent people—there are similar formulations about sons.[3] Analogous laws sometimes do embody conceptions of women as property, whose lives and bodies are controlled by male relatives: a law on rape declares that the *wife* of the rapist is to be raped by whomsoever the *father* of the victim chooses.[4]

A few centuries later, this literal construal of exact retribution has vanished. Now it is the perpetrator of the deed who must pay in the manner, and to the extent, of the damage inflicted: his or her life must be exacted to pay for the life of the victim.[5] A transition in ethical practice (not only in law) has occurred: where it was previously supposed that, when harm has been done to a member of one family, it is right to inflict the same injury on the corresponding relative in the family of the perpetrator, it is now the doer of the deed who should suffer. More than two mil-

2. James Pritchard, ed., *Ancient Near-Eastern Texts* (Princeton, NJ: Princeton University Press, 1950), 170, 175.

3. Ibid., 176.

4. Ibid., 185.

5. Of course, not all relatively ancient societies maintained the lex talionis. Nordic and Saxon groups developed the notion of "wergeld," a monetary payment compensating for lives taken. In the ancient Near East, however, the idea that murderers must forfeit their lives remained central. Witness the Hebrew Bible.

lennia on, we may demur at the thought that this is the final word on the matter, but it is hard to resist seeing the change as an improvement. We envisage cases (the overwhelming majority?) in which the corresponding relative knew nothing of the crime, cases where he or she was a child or even a friend who mourns the victim. However that may be, if the relative was not involved in the killing, justice miscarries if the relative loses his or her own life while the murderous relation goes free. Even if the perpetrator is punished "through him or her," that fails to support the practice, for the relative cannot be treated as part of the machinery of punishment, as if his or her life were not important to him or her as well as to the perpetrator. When societies go after the criminal directly, how can it not be a progressive step?

Great myths and poetry of early civilizations celebrate figures whose recorded deeds express their devotion to an ideal of honor and greatness overriding considerations that move later ethicists. Prominent examples are Homeric heroes.[6] We do not need to know if the *Iliad* has a historical basis; the crucial question is whether the ethical attitudes expressed are those prevalent in some Homeric past. One basis for supposing they are is the improbability of oral presentations of a clearly defined ethical perspective, popular across many generations, if the ethical ideas failed to reflect the actual outlook of the audience (or an audience hearers could identify as part of their history).

One shift in the period between Homer and Solon replaced the emphasis on personal honor as the principal ethical end with the idea of a contribution to the common good.[7] The Homeric hero's wartime life was directed toward acquiring personal glory; his prowess might be embodied in trophies (often given away in acts that simultaneously marked the hero's generosity and his previous exploits).[8] To appreciate

6. Others include the noble warriors from German, Norse, and Japanese traditions.

7. This is by no means the end of the idea of honor in Western ethical traditions. The concept recurs again and again, in chivalric codes in the Middle Ages, in illuminating passages in Shakespearean plays, in standards for eighteenth-century gentlemen and ladies, in the military ideals of affluent nations in the nineteenth and early twentieth centuries.

8. For clear presentations of the central features of the heroic code, see Walter Donlan, *The Aristocratic Ideal in Ancient Greece* (Lawrence, KS: Coronado Press, 1980), chap. 1; Moses Finley, *World of Odysseus* (Harmondsworth, UK: Penguin, 1980); chap. 5; and Joseph

the transition, juxtapose a passage from the *Iliad* with Thucydides' later "account" of Pericles' funeral oration.[9] Hector responds to various pleas not to engage Achilles in single combat by affirming the demands of honor. He knows his death would spell disaster for his city (and his family), but he cannot accept the dishonor resulting from refusing the challenge.[10] By contrast, when we read Thucydides' "Pericles," the common good comes first. "Pericles" says of the fallen:

> Some of them, no doubt, had their faults; but what we ought to remember is their gallant conduct against the enemy in defence of their native land. They have blotted out evil with good, and done more service to the commonwealth than they ever did harm in their private lives.[11]

These words are meant to honor a group, not an individual, and they do so by highlighting individual devotion to the good of the group.

Between the time recorded in the *Iliad* and the events commemorated by Thucydides, Greek warfare had changed profoundly. Military actions were now dominated by the organization of armed troops into the phalanx. (Men bearing heavy armor and a large shield were arrayed shoulder to shoulder and marched forward together, presenting long spears.) Success in battle depended no longer on the strength, endurance, and skill of an outstanding individual—an Achilles, a Hector—but on disciplined maintenance of one's place in the line. Conduct routine in the *Iliad*—Achilles' refusal to participate, Hector's rejection of the counsels of prudence, Diomedes' private treaty with his guest-relative Glaukus—now appears selfish, irresponsible, capricious, and quirky. The predominance of honor gave way to the virtues of moderation, self-discipline, and loyalty.

Bryant, *Moral Codes and Social Structure in Ancient Greece* (Albany: State University of New York Press, 1996), chap. 2.

9. Thucydides clearly warns that he reconstructs speeches by combining the sense of what was said with the thrust of what it would have been appropriate to say (*Peloponnesian War* [Harmondsworth, UK: Penguin Classics, 1972], 47).

10. See Finley, *World of Odysseus*, 115–17, for an excellent discussion of this episode.

11. Thucydides, *Peloponnesian War*, 148.

Pace Nietzsche, the substitution of solidarity in pursuit of a common project for individual ventures dominated by the thirst for honor is, at least in some respects, a step forward. Sacrifices undertaken in pursuit of honor often appear irresponsible, even absurd—Hector's decision has foreseeable consequences dreadful for him and for those about whom he cares. The transition can be viewed as restoring a healthier type of normative guidance, one closer to the early stages of egalitarian deliberation about the character and promotion of shared ends, even a correction of spectacularly destructive altruism failures.[12]

Many people conceive Christianity as transforming the ethical framework of the Greco-Roman world. What they probably intend is: some features of the ethical attitudes of most social groups identifying themselves as inspired by Jesus were absent from most other groups living under the aegis of Rome. What might these features be?

Obvious answer: the growth of Christian belief increased compassion in the ancient world. Jesus enjoined his followers to forgive their enemies and love their fellows. His influence reformed brutish Roman institutions. According to an eminent Victorian:

> No discussions, I conceive, can be more idle than whether slavery, or the slaughter of prisoners in war, or gladiatorial shows, or polygamy, are essentially wrong. They may be wrong now—they were not so once—and when an ancient countenanced by his example one or other of these, he was not committing a crime.[13]

There was no simple impact of Christianity on the ancient world, even with respect to the conduct of war, the abolition of slavery, the

12. Here it is worth recalling Mill's insight that attention to the consequences does not rule out self-sacrifice but simply demands that the sacrifice be worth something (John Stuart Mill, *Works*, Vol. 10 (Toronto: University of Toronto Press, 1970), 217).

13. W. E. H. Lecky, *History of European Morals from Augustus to Charlemagne* (New York: Braziller, 1955), 1:110. For Lecky's advocacy of the idea of ethical progress, see also vol. 1, 100–103, 147–50; vol. 2, 8–11, 73–75. The apparent relativism of Lecky's formulation is misleading rhetoric: he does not literally think that what is right (or wrong) has changed, but that what is *taken to be* right (or wrong) is altered. This is plain from his confidence in the Victorian values that have emerged.

character of public spectacles, or views about marriage. The decades
and centuries after Constantine were marked by frequent acts of violence
carried out by Christians and commended by their leaders (the savagery
of the Crusades is prefigured in quarrels about orthodoxy erupting from
the fourth century onward); nor did Christianity play any straightforward role in replacing chattel slavery with serfdom and villenage.[14]
Nor is the priority of love and forgiveness consistently upheld in all canonical Christian writings.[15] The evangelists attribute to Jesus such dicta
as "I came not to bring peace but a sword" and also describe the disciple
commanded to forgive others a very large number of times (490), causing
the death of members of the movement who failed to contribute all their
goods.[16]

On Lecky's account, Christianity introduced a progressive shift, centered on adopting a new ideal of altruism. Recall the dimensions of
altruism: intensity, range, scope, recognition of consequences, and empathetic understanding (§5). Even without taking into account further
complexities (those noted in §21), the ideal can be formulated in several
ways. In the extreme version, one commending "golden-rule altruism"
with respect to every person and every context, and demanding complete
accuracy about people's preferences,[17] it would be quite impossible to
follow. Any community trying to adopt it would face difficulties when
individuals' initial preferences for indivisible goods are incompatible
(and in widespread "Alphonse-Gaston" situations where both parties are
moved by the wish to abnegate their desires in favor of the other). Any

14. For an insightful discussion of issues about slavery, see Moses Finley, *Ancient Slavery and Modern Ideology* (London: Chatto and Windus, 1980). Historians from Gibbon on
have noted the savagery of the conflicts among early Christian sects. A lucid account of the
Crusades is offered by Steven Runciman, *A History of the Crusades*, 3 vols. (New York:
Harper and Row, 1964–1967), see esp. his narrative of the massacres that attended the "triumph of the cross" in Jerusalem, 1:286ff.).

15. As §18 noted, this theme is anticipated in Mesopotamian texts predating the Gospels
(see n. 20).

16. See Matthew 10:34–39 (also Luke 12:49–53), Matthew 18:21, and Acts 5:1–11.

17. I characterize the ideal in terms of preferences rather than interests because a Christian formulation in terms of interests would adopt a very special notion of the interests of
individuals, one that does not obviously translate into a judgment that the ideal would
mark a direction of ethical progress.

viable version requires a system of principles explaining when the needs of one individual are more urgent than those of another, thus assigning the roles of altruist and beneficiary. We can conceive the appealing parts of Jesus's message as articulating a part of this system, by identifying the urgent needs of those often excluded from consideration—those beyond the range of altruistic dispositions.

The popular thesis that Christianity represents an ethical advance is most plausible when the movement is viewed as promoting altruistic responses to marginalized people whose most basic desires have previously not been met. Victorian confidence about the ethical progress made by Christianizing the Roman Empire supposes that sort of expansion. It should allow for variation across regions and across periods in the newly Christian world, for different advances made by groups whose altruism was extended to different targets and to different extents.[18] Instead of thinking of a definite Christian ideal, we would do better to conceive of a general trend, for which Christianity provides a forceful expression.

In none of these instances do we have any sources revealing how and why people made the apparently progressive shifts. For them we need to move closer to the present.

§24. Second-Sex Citizens

During the last two centuries, in the countries of Western Europe and North America, there have been important changes in the civil status of women and in women's abilities to gain access to positions and privileges, previously viewed as an exclusively masculine domain. This shift has not proceeded at the same pace across all sectors of the societies in question, nor has it eradicated earlier attitudes opposing women's entry into spheres from which they were previously blocked, nor has the movement finally attained the goals for which many of those involved in it

18. The expansion of altruism in this way is hardly the exclusive province of Christianity. As noted, the injunction to love and forgiveness, even toward enemies, appears in Babylonian literature several centuries before Jesus, and the ancient world contained groups of non-Christians (for example, Jews and Stoics) whose ethical codes extended the altruism of prior traditions.

have striven.[19] Nevertheless, things have changed: it no longer seems appropriate for the leading English jurist to strip his daughter naked, to tie her to a bedpost, and to whip her until she will agree to marry the psychologically disturbed nobleman whom he has selected as her husband; or for a widow to give up her children to her husband's family, even though, by the lights of the surrounding community, the mother, not her in-laws, belongs to the orthodox church; or for women to be denied any education; or, when education is grudgingly allowed, for them to be debarred from receiving degrees, despite the fact that one of their number shows herself superior to her male contemporaries on the most prestigious mathematical exam of the day.[20] The list of horror stories from the past could be enormously extended. The plausibility of ethical progress in this domain is signaled by the reactions of many citizens of contemporary democracies, who not only firmly believe these practices were utterly unjustified, but also cannot conceive how reflective people could ever have permitted them.

The character of the advance is twofold. First, rules preventing women from playing coveted roles in their societies, from having access to particular institutions, from possessing things men around them wanted to acquire, and from exercising certain kinds of choice were rejected as ethically wrong. Second, the presence of women in roles and institutions traditionally held as male preserves has led to improvements in those roles and institutions.[21] The first type of change was consolidated earlier,

19. The writings of Catherine MacKinnon serve as important reminders of what a significant number of women (as well as some men) think remains to be done; see her *Feminism Unmodified* (Cambridge, MA: Harvard University Press, 1987) and *Towards a Feminist Theory of the State* (Cambridge, MA: Harvard University Press, 1989).

20. See J. Morrill, ed., *The Oxford Illustrated History of Tudor and Stuart Britain* (New York: Oxford University Press, 2001), 97, for the account of Sir Edward Coke's coercion of his daughter; George Walker, although a Quaker, was given custody of his children, despite the fact that his ex-wife, Ann, was Anglican—the law of Virginia ranked patriarchy ahead of orthodoxy (see Julia Cherry Spruill, *Women's Life and Work in the Southern Colonies* [New York: Norton, 1972], 345); in 1889, Charlotte Angas Scott, a student at Girton, obtained the highest score in the Mathematics Tripos at Cambridge, although she was not able officially to register for a degree.

21. Many people would view the impact of women on life in the professions from which they were so long excluded as a good thing. Within academic discussions, for example, in-

and some view the second as more controversial. I take both as instances of ethical progress.

How were the advances made? In the ancient examples it is impossible to identify psychological processes through which individuals, or groups, made ethical discoveries. Here, however, there is material to which investigators can turn in hopes of picking out the new perception or new piece of reasoning that fueled ethical evolution. A sequence of texts, retrospectively inspiring, leads from the writings of Mary Wollstonecraft at the end of the eighteenth century, to the documents of the nineteenth-century American feminist movement, to the classic essay co-authored by John Stuart Mill and Harriet Taylor, to the fiction of Charlotte Perkins Gilman and the social commentary of Virginia Woolf, to Simone de Beauvoir, to Betty Friedan, Catherine MacKinnon, and their successors.[22]

As with scientific revolutions, the triumph of a radically different perspective proves far more complex than might have been supposed.[23] Once the revolution is over, the confident insistence on male privilege seems monstrous in its blindness. How could Sir Edward Coke have tied his daughter to that bedpost, or Sophie Jex-Blake's father have hampered

clusion of women has sometimes fostered a more cooperative approach to research, and established this as a rival model for the aggression of male-male competition.

22. Wollstonecraft, *A Vindication of the Rights of Woman* (New York: Modern Library, 2001); Alice Rossi, ed., *The Feminist Papers* (New York: Bantam, 1977); John Stuart Mill (and Harriet Taylor, whom I include as coauthor here), *On The Subjection of Women* in Mill *On Liberty and Other Essays* [Oxford: Oxford University Press (World's Classics), 1998]; Charlotte Perkins Gilman, *Herland* (New York: Pantheon, 1979), and *The Yellow Wallpaper* (New York: Routledge, 2004); Virginia Woolf, *A Room of One's Own* (New York: Harcourt, 1957); Simone de Beauvoir, *The Second Sex* (New York: Vintage, 1974); Betty Friedan, *The Feminine Mystique* (New York: Norton, 1963); Germaine Greer, *The Female Eunuch* (New York: Bantam, 1972); Catherine MacKinnon, *Feminism Unmodified* (Cambridge, MA: Harvard University Press, 1987), and *Towards a Feminist Theory of the State* (Cambridge, MA: Harvard University Press, 1989); Donna Haraway, *Simians, Cyborgs and Women* (New York: Routledge, 1991); and bell hooks, *Ain't I a Woman* (Boston: South End Press, 1981). Although I mainly focus on Anglophone texts, there are many other important sources—for example, the response of Olympe de Gouges to the Declaration of the Rights of Man and the Citizen, produced by the (male) leaders of the French Revolution.

23. The intricacies Kuhn discerned in major scientific debates (*The Structure of Scientific Revolutions* [Chicago: University of Chicago Press, 1962]) are even more apparent in the ethical case.

her energetic wish to engage in medical practice? The autocratic men in these stories resemble the cartoon figures of old narratives in the history of science, the simple-minded Aristotelians who do not understand the brilliance of Galileo's arguments, and who maybe even refuse to look through his telescope. Investigating more closely, we find not one sequence of texts, with a compelling set of insights demolishing unsupported prejudice, but two—and each connects with particular parts of past ethical practice.

The heart of the feminist insight is a *factual* claim: the social practices prevalent in society (that is, the society in which the feminist author writes) confine the desires of women to a narrower range than would be achieved under different forms of socialization. Later, that claim is expanded—"and the same goes for men, too." The claim strikes directly at the conservative case. In many societies, from the ancient world to the present, women are assumed not to want certain kinds of possessions and positions, supposed to be incapable of particular kinds of choices. If they occasionally do, by some quirk, express a desire for the goods or offices, or want to make the choices, these preferences diverge from their interests. Conservatives see no sense in which society fails to respond to the wishes—or to the *proper* wishes—of its female members. To the extent that the society is good at socializing young girls, the "deviant" wishes will not arise with any serious frequency, and the theses about lack of desire and lack of ability will rarely be challenged. Sometimes the theses are buttressed further by assertions about the divine will.

Once entrenched, attitudes that certain types of female desires are deviant, and thus not to be endorsed, are difficult to displace in societies skilled at socializing the young. The rare girls and women who voice "deviant" wishes can be dismissed as in need of correction; they rarely have the chance to challenge common views about their incapacities. Under such conditions, initially inexplicable attitudes no longer appear monstrous—although the extreme case of whipping a young woman into submission is hard to view as anything but pathology. Fathers who discourage their daughters from public life are profoundly wrong about the aspirations they check, but their society has not only drummed into them the impropriety of those aspirations but also made it hard for them to acquire evidence about what would happen if the wishes were fulfilled.

One way to counter paternalism is to show that hopes expressed occasionally by a small number of women would be far more widespread if society did not so efficiently smother them. How can that be done? Wollstonecraft's *Vindication of the Rights of Woman* often seems timid to contemporary readers, who misread its clever rhetorical strategy. Wollstonecraft argues for a limited goal—allow the education of women—precisely because she can connect that goal with improved fulfillment of roles traditionally assigned to women. Her conservative opponents, whether they maintain that the tasks of bearing and raising children, and supporting husbands, are divine provisions for women, or whether, like Rousseau, her principal foil, they emphasize proper nurturing of the (male) citizen, are committed to allowing women access to whatever bests fit them for the roles of wife and mother. Hence, Wollstonecraft can argue for her proposed change by showing the superior ability of educated women to discharge "their" roles. She highlights the point in a passage, often embarrassing to contemporary readers: the fates of uneducated and educated wives, and of their children, are contrasted; the comparison culminates in a vision of the educated widow; having successfully raised her family ("her work is done"), she ascends to rejoin her husband in heaven. Because her opponents are committed to take the conventional role seriously, the rhetorical effect is devastating. How much does this gain? Surely the educated woman will be confined to the domestic sphere?[24] Opening the door to education, as Wollstonecraft probably saw, weakens the power of traditional systems of socializing girls and young women and thus increases the chance women will express desires for broader roles in society. It is a crucial first step in normalizing those desires.

On the Subjection of Women goes further, replacing the argument that women's education is needed to develop better wives and mothers with an appeal to individual freedom. Education can be viewed, as it is by Mill and Taylor, as a crucial device for men and women to formulate what is

---

24. Late in *Vindication*, in chap. 9, Wollstonecraft does venture a little further, advertising to the possibility that women may do some kinds of work ("keeping a little shop," participating in medical care). I read this as a clever signal to sympathetic readers (the ones who have come this far with her), to the effect that the changes for which she officially campaigns are only beginnings.

central to their lives, to "find their own good in their own way."[25] They call for social experimentation, both as a means for providing the young with potential models from which they can assemble their individual conceptions of how to live, and as the proper expression of what people want, not to be confined unless it does harm to others.[26] Later, when the desires of educated women to participate in public life—and, in some cases, to change the character of public life—have become even more widespread, Woolf documents the ways in which those desires continue to be resisted.[27]

Why does confinement continue? Woolf's own quotations from the oppressive men who rein in their daughters reveal the structure of conservative thought. Suppose the step recommended by Wollstonecraft is taken: a society of educated women contains wives and mothers who discharge their roles with unprecedented success. Not only would broadening the activities of educated women require new arguments, but, *according to the case already made for educating women*, it would likely be counterproductive. If the emphasis is firmly on improving the conduct of wives and mothers, pitching women into the public world appears a bad idea, one likely to produce weakened marriages and neglected children. Conservatives protest the Millian insistence on the primary value of individual development, on formulating freely one's own plan of life—by appealing to higher, divinely prescribed, goals for human existence, by emphasizing the health and flourishing of individuals or of society—*but they can even adapt their reasoning to the framework their opponents take for granted*. Desires are to be honored only if they do no harm to others. If women are given access to public life, they will do harm—their husbands and children will suffer. Women's desires for time-consuming careers, for prominent positions in society, are viewed as altruism failures.

25. See John Stuart Mill, *On Liberty* [Oxford: Oxford University Press (World's Classics), 1998], chap. 1.

26. John Stuart Mill, *On the Subjection*; plainly the essay reinforces and is reinforced by the central ideas of *On Liberty*.

27. Virginia Woolf, *A Room of One's Own* (New York: Harcourt, 1957), and especially *Three Guineas* (San Diego: Harcourt Brace Jovanovich, 1966). A crucial move in the latter is the suggestion that public life be transformed by the pressure of women who form a "society of outsiders."

Married women without children engage less frequently in such failures, but female pursuit of a career can be fully exempt from them only if the woman is free of family commitments. Since the desire for a family is central for women, the Mill-Taylor recipe will produce wrongdoing or unhappiness.[28]

So begins a (new) round of catch-22. From the reformers' perspective, once the public aspirations of women are no longer viewed as pathological, the lack of response to *them* constitutes a class of altruism failures; the counterpart desires of husbands and fathers are endorsed at the cost of suppressing women's wishes. Principles of fairness, shared by conservatives and reformers, oppose constant sacrifices by one spouse to benefit the other. Conservatives see asymmetries here—women have special talents and abilities expressed in nurturing the family; their happiness is centrally bound up with the family's flourishing. The obvious counter: this is a product of existing conditions of socialization, of the particular way the institution of marriage has been framed, and things could be done differently. We should experiment. But, conservatives insist, experiments are properly canceled if they risk great damage. Framing the roles of husbands and wives differently would, given "the natural desires of the sexes," damage a valued institution (marriage) and cause frustration and unhappiness for parents and children alike. An obvious charge: conservatives beg the question. The charge is met with a countercharge: tu quoque. To assume the experiments will not rub against human nature is already to presuppose the desires of men and women to be adaptable, and that is equally to beg the question. Each side must argue from its preferred ideas about the plasticity of human preferences.[29]

28. Arguments of this kind survive into the present. Their persistence is a product of the continued inability to solve broader problems about the provision of public goods and the education of the young. Resolution is hard because the issues are so entangled. See Chapter 10 (§60).

29. The line of reasoning attributed to conservatives here is still present in many societies, and in many corners of societies that officially endorse women's entitlement to seek demanding careers. Because the problem of combining work and family life remains unsolved, for a large number of women and for a smaller number of men, issues about reframing the institution of marriage remain. These combine with other questions about the forms of the division of labor in contemporary societies, about the pressures that division of labor exerts, about the distribution of resources, and about the provision of public

Brief rehearsal of this dialectic exposes important aspects of the ethical transition. The Mill-Taylor point, that women's capacities are unknown because potentially revealing experiments in socialization have never been tried, is indisputable; the conclusion that those experiments *should be tried* is the locus of controversy, yielding the subsequent impasse. How then was this revolution actually resolved?

Through demonstrations and demonstration. The protests of suffrage movements, including the willingness of women to sacrifice their lives to reveal how strongly they wanted full participation in society, made their aspirations impossible to ignore. Female labor during the Great War demonstrated women's capacities, and also the possibility of combining work with the nurture of children. It is no accident that the United States and the United Kingdom both granted rights to vote after the end of the war. Later contributions of women in the Second World War, and their withdrawal into a more traditional domesticity in the immediate postwar period, provided the background against which the women's movements of the 1960s could uncover suppressed and unsatisfied yearnings. The centrality of consciousness raising to feminism recalls the advance already made by Wollstonecraft: the first step is to reveal that certain wishes are widespread, not therefore easily dismissed as pathological, and that they remain unexpressed because of the smothering effects of social expectations. The women of the 1960s who attended group meetings, sharing aspirations and experiences, could look both to the demonstrated capacities of their predecessors at times of national need, and to the partial fulfillment of desires they recognized in themselves. Their voices could not be ignored. The impasse was broken. Ethical progress was made.

---

goods. The resultant entanglements make debates among conservatives and reformers even harder to resolve. Where women's lives are able to combine demanding work and family most smoothly, this is frequently achieved at the cost of deferring the burden to other women whose choices are far more restricted (women of lower socioeconomic status who serve as caregivers or housekeepers). Wollstonecraft's assumption of the presence of servants in the domestic arrangements she envisages is not quite the anachronism it initially appears—nor is Woolf's reliance on the idea that someone else will do the shopping and the cooking. (I am indebted to Martha Howell for presentations on Wollstonecraft that have helped me to see these aspects of her problem more clearly.)

What was discovered? Factual knowledge advanced: people learned that, under different conditions of socialization, women wanted things traditionally denied to them; that they found satisfaction in attaining some of these things; that fulfillment of the wishes did not thwart desires previously seen as central to female nature—public life combined more or less satisfactorily with family life.[30] Increased factual knowledge proliferated desires for access to public life, fostering acceptance of the desires as prevalent and no longer pathological. Recognition of the suppression or frustration of those desires aroused sympathy, recruiting male as well as female allies for the reform movement. Like the early elaboration of normative guidance, in which particular altruism failures cause too much trouble, the increase of sexual egalitarianism occurred partly because, in the end, traditionalists wanted a quieter life.

§25. Repudiating Chattel Slavery

If there is one example in which the attribution of progress is almost incontrovertible, it is the abolition of the "peculiar" institution, chattel slavery.[31] Opposition to slavery intensified in Britain in the late eighteenth century and in America in the first half of the nineteenth, culminating in the Emancipation Proclamation and the Civil War. Ultimately the view that slavery is ethically permissible was replaced by the denial of that claim. How was that advance accomplished?

Seventeenth- and eighteenth-century northern Europeans, whether resident in the ancestral countries or dispersed among the colonies, attempted to embed slavery within their ethical codes. Apologists drew concepts and distinctions formulated centuries earlier in efforts to justify ownership of human beings. One traditional defense distinguished between people who are permissibly enslaved and those who are

30. The "more or less" enters here as a reminder of the difficulties already remarked (n. 29), of the fact that special social circumstances are needed for the combination to work smoothly. See §60.

31. I think it can be opposed only by arguing that the notion of ethical progress cannot be given a clear sense. I am grateful to Edie Jeffrey for reinforcing my conviction that this example is indispensable for any account of ethical progress, and for giving me good advice about how to investigate it.

not—formulated in the Hebrew Bible by differentiating the chosen people from their captives in war, by Aristotle in terms of variants in individual nature, and by the medieval church through separating the faithful from infidels. Colonial Christians added further support from the scriptures. The Pentateuch testifies that the patriarchs had slaves (and had sexual relations with female slaves); the letter to Philemon endorses slavery; further, the particular people enslaved in North America (people of African descent) are descendants of Ham (or, in some versions, Canaan), inheritors of a biblical curse. Protestant Christianity also contrasted the liberty of the soul to attain to God's grace with mere bodily liberty. On this basis, some claimed, slave traders were doing their captives a favor.[32]

The slave-owning cause constructed its own account of the trade. The slaves' native situation in Africa was portrayed as a state of Hobbesian nature, dominated by strife, bestial practices, and utter ignorance. After transporting the unfortunate people across the Atlantic, kindly slave owners provided food and shelter (as well as paternal affection) in exchange for toil. Even more important, slaves were given the opportunity to hear the true religion and gain spiritual salvation.

All this is rubbish, but it is impossible to understand Christian acceptance of slavery without recognizing the self-serving interpretation. In 1700, Samuel Sewall published a pamphlet proposing an analogy between the slavery of the colonies and the (unlawful) servitude of Joseph. His tract drew a response from John Saffin, who, in 1701, addressed Sewall's suggestion that "we may not do evil that good may come of it" by writing, "It is no Evil thing to bring them out of their own Heathenish Country, where they may have the Knowledge of the True God, be Converted and Eternally saved."[33] Five years later, Cotton Mather saw African slaves as providing religious opportunities for colonists:

32. In *The Problem of Slavery in Western Culture* (Ithaca, NY: Cornell University Press, 1966), David Brion Davis provides an illuminating account of all these apologist strategies and their relationship to ancient and medieval thought.

33. John Saffin, "A Brief Candid Answer to a Late Printed Sheet Entitled *The Selling of Joseph*" (1701), in *Against Slavery: An Abolitionist Reader*, ed., M. Lowance (New York: Penguin, 2000), 16.

The State of your *Negroes* in the World, must be low, and mean, and abject; a State of servitude. No *Great Things* in this World, can be done for them. Something, then, let there be done, towards their welfare in the *World to Come. . . . Every one of us shall give account of himself to God*[34]

In the middle eighteenth century, the influence of the argument diminished because the efforts to propagate the Gospel were so obviously unsuccessful. Slaves preferred to spend their Sundays dancing, trading, and resting—and, as David Brion Davis notes, they did not flock to a religion "which sanctioned their masters' authority, which enjoined them to avoid idleness and to toil more diligently, and which promised to deprive them of their few pleasures and liberties."[35] Colonists concluded, however, that slaves were incorrigible. The collapse of one line of proslavery argument buttressed another.

Saffin's response to Sewall already claimed that Africans and Europeans were distinguished in moral and intellectual temperament: his tract closes with a piece of doggerel attributing innate vices (cowardice, cruelty, libidinousness, etc.) to the black races.[36] The judgment survived into the nineteenth century. As late as 1852, Mary Eastman could write a response to *Uncle Tom's Cabin* in which she assembled all the main parts of the "difference" argument: Africans are descendants of Ham, cursed by God, with traits of character requiring firm discipline by wiser (and benevolent) people of European ancestry; slaves are no more appropriate bearers of freedom and self-government than wayward children.[37] Ideas like these were current, not only among literalist Christians but also in Enlightenment circles. Although Montesquieu, the most insightful early critic of slavery, punctured the appeal to innate differences,[38]

34. Cotton Mather, "The Negro Christianized: An Essay to Excite and Assist That Good Work, Instruction of Negro-Servants in Christianity," in Lowance, *Against Slavery,* 19.

35. Davis, *Problem of Slavery,* 218; see also 211–22.

36. Lowance, *Against Slavery,* 17.

37. See Mary Eastman, *Aunt Phillis' Cabin: Or Southern Life As It Is,* excerpted in Lowance, *Against Slavery,* 296–300.

38. See Charles Montesquieu, *Spirit of the Laws* (Cambridge, UK: Cambridge University Press, 1989), for Montesquieu's famous remark that Africans cannot be humans

eighteenth-century speculative anthropology inspired Voltaire, Hume, Buffon, and their intellectual descendants to advocate African inferiority. Adam Smith was a rare dissenter, but he was soundly rebuked by the Virginian Arthur Lee, who drew on his extensive experience of black slaves to set Smith straight.[39]

General considerations about racial hierarchy were coupled with claims about the behavior of Africans, both in their native countries and in their state of servitude. A body of literary attempts to depict the nobility of enslaved Africans (Aphra Behn's *Orinooko* is a prominent representative) was countered by a far larger volume of writings from the people (slave owners) who claimed to know the subject best. Achievements of individual slaves were systematically undervalued. Expressions of a conviction that black people are doomed to lesser accomplishments (also unpleasing, if not disgusting) are even found in the words of two of America's most high-minded presidents. Jefferson wrote:

> Whether the black of the negro resides in the reticular membrane between the skin and the scarf-skin, or in the scarf-skin itself; whether it proceeds from the colour of the blood, the colour of the bile, or from that of some other secretion, the difference is fixed in nature, and is as real as if its seat and cause were better known to us. And is this difference of no importance? Is it not the foundation of a greater or less share of beauty in the two races? Are not the fine mixtures of red and white, the expressions of every passion by greater or lesser suffusions of colour in the one, preferable to that eternal monotony, which reigns in the countenance, that immoveable veil of black which covers all the emotions of the other race? Add to these, flowing hair, a more elegant symmetry of form, their own judgment

because if they were it would follow that we (Europeans) are not Christians. His ironies were unappreciated. For some uncomprehending reactions to Montesquieu, see Davis, *Problem of Slavery*, 403. It is also worth noting that one of Montesquieu's most important arguments against slavery imagines that the roles of slaves and masters are determined by lot; for what seems to be an anticipation of Rawlsian appeals to ignorance of social position, see the addenda to *Spirit of the Laws*.

39. Arthur Lee, *An Essay in Vindication of the Continental Colonies in America, from a Censure of Mr. Adam Smith in his "Theory of Moral Sentiments"* ("Printed for the author" London, 1764).

in favour of the whites, declared by their preference of them as uniformly as is the preference of the Oranootan for the black women over those of his own species.[40]

Jefferson goes on to assess the character and accomplishments of the slaves he knew: their passions are transient and instinctual, they have scant power of reason, little imagination, virtually no artistic skill; Jefferson allows that their moral lapses (lying, stealing, and so forth) can sometimes be traced to the difficulties of their situation, but even virtues are transmuted into defects—the courage of African blacks is seen as absence of forethought. Although he concludes that "the opinion that they are inferior in the faculties of reason and imagination, must be hazarded with great diffidence," his preceding catalog reveals Jefferson not only hazarding it but showing little diffidence about the constituent claims. Decades later, Lincoln echoed Jefferson's judgment, averring a "physical difference between the white and black races which I believe will forever forbid the two races living together on terms of social and political equality."[41]

The transition from an ethical practice that permits slavery—either as unproblematic or as problematic but, on balance, acceptable—to one identifying it as a patent ethical wrong surely looks progressive. How was it accomplished? A collection of counterarguments systematically dismantled the justificatory attempts of apologists. They dissect the evidence for taking black Africans to have inherited some biblical curse; they note other ways of bringing the African soul to grace than subjecting the African body to the middle passage, the slave auction, unremitting toil, sexual, abuse and the lash; they display the accomplishments of individual slaves, or ex-slaves, whose words and works refute theses of innate racial difference.[42] The overall abolitionist campaign consisted in

40. Thomas Jefferson, *Notes on the State of Virginia,* in *The Portable Thomas Jefferson, ed.* Merrill D. Peterson (New York: Penguin, 1975), 186ff. It is interesting to ask how the man who wrote these words conceived his relationship with Sally Hemmings.

41. Cited in the "General Introduction," to Lowance, *Against Slavery,* xxiv. Perhaps, as the editor notes (xxv), Lincoln was simply bowing to political pressure.

42. The writings and speeches of Frederick Douglass are notable examples of this last strategy. In his "General Introduction" to *Against Slavery,* Mason Lowance tells an

destroying all the devices deployed to avoid applying to people of African descent the same attitudes and principles routinely accepted as governing peaceful interactions among the civilized—it tore down the distortions allowing Europeans to view Africans as utensils rather than people.

Besides the negative side of the campaign, there were also positive discoveries. A few courageous visitors to the parts of the African interior from which slaves were drawn were surprised to discover communities with different customs, but with stable social relations and, above all, familiar human needs and feelings. As more was learned about slave recruitment and the character of transatlantic voyages, many of the vaunted benefits conferred by enslavement were disclosed as a farrago of nonsense. Factual discoveries, integrated with strenuous readings of the scriptures, allowed slaves at last to be seen and to become targets of sympathy. Audiences eventually responded to the eloquence of Douglass and others, black and white, who cataloged slave suffering, but their applause depended on earlier advances, made on a more abstract theological basis.

Although they did not consistently condemn slavery, American Quakers were often especially concerned with the problem and sometimes moved to argue for abolition of the institution. Arguably, *the* pioneering abolitionist was John Woolman, whose *Journal* records how he reached his position. Woolman's public campaign culminates in his *Some Considerations on the Keeping of Negroes* (first published in 1754), in which he argues that "*Negroes* are our Fellow Creatures." Woolman's defense linked slave suffering to that experienced by the outcasts who excited Jesus's sympathy.[43] His spiritual odyssey depended as much on his reflections on the New Testament as on his experiences of slavery. There

---

interesting story of one of Douglass's presentations: "Once during the opening moments of a lecture in London, an audience expressed hostile disbelief in his past as a chattel slave because his oratory and elocution were so powerful. (It was well-known that slaves were held in illiteracy and ignorance as a means of control.) Without speaking another word, Douglass promptly stripped off his shirt and turned his flayed back to the incredulous audience to show the scars of his floggings" (*Against Slavery*, xxx).

43. Excerpts from *Some Considerations* appear in Lowance, *Against Slavery*, 22–24; see also John Woolman's *Journal* (New York: Citadel, 1961).

never seems to have been some perception of the ethical standing, the worth of the slaves; what troubled Woolman was the conflict between the institution and his Christian duty.

The early pages of the *Journal* explain how the sixteen-year-old Woolman "began to love wanton company," how a period of self-indulgence was punctuated by ever-longer intervals of remorse, and how eventually he "recovered" and came back to "live under the cross".[44] As his own ability to resist temptations increased, he began to be troubled by the backslidings of others, and to be "uneasy" when he did not remonstrate with them; uncharitably, we might describe the twenty-three-year-old as a bit of a busybody. The first (mentioned) concern about slavery arose when he was asked to perform a task:

> My employer, having a negro woman, sold her, and desired me to write a bill of sale, the man being waiting who bought her. The thing was sudden; and though I felt uneasy at the thoughts of writing an instrument of slavery for one of my fellow-creatures, yet I remembered that I was hired by the year, that it was my master who directed me to do it, and that it was an elderly man, a member of our Society, who bought her; so through weakness I gave way, and wrote it; but at the executing of it I was so afflicted in my mind, that I said before my master and the Friend that I believed slave-keeping to be a practice inconsistent with the Christian religion.[45]

Shortly afterward, Woolman refused to sign a similar document for a young acquaintance, also "of our Society."[46]

The language of this passage is telling. The woman sold remains anonymous. Perhaps Woolman did not know her—he kept his master's shop, and lived there alone, at a distance from his employer's house. Yet this bare characterization ("a negro woman") typifies the entire *Journal*. Slaves appear in it only under the most abstract descriptions, never perceived as individuals. Woolman provides no extended portrait of their

---

44. Woolman, *Journal*, chap. 1; quotes from 4, 5, 8.
45. Ibid., 14–15.
46. Ibid., 15.

behavior and capacities, one that might persuade others to see them as people. Similarly, the concern with his own spiritual standing, evident in the hesitations over the bill of sale, is omnipresent. When he discloses his discomfort, to his master and the purchaser (and to his readers), he claims an inconsistency between Christianity and slavery. The nature of that inconsistency is alluded to by the characterization of the woman as a "fellow *creature*."

What led Woolman to draw up the document? He clearly saw it as an action commended by his duty of obedience (the Christian servant obeys his Christian master), and yet he knew slaves often suffer (that was surely the initial cause of his "trouble"). He temporarily suppressed doubts—the buyer was elderly and a Friend, qualities likely to prevent sexual and other forms of abuse. As he reflected, however, he recognized his participation in an institution prone to inflict suffering on "fellow-creatures," and, although the chances of serious abuse in this case seemed remote, they were not zero. Once the document had been signed and the woman "conveyed," there was no guarantee she would not be maltreated. He would have been partly responsible.

Perhaps I overinterpret the passage. But this reading accords with Woolman's subsequent discussions of his growing opposition to slavery. He is constantly concerned that he is infected by living among those who support themselves by slavery—interested, too, in saving them from corruption.[47] At times the spiritual plight of slave owners troubles him, and his reprimands have the character of the sober young man who intervened to save his acquaintances from "wantonness." Moreover, more thoroughly than his predecessors, he takes seriously the Christian apology for slavery, quoting scripture to rebut the characterization of slaves as inheriting the curse laid upon Ham, and urging priority for the official aim of redeeming these "lost people":

> If compassion for the Africans, on account of their domestic troubles, was the real motive of our purchasing them, that spirit of tenderness being attended to, would incite us to use them kindly, that, as strangers brought out of affliction, their lives might be happy among us.

47. Ibid., 22, 39, 53.

And as they are human creatures, whose souls are as precious as ours, and who may receive the same help and comfort from the Holy Scriptures as we do, we could not omit suitable endeavors to instruct them therein.[48]

The case Woolman makes to his slave-owning interlocutors, and to his readers, lies within the abstract framework of Christian duty. His goal is to remove the blemishes from the Christian community, whether they are individual propensities to "wantonness," the "burdensome stone" of slave ownership, or the traffic in "impure channels," which distresses him during his visit to England.[49]

Woolman made a large and important progressive step. It is hard not to admire his rejection of slavery, or the courage and perseverance displayed in his many attempts to persuade others. *His reasons, however, are not those of any contemporary secular ethical framework.* As in earlier instances, progress is not achieved through some clear new ethical insight. To be sure, there are genuine cognitive accomplishments, consisting in the recognition of previously masked facts; Woolman and his successors appreciate that the view of the African lineage as cursed is groundless, that conditions in "savage Africa" are hardly ameliorated by shackling and confining human beings, separating them from their kin, and beating and raping them; his successors come to see that slaves (and ex-slaves), with little opportunity and virtually no motivation, can do remarkable things. Later abolitionists, building on Woolman's advance, recognize how altruistic dispositions, shaped in the prevailing ethical practice, are confined from any extension to black Africans only because the pertinent people are kept out of sight, portrayed from a distance as brutish and incapable of "superior" feelings. The course of the change in attitude, and the consequent growth of sympathy for slaves, is unsteady and incomplete, even in the late eighteenth and nineteenth centuries (witness Jefferson, and perhaps Lincoln). It comes about at all only because profoundly devout men and women wrestle with problems of scriptural interpretation, eventually producing the possibility of seeing the sufferings

48. Ibid., 54; see also 53–56.
49. Ibid., 54, 212.

as inflicted on *real people*. Finally, the men and women routinely bought and sold are no longer anonymous, no longer undifferentiated "fellow creatures," but fully, individually, and equally, human.

§26. The Withering of Vice

My final example represents an entire class of transitions occurring in the secularization of (some) ethical codes. In these episodes, conduct previously regarded as wicked, depraved, or sinful comes to be seen as ethically permissible, and perhaps appropriate for some people. Ethical concepts prominent in earlier discussions are abandoned or refined. So citizens of many affluent societies no longer condemn those who express sexual desires for people of the same sex. Sometimes the shift is only partial: homosexuality is no longer a vice, but still something regrettable—a sickness, a defect, an incomplete form of sexual fulfillment. When the transition is thoroughgoing, same-sex preference simply becomes the way in which some people give direction to their sexual desires, neither intrinsically better nor worse than heterosexuality. Terms previously used to characterize those drawn to their own sex are rejected as prejudiced, confused, and uncharitable; even broader notions—"vice," "sin"—come to seem askew. A more egalitarian view prevails. Homosexual intercourse, like its heterosexual counterpart, can be loving or exploitative, tender or cruel, deeply expressive or a shallow pleasure. Homosexual relationships can vary along all the dimensions of heterosexual ones.

Ethics is not all about regulating sex. Nonetheless, probably from the beginning, ethical codes have appraised various sorts of sexual activity, allowing some, forbidding others. When homosexuality is no longer characterized as a vice, the framework of appraisal is modified. Instead of focusing on the sexes of the partners (or on the anatomical organs brought into contact), actions are judged on other grounds: whether they are coercive, exploitative, in violation of prior promises, and so forth. In consequence, people who had fought to curb desires that often arose with great violence within them, people who were compelled to seek transient expressions of their sexual passions in clandestine and unsatisfactory encounters, people who constantly feared exposure of their secret lives, people whose central love for someone else could never be fully developed in arrangements that openly expressed it, are succeeded by similar people for whom all these problems are overcome. It is hard not to view that as ethical progress.

From an older perspective, one still surviving in some societies and in some groups even within countries that have made the transition, any tolerance of "deviant" sexuality is a sign of corruption, a mark of ethical decay. That perspective relies on two major claims, emphasized differently in different versions. First, homosexual desires are genuinely deviant, unhealthy eruptions within degenerate people, who should be encouraged to suppress them in favor of more salubrious (heterosexual) inclinations. Second, these desires—or, at least, the expression of them in homosexual behavior—are forbidden by divine command. Accepting same-sex preference rests upon establishing facts about the prevalence of homosexual desires and about the consequences of expressing them, as well as undermining the thought that satisfaction of these desires is forbidden by the deity.

As in the case of women's aspirations, discussed in §24, the normality of the desires is difficult to recognize in a society where they are seen as deviant. When homosexual acts count as a form of vice, when those who engage in them are reviled, mocked, and even prosecuted, the society will lack reliable statistics about same-sex desires and their behavioral expression. There will be little public knowledge of the character and consequences of homosexual relationships. Finally, those relationships will be profoundly and adversely affected by pressures to keep them hidden: not only will men and women struggle to find ways of meeting potential partners, forced to seek love furtively in squalid places, but they are also likely to absorb the social condemnation of what they do, feeling shame and guilt even while they achieve some temporary satisfaction. All this supports a public image of homosexual activity as infrequent, deviant, insalubrious, and stripped of all positive traits associated with the expression of love.

In part, this picture was rectified through the scientific study of sexual behavior, from sexologists of the late nineteenth century to psychologists and sociologists of later decades.[50] Whether or not its methods and

---

50. Even studies of sexual behavior that regard homosexuality as defective can play a liberating role—just as Wollstonecraft's apparently limited plea for female education opened the way to broader expression of women's aspirations. Freud's recognition of

data were completely reliable, Kinsey's famous report played a large part in undermining the repudiation of homosexuality as deviant.[51] If men and women were engaging in homosexual contact at the rates Kinsey claimed, the effects of the behavior could hardly be so terrible.

Also important was a related shift in ethical practice, acceptance of the wrongness of treating private consensual homosexual acts as criminal offenses. Against the background assumption that the law should intervene only to prevent conduct causing harm to others, increased understanding of facts about homosexuality induced many countries to repeal their (frequently harsh) statutes.[52] These legal steps neither modified the common evaluation of homosexual acts as immoral (as vice) nor removed the stigma associated with homosexuality.[53] To declare oneself a homosexual was an act of great bravery when same-sex acts were criminal, and it continued to require courage even after legalization, when "only" scorn and derision remained. Coming out was still hard to do.

Yet, just as consciousness-raising was crucial to full public recognition of the prevalence and extent of women's aspirations for public roles, so acts of coming out presented a different picture of homosexuality. Individuals who had previously seemed "normal" and "respectable" suddenly exposed the "darkness" and "squalor" of their private lives.

---

widespread homosexual wishes, even though associating them with incomplete development, modified prevailing ideas about their frequency.

51. Alfred C. Kinsey, Wardell B. Pomeroy, and Clyde E. Martin, *Sexual Behavior in the Human Male* (Philadelphia: W.B. Saunders, 1948); Staff of the Institute for Sex Research, Indiana University, Alfred C. Kinsey, Wardell B. Pomeroy, Clyde E. Martin, and Paul H. Gebhard, *Sexual Behavior in the Human Female* (Philadelphia: W.B. Saunders, 1953); E. O. Laumann, John H. Gagnon, Robert T. Michael, and Stuart Michaels, *The Social Organization of Sexuality* (Chicago: University of Chicago Press, 1994).

52. This depends on a prior ethical shift, adoption of the Millian conception of law (classically expressed in *On Liberty*). During the 1960s and 1970s, that conception combined with increased factual knowledge to produce a cascade of liberalizing reforms in European countries and in some parts of North America (Canada and some states in the United States, with Illinois leading the way). Denmark (1933) had taken the step much earlier, and, interestingly, the focus on the private may have inspired France (which had no antihomosexuality law) to institute a law against *public* displays of homosexual affection.

53. As Mill so clearly sees (in *On Liberty*) the effects of social stigma can be just as confining as those of the criminal law.

There were so many of them that standard assumptions about normality and respectability had to shift. The varieties of homosexual relationships became visible, and so too did the ways in which social attitudes blocked attainment of positive features people, whatever their sexual preference, want in their connections with others. As homosexuals resisted invasion of their lives, at Stonewall and after, the initial reaction to "deviants" who opposed the forces of "law and order" gave way to sympathy for people prepared to fight for the right to love whom and how they chose.[54]

Making the realities of homosexual desire and homosexual life visible was one part of the revolution. The other consisted in weakening the force of the idea that this is a form of sexual behavior proscribed by God. As some societies, notably in Europe, experienced a large decrease in the proportion of their citizens who accepted the authority of particular religious texts (the Hebrew Bible, the Old and New Testaments), justifying injunctions against behavior by appeal to the authority of the scriptures became increasingly suspect. Even among the devout, however, emphasizing a ban on "men lying with one another" came to appear curiously selective. Socially liberal theologians pointed out that the prescription occurs in a lengthy catalog of rules, almost all of which are disregarded by Christians and many of which are neglected by Jews. They commended the central scriptural doctrines, the ones enunciated again and again, illustrated with famous parables. If the will of the deity is to be honored, we should focus on what is centrally on his mind.

As with the examples of the previous two sections, it would be folly to claim that progress has gone as far as it can. Purging evaluation of sexual activity from any consideration of the sex of the partners—attending to the relevant qualities of homosexual and heterosexual relations alike—remains incomplete. The withering of vice depends on achieving a more selective, and more sophisticated, view of divine commands. It might have been accelerated by a deeper skepticism about the whole idea.

54. This transition can be traced in newspaper responses to the nascent Gay Pride movement; see in particular the reports in the *New York Times*, in the immediate aftermath of Stonewall and in subsequent years.

§27. The Divine Commander

Several episodes of previous sections would have gone differently—advancing further or faster—if there had been another revolution. From Plato on, philosophers have scrutinized the idea of grounding ethical codes in the commands of a deity. Although the arguments presented are powerful, they have failed to dislodge the idea, still popular around the world.[55] I shall later consider why this might be. First, however, the arguments.

Plato offers a dilemma. Either there is an independent standard for assessing the commands issued by the deity[56], or there is not. If there is, divine commands can be appraised as good or bad, so we can justify our following them if they are good; but creating this possibility simultaneously displaces the deity as the source of ethics; there is a fundamental measure of ethical goodness (rightness, virtue) prior to the divine acts of commandment. If, on the other hand, there is no prior source, we can no longer appraise the deity as good, nor see the commands as anything but arbitrary expressions of will; in consequence, the injunctions no longer have ethical force. Kant recapitulates the point succinctly, claiming there have to be prior sources for the moral law, because, without them, we could no longer recognize the "Holy One of the Gospels."[57]

After the twentieth century's spectacular organization of social machines for the brutalization and massacre of human beings, we should be sensitive to the ethical status of following orders. In Russia and Rwanda, Johannesburg and Jerusalem, defendants have sought to excuse themselves by claiming they were merely following orders issued by authorities. To judge them guilty, as courts and citizens do, presupposes that

55. I simplify. Some religions suppose transcendent beings are impersonal and lack wills. In these instances, one should speak of prescriptions to "align" oneself with the transcendent forces. Adopting more circumspect language here would be clumsy and obscure the lines of argument.

56. In Plato's *Euthyphro,* the divine source is represented as plural—ethics is a matter of what the gods love—and Socrates has a preliminary bout with Euthyphro in which he takes advantage of the possibility of the gods having divergent tastes. This is a flourish on the main line of reasoning.

57. Kant, *Groundwork of the Metaphysics of Morals,* a good English translation is that of Mary Gregor (New York: Cambridge University Press, 1998) (Akademie pagination 408).

the ethical characteristics of commanders affect the ethical status of those who obey. You can follow orders issued by a wrongdoer without yourself doing wrong—your boss, who is, unbeknownst to you, an embezzler, tells you to type an apparently innocuous document; but if you are in a position to detect the commander's corruption, you should resist the command. Often, the order given should cause doubts about the character of the person who issued it, as when you are told to herd the prisoners into the gas chamber.

Suppose ethics is really founded on a divine command: there is no prior source of goodness and badness, rightness and wrongness, except the will of the deity. You hear orders enunciated by the deity's representative, or read them in a sacred text. Should you follow? There is no independent standard by which you can judge the command. Issuing a different set of prescriptions would be neither better nor worse—the actual list reflects an arbitrary choice. You might obey, in the way you drive on the right in many countries. Equally, you might resist. The deity has commanded obedience—but he might equally have ordered disobedience. Why comply with actual orders, rather than those he might have given? Nothing demands that, except another order, and following *that* order has no higher backing than the command to comply with it.

In fact, however, your situation is worse, for sometimes the deity commands people to harm others. He orders a man to kill his son, declares a geographical region must be taken by force from those who inhabit it and most of the residents slaughtered, and insists that we put to death (or at least expel from the community) any men among us who are found "lying together".[58] You feel uncomfortable about following orders like these, but you find them in the sacred text and go along. In obeying are you so different from the functionaries who did their jobs in the machinery of death?

You have independent evidence about the will of this deity. Apparently he demands complete subordination and service: special places are to be erected for his worship and adoration; his will is to be carried out

---

58. Leviticus 18 and 20 suggest different punishments for the "lying together." Similarly, there are variations in what Canaanites are to expect, although, at best, only the young women will survive.

in everyday life. Those not prepared to follow the commands are to be punished, and the punishment will be eternal and infinitely agonizing on all possible dimensions.[59] Knowing this, you might be cowed into submission, just as some of the underlings of twentieth-century dictators did what they did out of fear of retribution. The deity is very powerful, the author of the whole show. Sheer power, however, has no bearing on whether you ought to follow his caprices.

Consider your predicament more carefully. You recognize the deity has commanded a large number of things, some of them apparently wasteful and expressing his narcissism (demanding elaborate forms of worship), some of them apparently breathtakingly evil. You do not know if there is an independent ethical standard by which the commands and the commander himself can be measured. If there is, your own independent judgment suggests some of the actions are radically at odds with it. If there is not, you are simply being ordered to satisfy a caprice, one alienating you completely from your human sympathies. Compliance is, at best, ethically neutral and quite possibly ethically incorrect. Hence, you should surely not follow the order.

An obvious response: who are you to judge? You are a thoroughly finite being whose knowledge is puny. But you should be clear on just what sorts of knowledge are pertinent to your predicament. If there is no antecedent ethical standard, no sense can be given to the idea that the deity knows more about what ought to be done than you do. Moreover, there is no sense in which satisfying his caprices is better than responding to your own human sympathies. He is more powerful than you are and knows more facts (perhaps all the facts) about the universe he created. Nevertheless, without an independent standard, following the orders of the more powerful and factually knowledgeable cannot count as better than following the orders of the weaker and more factually ignorant. On the other hand, if there is an independent standard, perhaps the deity has an access to it that his finite creatures do not: he has greater

59. See a posthumous essay of David Lewis, "Divine Evil" (in *Philosophers without Gods*, ed. Louise Antony [New York: Oxford University Press, 2007]; I completed this essay from an outline left by Lewis at his untimely death). The essay considers various possible ways for Christians to avoid supposing their God causes suffering on an infinitely vaster scale than any of the world's most celebrated human evildoers.

ethical knowledge and transmits it to us in his decrees. When our own judgment suggests the commands are hideously evil, we should wonder if our ethical knowledge is partial, and if the deity sees things more clearly than we do. To follow the orders, however, requires more than the bare possibility of the deity's superior insight. We must either have evidence for thinking the commander has special access to the standards of ethical correctness, or we must simply take this on trust. The former option is not available unless we use our own judgment about what the standards are, and, if we do so, the fact that the deity commands things that are, by our lights, horrible tells against the hypothesis of special access. In the end, then, the suggestion must be that we simply have to have faith in the deity as a source of ethical insights.

This is the best way to think about the divine commander. According to it, ethical standards are not created by the deity's fiat, but the deity has superior knowledge of those standards and communicates the knowledge to us (or to a few of us); we should trust that this is so and consequently obey. *We are now exactly in the position of the functionaries who defended their participation in acts of massacre and genocide.* The defendant speaks: "My job was to follow the orders. Although I felt uneasy about some of these orders, it was not for me to question them. For I trusted they were given by a leader who saw the whole situation far more clearly than I could ever do. I had faith in the leader, faith in the superiority of his judgment to my own, and faith in the rightness of not letting my own doubts intrude. That's why I obeyed." The defense is no more adequate in the context of following divine commands than when the one in charge is a human dictator.

Conceiving an unseen enforcer is a useful technique for socializing members of the group in the ethical code, and thus valuable in cultural competition (§17). The intellectual problems of viewing ethics as an expression of the divine will have been articulated by Plato and his successors, but the arguments fail to dislodge the thesis that the precepts of the group articulate the commands of the local deity(ies). Why is that?

The answer returns us to a central question of this chapter: Is the mere-change view acceptable? Ordinary thought about ethics accepts the possibility of ethical progress and seeks an independent standard against which ethical practices can be appraised. *What could that be, other than the will of some greater being?* Abstract philosophical substitutes are hard to grasp, or to fit to prominent examples of ethical advance.[60] So, for all its flaws, the picture of the divine commander survives.

Understanding the ethical project, its origins, its evolution, and the historical episodes supporting a conception of ethical progress can free us from the choice between unconvincing philosophical abstractions and problematic religious foundations. Showing that will be the work of the rest of this book.

60. The next chapter will defend this claim. I suspect that many people have an inchoate appreciation of it.

PART TWO

A Metaethical Perspective

CHAPTER 5

Troubles with Truth

§28. Taking Stock

The history of Part I aims to provide considerations for freeing us from unsatisfactory conceptions. We turn now to metaethical liberation, focused on questions about the possibility of truth and knowledge in ethics. The first goal is to show how standard accounts of possibilities of ethical truth and knowledge, explanations put forward to resist the "mere-change view" and its kin, fit the history poorly. Recognizing that will prepare for a positive proposal in Chapter 6.

The examples of the last chapter show apparent ethical progress. A handful of instances do not, however, portray the historical unfolding of ethical practice as a story of constant advances. Far from being prevalent in history and prehistory, progressive transitions might be quite rare. Given the large number of normative traditions, and the long period of time through which they have evolved, the total number of changes in ethical practice is vast. Most are unknown, and, for times and places at which ethical change can be studied, little has been done to understand the character of ethical *practice* and the ways in which it has evolved. We have "histories of ethics," attending to the *theories* philosophers and religious thinkers have constructed, but few studies systematically exploring the ethical practices of societies and their modifications.[1] Consequently, little is known about the prevalence of progress in ethical practice.

An *unsystematic* review prompts tentative skepticism about the steadiness of ethical progress. Abolishing chattel slavery was a progressive step. Yet it was preceded, three centuries earlier, by the reinstitution of a practice—the buying and selling of people—that had been rare in Europe for a long time; and it was followed, within a few decades, by practices of discriminating against the newly freed slaves and their descendants. Any Christian advances in spreading compassion in the Roman world gave way to sectarian warfare: the Romans who exclaimed at the love Christians displayed toward one another would have rethought if they had witnessed the appalling bloodbaths of the controversy over Arianism. Perhaps each progressive change has a regressive counterpart?

To resist the mere-change view is not to defend the prevalence of ethical advances. It is to suggest the *possibility* of progress. If that possibility makes sense, there must be conditions marking off the progressive transitions from others (regressive or simply nonprogressive). What constraints govern improvement of ethical practice?

To investigate the viability of a notion of progress, it is helpful to start with the impulse leading to judgments about progressiveness (judgments the examples of the last chapter aimed to elicit). Assume people today

---

1. Even Lecky's (Victorian) study of a long period in the history of ethics concentrates more on theoretical ideas than on the actual practices of groups of people. For the ancient world, the writings of Kenneth Dover (*Greek Popular Morality in the Time of Plato and Aristotle* [Oxford: Blackwell, 1974] and *Greek Homosexuality* [Cambridge, MA: Harvard University Press, 1978]); Moses Finley (*World of Odysseus* [New York: Viking, 1978]); Walter Donlan (*The Aristocratic Ideal in Ancient Greece* [Lawrence, KS: Coronado Press, 1980]); and Joseph Bryant (*Moral Codes and Social Structure in Ancient Greece* [Albany: State University of New York Press, 1996]) point in appropriate directions, as do the discussions of Peter Brown, *The Body and Society* (New York: Columbia University Press, 1988) and Jerome Carcopino, *Daily Life in Ancient Rome* (New Haven: Yale University Press, 2003), on the Roman world. I know of no sources for other places and later periods (collectively) as good in elaborating the ethical lives of ordinary citizens. Many historical works offer particular insights, but, without a focus on transitions in practice, ethical change is hard to study seriously. Hence my (amateur) efforts in the previous chapter.

agree in repudiating chattel slavery. They are repelled by the buying and selling, the harsh treatment, the division of families. Because they judge ethical progress to have been made when slavery was abolished, they do not want to go back.[2] When they contemplate the world *before* the reintroduction of slavery and the world after slavery returned, they prefer the earlier state to the later (in this respect, at least).

A first, very simple, subjective criterion for progress can be based on these reactions: a change in an ethical code is progressive just in case those who live after the change prefer life in the later world to life in the earlier one. Is the subjective criterion adequate as an account of ethical progress? There are ample reasons to worry. Desires expressed in attributions of ethical progress are purely contingent. Were the individuals who make these judgments to be placed within a rival tradition, one making different transitions in ethical practice (typically transitions running counter to those actually preferred), they would be likely to endorse incompatible judgments of progressiveness.[3] Human beings may be malleable enough to be brought—by the right, or the wrong, systems of training—to issue radically different verdicts on the same situations. Even when many different traditions agree in their modifications of ethical practice and retrospective endorsements, history might easily have gone differently, so consensus judgment is thoroughly contingent.

The skeptic issues a challenge: "You can call this 'progress' if you like, but this is no more than a way of comforting yourself; you have socially shaped preferences for the kind of life you now lead, in contrast to the lives your forebears led; you dignify these preferences with a label, but it is nothing more than an honorary title, masking the contingencies of the events and of your retrospective judgments." If things are as the challenger says, there are two possibilities. Either nothing more can be done to elaborate or defend the subjective criterion, and it is useless for

2. Or, at least, without the publicly tolerated institution of the seventeenth and eighteenth centuries. They may also be indignant about contemporary practices (illegal practices) of human trafficking—and believe that stronger measures, both legal and ethical, are needed to combat them.

3. In fact, actual people who lived through progressive transitions made incompatible assessments. Not all those who lived through emancipation preferred the world it produced.

opposing the mere-change view, or the subjective criterion can be developed to meet the challenge. Start with the latter option.

A point must be conceded to the skeptic. Often in the history of ethical practice, a modification is made at one stage, only to be reversed later. Decisions to afford certain groups of people protections and opportunities are sometimes unstable: consider attitudes towards Jews in Western societies, from the Middle Ages to the present; similarly, communities have changed their minds about the fair distribution of wealth or the responsibilities of the rich to provide for the poor. Pending a revisionary reconstruction of these apparent oscillations, mere postrevolutionary satisfaction looks too weak to suffice for ethical progress. The constraint allows progress to occur whenever people are immediately happy with a change. Modifications in contrary directions can count equally as progressive.

An obvious remedy suggests itself. Not only do progressive changes require postrevolutionaries to prefer life under the new dispensation to life under the old, but the preference must also endure. As the ethical project evolves, later generations must endorse the preference, looking back on the revolution as a progressive step. So strengthened, the criterion avoids the simplest difficulties, posed by rapid oscillations, but it needs to fix a period to serve for certifying past progress. How long must the preference endure? Any prescription seems arbitrary and, furthermore, subject to the obvious possibility of reversals occurring at a slower pace: for any chosen span of time, a preference could endure throughout that span, only to be reversed later.

Very well. Demand something yet stronger: the desires must be not only stable throughout some period after the revolution, but stable "in the limit."[4] Now the subjective criterion answers the worries about oscillations besetting weaker versions, but the notion of the "limit" of ethi-

4. Here, there are obvious connections with ideas of classical pragmatism, specifically to the account of truth often attributed to Peirce in Nathan Houser and Christian Kloesel, eds., *The Essential Peirce*, vol. 1 (Bloomington: University of Indiana Press, 109–123) and apparently also present in James. In the ethical context, the idea surfaces in Jamesian references to "the last word of the last man." William James, "The Moral Philosophers and the Moral Life," in *The Will to Believe and Other Essays* (Cambridge, MA: Harvard University Press, 1979).

cal practices needs explanation. Should this notion be tied to the *actual* course of human history, or should it be understood in terms of some *idealization* of our future? The *actual past* includes occasions on which apparently progressive ethical transitions have been reversed—for example, episodes in which particular groups have been protected by the ethical code, only later to be excluded again. To judge the inclusions as progressive is to diagnose the reversals as events of terrible social blindness. At some periods in human history, people—socially blinded people—have preferred to live in a world produced by a *regressive* transition, and there are potential causes that induce social blindness in any group. What guarantees the absence of these causes in the limit of the extension of our ethical practices? Understanding the criterion in terms of the actual course of human history, it cannot be expected to judge progress correctly.

Again, there is an obvious remedy. Require the preferences to be stable, in the limit, in a history that proceeds under "ideal" conditions: a transition within ethical practice is progressive if and only if, in a future proceeding from the transition and modifying practice under ideal conditions, there is a stage after which the preference for life after the transition is never reversed. This version is superior to any previously considered, but it contains, at its core, an unexplained notion, that of the ideal conditions these futures must satisfy. The ideal conditions depend on certain kinds of perturbing forces being absent, the variety of causes producing the forms of blindness of our ethical past. Historians are divided on how to account for socially induced ethical blindness, even in the most well-studied episodes.[5] Because we understand so little of earlier modifications in ethical practices, let alone the potential causes of regressive steps in it, the causal factors diverting the *past* course of ethical evolution from the ideal state cannot be specified. Nor can we hope to identify further factors possibly affecting our descendants. Our only purchase on the idea of the ideal unfolding of ethical practice is through

---

5. There is a large and sophisticated literature devoted to questions about how German citizens could have avoided seeing what was done to the "outcasts" under the Nazi regime, but it is hard to draw from it any definite catalog of all the perturbing forces potentially at work.

the distortion of human preferences: the perturbing forces are just those leading people to prefer to live in the world produced by a regressive transition. That specification of the "perturbing forces" would make the subjective criterion quite hopeless. Transitions count as progressive just in case the worlds they produce are preferred by people unperturbed by forces generating misjudgments about progressiveness.

Even in its best version, the subjective criterion is not good enough. It reduces to banal circularity. Furthermore, it introduces epistemological problems. We judge the transitions of the last chapter as progressive. Are we committed to a prediction about the future course of ethical evolution, actual or ideal? If progress was made, preferences for life after slavery or life with greater female equality will be shared by our descendants—so long as they are not "improperly swayed" by unspecifiable forces. How can we assess that? The judgments of progress are grounded in aspects of the episodes themselves, not because we anticipate the eventual stability of human desires (assuming no "perturbations"). Any confidence we have about the future, if it proceeds "properly," rests on thinking there is something about the transition to which people who come after us will continue to respond (insofar as their vision is not distorted). Their desires, like ours, are secondary, symptoms of the progressiveness of the transitions, not constitutive of it.

§29. Prima Facie Problems

The most obvious alternative to the mere-change view introduces an idea of ethical truth. Progress occurs when an ethical practice substitutes truth for falsehood. More generally, you might think of ethics as a form of inquiry subject to external constraints, constraints beyond the contingent preferences people have. In the rest of this chapter, that general thought will be confronted with the past evolution of the ethical project.

We have examples of apparently progressive changes during the past few millennia, and some larger modifications, about which we know much less, occurring in the more distant past. The examples invite obvious questions: Where exactly do historical actors bump up against the external constraints and acknowledge their force? How did the constraints play a role in the larger changes of the Paleolithic and early Neolithic? I offer blunt answers, to be defended more carefully in this and the two following sections: the transitions recorded in the historical record, as well as those hypothesized for the prehistoric past, are best conceived as "local adaptations," not episodes of ethical discovery.[6] The changes appear to be responses to difficulties of the social situations in which individuals and groups find themselves. "Moments of ethical insight" are elusive. My first aim is to show these answers to be prima facie plausible.

The historical figures who figure in ethical transitions, the vast majority of them unidentifiable as individuals, do not start from some situation in which they lack ethical convictions, follow a process of reasoning or observe some facet of reality, and thereby arrive at a well-grounded belief in an ethical judgment. Actual historical agents (and their prehistoric counterparts) were born into societies and socialized from early childhood. They acquired practices of expressing ethical evaluations, an extensive repertoire of ethical concepts, and dispositions to accept a body of ethical statements, most of which they never questioned. For the revisionary historical actors who stand out relatively clearly—Mary Wollstonecraft, John Woolman—what occurs is a *change* in ethical conviction: ethical beliefs transmitted within the society and shared by everybody around are rejected in favor of claims incompatible with them.

The psychological processes these people seem to undergo differ radically from the forms of evidence conjured in typical philosophical accounts of ethical justification. There are no special abstract forms of reasoning, nothing reported as a moment of "perception" or "intuition." Reformers take up the ethical project as framed in their culture, making proposals on the basis of empirical information they find salient.[7] In light of the background ethical views inculcated in the society, women act well by performing particular tasks as wives and mothers, while black

---

6. Here I recapitulate themes presented in a pregnant passage from John Dewey, *Human Nature and Conduct* (Amherst NY: Prometheus Books, 2002) 103.

7. Their justificatory achievements and the limits of what they can support will become fully clear only after the account of ethical method in Chapter 9. For the time being, though, I emphasize only the differences from the views of those who treat ethics in terms of conformity to external constraints.

slaves need to be brought to godliness. Wollstonecraft recognizes *facts* about the world: uneducated women often fail at some of the allegedly important tasks. So too does Woolman: slaves are not made more God-fearing by being treated as they usually are, and their owners do not exemplify standards of godliness. Recognizing these facts, the reformers construct arguments bearing on prevailing ethical assumptions. The discoveries they make about the world are not at all mysterious. It may have taken an acute observer, one sensitized by his or her own worries about personal salvation, to see how socially accepted behavior affected slaves and slave owners, but ordinary forms of observation can deliver the kind of information Woolman acquired and used. The forms of reasoning are of types well understood from other contexts: if women should perform particular tasks as wives and mothers, and educated women do a superior job (by current standards), women should be educated. To have uncovered the facts, seen their salience, and made the case were significant achievements, but they did not turn on some novel apprehension of "ethical reality."

Compare episodes in which prevalent scientific beliefs are changed. Historical studies have brought home the difficulties attending the disclosure of radical novelties: seeing the swaying incense burner as a pendulum requires a shift of perspective hard to achieve.[8] Occasionally, however, the unexpected forces itself upon the observer; Röntgen cannot overlook the fluorescence on the screen. What analogs can be found in episodes of ethical change? What psychological processes go on in innovators, and how are the constraints on ethical progress registered in their thinking or feeling? How exactly do reason, intuition, and perception work to generate new ethical insights? Woolman and Wollstonecraft saw aspects of the world those around them had not brought into focus; ancient aristocrats probably learned that, without a phalanx of hoplites, their city-states were indefensible. Once those features were widely

8. Here the loci classici are Thomas S. Kuhn's *Structure of Scientific Revolutions* (Chicago: University of Chicago Press, 1962 and 1970) and N. R. Hanson, *Patterns of Discovery* (Cambridge, UK: Cambridge University Press, 1958). The example of the pendulum is Kuhn's, and seems to me one in which his invocation of a "gestalt switch" is most convincing.

appreciated, continuing ethical practice in the traditional ways became *socially problematic.*

Consider what histories supporting philosophical talk of subjects "perceiving" or "intuiting" good or bad, right or wrong, would be like. Woolman might simply have "seen" the injustice of treating an individual slave in one of the usual abusive ways; Wollstonecraft might have "intuited" the wrongful oppression to which members of her sex were subjected. Reflecting on the crucial processes, they could have offered pointers to help others attain the insights vouchsafed to them. Although their efforts to persuade are extensive and passionate, neither attempts anything of the sort. They are radically unlike the scientific figures who undergo transformative observations. Galileo teaches his readers to see the swinging censer differently; Röntgen shows the fluorescent screen.

What could encounters with external constraints be like? Some philosophers have supposed we can have contact with something deserving the name of "ethical reality," and that people have psychological capacities enabling them to arrive at well-grounded ethical judgments.[9] Imagine you come across some boys dousing a cat with gasoline and igniting it. You judge the action to be wrong. The judgment is immediate, tempting you to say you *see* the wrongness of what the boys do, just as you see the unfolding event. In the same way we can gain knowledge of cats, gasoline, and boys through perception, we can also learn about the goodness or badness of states of affairs, the rightness or wrongness of actions. Or perhaps judgment is mediated by a feeling of repugnance, a violent antipathy to what occurs, so the negative reaction forms the basis for our ethical assessment. The perception puts us into an affective state, and being in that state warrants our ethical judgment.

There is a better explanation. You make the judgment, and if you make it immediately, you do so because your society has inculcated psychological propensities to apply the vocabulary as you do. If you feel

9. This position has been most clearly and precisely articulated by Nicholas Sturgeon; particularly valuable is his exchange with Gilbert Harman. See Nicholas Sturgeon, "Moral Explanations," in *Morality, Reason, and Truth,* ed. David Copp and Dean Zimmerman (Totowa, NJ: Rowman and Littlefield, 1985), and Gilbert Harman, "Moral Explanations of Natural Facts," *Southern Journal of Philosophy* 24 (1986): 57–68. In the following text, I use an example extensively discussed in this exchange.

particular emotions—horror, repugnance—it is because your society has connected affective states with cognitions and volitions and has marked out certain types of sentiments as "ethical," rather than merely idiosyncratic responses. In either version, your psychology has been shaped to generate, on perceiving particular cues (the cat's agonized squeals and squirming), an immediate, or relatively immediate, judgment.[10] Background ethical practices of your society, communicated to you in your early socialization, underlie your judgment. For those who propose *modifications* of practice, the requisite preparation for such direct responses is absent. Without something like a fogged photographic plate to subvert prior expectations, Woolman or Wollstonecraft would have no ability for direct assessment.

Socialization plays an analogous role in ordinary observation and in the refined scientific versions of it. The technician observing the bubble chamber or the sequencing machine arrives at judgments about types of particle collisions and characteristics of the DNA. Here too, immediate responses depend on prior training: this person does not have any peculiar ability to "see" subatomic particles; she differs from others only in having been taught to respond to particular types of tracks in prescribed ways. Her training enables her to gain knowledge of events involving subatomic particles, but it does so only in virtue of historical events through which these initially inaccessible aspects of reality were detected and ways discovered to make them manifest. The knowledge gained through current observation depends on the achievements of a previous group of observers, who were able to transform a world in which the remote particles were unknown to a world in which they could be detected by the contemporary devices. Historians and philosophers can reconstruct the processes of observation and reasoning behind our enhanced ability to observe nature. Observable phenomena provided a basis for relying on microscopes, techniques of microscopy were extended to reveal Brownian motion, recognition of Brownian motion led to knowledge of atoms and manufacture of devices for fathoming the properties of the atom, experiments with these devices led to the detection of the

---

10. In the version where your evaluation depends on appreciating your own emotional response, there is an intermediate step, but the process from perception to judgment remains relatively direct.

subatomic world, and so on. The history is long and complex, but our confidence in the quick judgments of the technicians who observe tracks in bubble chambers would be radically undermined if we thought it could not be told.[11]

Back to your horrified witnessing of the tortured cat. Like the technician at the bubble chamber, you make an immediate judgment—"There's a positron!,", "That's wrong!"—and you do so by exercising socially inculcated psychological dispositions. To view you in this way is not yet to cast doubt on the thesis that you are making contact with some external constraint and thereby arriving at new ethical knowledge. *Sustaining* the thesis, however, requires a story about the historical background to your belief, and to the dispositions ("techniques") instilled in you, one similar to the tale told for the detection of the subatomic world. Reconstructing history requires tracing a justified route—disclosing processes of reasoning, perception, intuition, or whatever—that led from the stage at which people looked on at squirming animals with indifference (making no ethical judgments about the actions producing the writhings) to the stage at which the judgments ("That's wrong!") became firmly accepted and the young became socialized to have propensities for making such judgments.

This particular ethical change did not figure in the catalog of the last chapter. Its history is obscure—even the epochs and societies within which it occurred cannot be identified.[12] We can, however, consider ethical revolutions for which we have historical evidence, and inquire

11. Historians and philosophers of science have paid great attention to this example, so it is reasonable to claim confidence here. See, for example, the writings of Mary-Jo Nye (*Molecular Reality* [London: McDonald, 1972]), Wesley Salmon (*Reality and Rationality* [New York: Oxford University Press, 2005]), Ian Hacking ("Do We See Through a Microscope?" in *Representing and Intervening* [Cambridge, UK: Cambridge University Press, 1985]), and Peter Achinstein ("Is There a Valid Experimental Argument for Scientific Realism?," *Journal of Philosophy* [2002] vol 99 pp. 470–495).

12. I suspect it is very ancient. An obvious conjecture is that it occurred first in connection with domestic animals—but that is no more than a conjecture. Some thinkers will maintain, with considerable reason, that the "revolution" of responding to nonhuman animal suffering is still far from complete. (A classic is Peter Singer, *Animal Liberation* [New York: Random House, 1975]). As we shall see, ethical conclusions about nonhuman animals raise difficulties for my own approach. For discussion, and an attempt at resolution, see §47.

whether in these instances a parallel to the story about the discovery of subatomic reality is available. The ways people like Wollstonecraft and Woolman arrived at and defended their proposed changes in ethical practice are absolutely critical, for these are the places where the constraining power of the external sources will emerge—if it ever does. The rejection of chattel slavery provides an especially good instance for these purposes because here our access to the psychological lives of those who might have made "ethical discoveries" is relatively good (although there is much even about John Woolman's psychological development that remains unknown).[13] Nothing in the recorded testimony of any abolitionist discloses any analog of the critical observation in which external constraints are apprehended differently (and correctly)—the analog of Röntgen's fluorescent screen. Woolman's *Journal*, the most revealing document we have, contains no mention of an occasion when he saw the plight of slaves in a new way (a sudden revulsion at the battered bodies before him, that—somehow—transmitted ethical insight). As I remarked (§25), black slaves do not appear as individuals in Woolman's narrative—nor do individualized slave owners whose "corruption" Woolman might suddenly "perceive."

Woolman does feel an emotion—an "unease"—on the pivotal occasion when his master asks him to draw up a bill of sale. As background to his experiences, the *Journal* presents his repudiation of his own past "wantonness," his renewed discipline, often expressed in reprobation of others. We are given a portrait of a young man who views the conventions of behavior, even among the Friends (the Quakers), as not always sufficiently scrupulous, and who minutely scrutinizes his own conduct to forestall possible lapses. The "unease" takes very explicit forms, exhibited in the reflections leading him to quiet it and perform the task assigned him: the buyer is elderly and himself a Quaker, thus decreasing the chances of sexual and physical abuse and increasing those of spiritual guidance for the slave. In the background are facts Woolman recalls as

13. It is also a good case for present purposes because it is one principal defenders of the idea of contact with external ethical constraints typically cite. Woolman is preeminent among early abolitionists for explaining the circumstances leading to his new ethical stance.

he ponders what to do—slaves are often beaten, female slaves are often sexually coerced, concern for the spiritual development of slaves is rare. He is suddenly asked to do something that makes him, if only tangentially, complicit with these common features of the slave-owning institution, and, for a young man so sensitive to his spiritual temperature, even that tangential involvement promotes unease. This psychological explanation fits the record we have far better than any (nebulous) process through which some (mysterious) external constraint manifests itself.

It is instructive to contrast the historical sources available in the ethical case with those in another instance in which philosophers want to explore possibilities of new (disputed) knowledge. In *The Varieties of Religious Experience*, William James canvasses the reports of those who acquired a new belief about "the transcendent," and his discussion is fully based on particular episodes in which people "saw" something strikingly new. Whatever scruples we may have about the reliability of the processes James's subjects underwent, reports of this kind are, in principle, the right kind of material from which a defense of new religious knowledge might be drawn. Historical documents describe the experiences and their apparent power. In the repudiation of chattel slavery, records of any similar experiences, moments of sudden revelation, are entirely absent.

If progressive ethical inquiry increases conformity with external constraints, the absence of episodes in which such constraints are recognized, even if only dimly, is thoroughly perplexing. When we reflect on other examples of ethical revolutions considered in the previous chapter, there is simply no evidence of times and places at which some sense of these constraints modified ethical practice, allowing for the institution of reliable techniques for everyday "observation" (ways of cultivating psychological dispositions enabling people to "see" particular states and actions as good or bad, right or wrong). For some transitions, the idea is absurd. Greek replacement of the ideal of heroic courage with that of solidarity has far more to do with the technology of fighting than it does with any moment of insight disclosing the ethical flaws of the Homeric hero or the virtues of the hoplite. Nobody would insist on moments of ethical insight, analogs to the scientific observations perturbing prevalent ideas about nature, if the grip of a background picture of ethics were not so powerful as to make it appear that such moments *have to have* occurred.

So far, a prima facie challenge for the thesis that ethics is a form of inquiry responding to external constraints. Defenders of the thesis must explain how those external constraints play some substantive role in the evolution of the ethical project. Otherwise invoking them is idle, a piece of comforting rhetoric easily excised. Yet the issue should not be left there, with a pointed invitation to tell a different narrative or to interpret the history of Part I differently from ways so far suggested. Diagnosis of the flaws of popular philosophical views is intended to fashion a better approach to ethical progress. Hence, the problems just posed should be examined more deeply, to reveal how serious the challenge is.

§30. Truth, Realism, and Constructivism

Interpreting ethical progress as consisting in the attainment of new truths has evident attractions. Although ethical codes are often distinguished by their rules and paradigms for behavior, the latter often encapsulated in inspiring stories, the codes of most societies employ special concepts to reformulate the content of commands as (what appear to be) descriptive statements: "Do X!" or "Act as Y did!" are accompanied by "It is right to do X" and "What Y did was good."[14] If such statements are true or false, the progressiveness of a step in which particular new commands were introduced (or in which old rules were modified or dropped) can be understood by considering the *descriptive counterparts* of the commands and rules (statements that evaluate the actions required or commended as right, or good, or virtuous), taking the descriptive counterparts to eliminate falsehood in favor of truth. Returning to episodes from the last chapter, the modification of the lex talionis is progressive because the statement "It is wrong to punish someone other than the

---

14. Anthropologists recognize cases in which the name of the group substitutes for the vocabulary used to recommend or command actions. For the purposes of this chapter, it does not matter whether the predicates used are "good"/"bad," or "right"/"wrong," or "virtuous"/"vicious," or even the names applied to a specific group and to those it counts as outsiders.

perpetrator" is true; and the abolition of chattel slavery is progressive because "Slavery is permissible" is false (as is the claim apparently espoused by Cotton Mather and others: "Slavery is a good thing").

The challenge posed in the last section can now be re-posed: how do you integrate an account of ethical progress as the substitution of ethical truth for ethical falsehood (or the accumulation of ethical truth) with the actual evolution of the ethical project? There are two main possibilities. Either there are episodes of ethical insight, occasions on which new ethical truths are discerned, or progressive transitions, those attaining new truth, have to be viewed as fortunate occasions in which blind stumbling turns out well.

The latter option proposes that ethical practice evolves in ways not requiring the apprehension of the constraints, but sometimes turning out to conform to them. When this occurs, the historical actors are unaware of the fact. They argue with one another and eventually reach peace on quite different grounds. For example, they take an important progressive step in extending some of their prescriptions to cover interactions with a neighboring band because they appreciate the possible benefits of trade (§19). Only much later thinkers, perhaps only those with a synthetic vision of the ethical project as a whole, can understand how they have unwittingly responded to the external constraints. Is this a satisfactory amendment of the original idea, one able to acknowledge the history of the ethical project and also to find a place for the idea of external constraint?

People arrive at true statements in a variety of ways, including ways providing them with no justification for their new beliefs. If ethical progress pervaded the history of our practices, it would be hard to credit such "sleepwalking." Where progress is systematic and sure-footed, as it is in mathematics and in the sciences, there would be genuine problems in integrating the thought of progress as consisting in the accumulation of truth with denying episodes of insight: it would be reasonable to wonder how human beings could, so consistently, be so lucky. With respect to ethics, however, where progress seems unsteady, no such worry about a happy series of coincidences arises. The sleepwalkers stumble along, often, indeed perhaps most of the time, lurching from error to error, but occasionally lighting upon new ethical truths. After the fact, matters become clearer, as expressed in the confidence of some judgments about ethical progress. Eventually those latter-day evaluations will have to be understood, recognized as knowledge—and that will demand some account of how the knowledge is gained. Providing the account poses difficulties of its own. For the moment, however, expand the menu of possibilities for meeting the challenge of the last section by supposing the external constraints are not apprehended by the participants, but only by much later thinkers, gifted with special insights into the character of ethical transitions.

Consequently, the challenge could be met in one of three different ways: by developing an account of ethical truth, in light of which episodes of ethical change could be reconstructed to show how individual participants apprehended these truths; by using the account of ethical truth to understand the capabilities of the later thinkers who can at last recognize the external constraints; or by proposing an account of ethical statements on which they are not seen as having truth-values.[15] Those who suspect the problems of the last section are only prima facie difficulties may maintain that the appearance is generated because the unanalyzed notion of *external constraint* is tacitly interpreted in crude ways, so the recognition of external constraints comes to seem much more mysterious than it really is. They may expect an account of ethical truth to clear up the trouble. The next step, then, is to understand the notion of ethical truth.

The approach to truth most obviously suited to a thesis about external constraints is that associated with the sciences and with everyday descriptions of the world around us. In these domains, progress can be conceived as the acquisition of significant truth, where truth is understood in familiar terms, as correspondence to reality.[16] The conception

15. This last possibility would then need to be elaborated with some alternative account of the external constraints, supplemented with a proposal as to how these constraints are recognized either by participants in episodes of change or by the clever folk who come later.

16. My *The Advancement of Science* (New York: Oxford University Press, 1993), as corrected by *Science, Truth, and Democracy* (New York: Oxford University Press, 2001), provides an approach of this type. Other accounts are possible, for example, approaches conceiving progress as problem solving. Those would be closer to the perspective I shall eventually take in the ethical case.

of correspondence does not presuppose peculiar composite entities, "facts," to which true statements correspond. Rather, to put the point pedantically, simple atomic statements, "Jakie helped Krom," for example, are true in virtue of the referential relations between logically constituent terms and parts of reality, together with the set-theoretical relations of inclusion. "Jakie helped Krom" is true because "Jakie" and "Krom" are singular terms, picking out particular objects in the world, and "help" is a two-place relational term picking out a relation (a set of ordered pairs), and the pair <Jakie, Krom> belongs to that set.[17] The truth of more complex statements is understood in terms of the conditions laid down by Tarski for the language of first-order logic.[18] This understanding of truth for scientific sentences combines well with an account of progress in terms of the attainment of truth and with the actual historical development of the sciences.[19] For humbler types of scientific statements, we find no great mysteries: Frans de Waal was able to recognize the truth of "Jakie helped Krom" because he could observe Jakie, Krom, and the helping behavior. When the entities whose properties are recorded in true statements are more remote, extra philosophical work must be done to show how access to these entities becomes possible: but that work can, and has, been done (as in the instance of subatomic particles, considered in the previous section).

Because this approach to truth lends itself so easily to the acceptance of external constraints, it is good to begin with it; other conceptions of

17. Here I rely on the Tarskian account of the truth of logically simple sentences of first-order languages (sentences that do not contain connectives or quantifiers). I abstract from issues about tense.

18. Alfred Tarski "The Concept of Truth in Formalized Languages," in *Logic, Semantics, Metamathematics* (Oxford University Press, 1956), 152–278; in his important essay "Tarski's Theory of Truth," *Journal of Philosophy* (1972), Hartry Field shows how Tarski generates the notion of truth from the notion of reference. The correspondence theory I espouse does not assume, with Field's article, that it is possible to reduce the notion of reference to some physicalist basis. See my essay "On the Explanatory Power of Correspondence Truth," *Philosophy and Phenomenological Research* 64 (2002): 346–64.

19. There is a line of thought, descending from Kuhn's *Structure of Scientific Revolutions*, that denies this claim. For defense, see my *The Advancement of Science* and *Science, Truth, and Democracy*. See also my "Real Realism: The Galilean Strategy," *Philosophical Review* 110 (2001): 151–97, and "On the Explanatory Power."

truth will occupy us later. How would a correspondence account apply to the ethical case? Recall the simple example: some boys douse a cat with gasoline and ignite it. Call this event *E*. "*E* is wrong" appears to be a true ethical statement (we might normally use less restrained and abstract vocabulary). On a correspondence account, its truth will stem from the fact that "*E*" picks out an event and "wrong" refers to a class of events, a class in which *E* is included. No trouble threatens with the first part of this condition; any anxiety must result from difficulties with understanding the reference of the ethical predicate.

How does "wrong" come to refer to a particular class of events? Distinguish two broad possibilities. On a *constructivist* approach to ethical reference, we stipulate that particular events are to be labeled as "wrong"; the boundaries of this class are matters for our decision, not fixed antecedently to the decision. On a *realist* alternative, there is a preexistent division between actions that are wrong and those that are not, and the only place for human construction or convention lies in decisions about which sounds or signs to use in marking the distinction. Within each of these broad approaches, further important distinctions should be drawn.

Consider, first, the realist alternative. In attempting to integrate ethical truth and its apprehension with the history of ethical practice, it can adopt one of two rival stances concerning property detection. The property marked out by a predicate may be *simple* and *irreducible*, in the sense that no elaboration of conditions necessary and sufficient for the presence of the property is required to explain how people detect that objects (at least some objects) have it. The property of being an approximately straight line looks to be of this type. Although geometry texts sometimes define a straight line as the shortest distance between two points, we do not acquire the belief that a path is approximately straight by investigating all the possible alternative routes between its endpoints and satisfying ourselves that this one is (among) the shortest. Other properties, the *complex* and *reducible* ones, need subsidiary conditions to play a role in their detection. You attribute the property of being cilantro to the green stuff in the grocery store by attending to the shape of the leaves, and, if you worry about possible confusion with the parsley, by sniffing it; detection proceeds by explicitly noting the conditions. Properties appearing simple and irreducible may turn out to be complex and reducible, however, not because subsidiary conditions are consciously noted, but because the explanation of their detection involves unconscious apprehension of those conditions. If color properties apply to macroscopic objects in virtue of the disposition of those objects to scatter light of particular wavelengths, our detection of these color properties will be explained in terms of our ways of responding to the characteristic patterns of scattering.[20] A realist approach to ethical truth may thus suppose the properties marked out by ethical predicates ("good," "wrong," and so forth) are simple and irreducible or complex and reducible; with respect to the latter alternative, the subsidiary conditions may be consciously noted, or unconscious apprehension may play a role in detection.

Much more could be said about these options, but brief explanations are sufficient for refining the challenge of §29. Consider next the possibilities for constructivism. The very simplest version of a constructivist approach to ethical truth takes the reference of ethical predicates ("wrong," for example) as simply a matter for conventional determination. Any group, or maybe even any individual, can decide, on any grounds or on none, to apply the predicate so it picks out any set of objects, states of affairs, or events. This is the Humpty Dumpty theory of ethical truth—with respect to the key ethical vocabulary, we, collectively or individually, are always the masters—and it allows any ethical transition whatsoever to be "progressive" from the viewpoint of those who make it, or, since all can be assimilated, abandons any contrast between progress and regress. It capitulates to the mere-change view.

More interesting versions of constructivism suppose there are conditions on the processes through which the reference of ethical predicates is determined. The set of events marked out as wrong consists of those events acquiring a special status if particular procedures were followed.

---

20. I do not suppose this is the correct account of color properties. It is a popular one, and hence useful for introducing an option a realist about ethical truth may try to exploit. The example may inspire some to retract the supposition that the property of being an approximately straight line counts as simple and irreducible: perhaps the only such properties are geometrically more basic, or perhaps there are no such properties at all. I need not quarrel with such reactions. Since I shall conclude that realist truth about ethics cannot be sustained in any of its guises, it is best to be inclusive about the possibilities allowed.

So, for example, there might be some form of reasoning each individual can undergo that divides actions into a number of types; or there might be some hypothetical social process, in which groups of people can engage, generating distinctions among events; or there might be some actual social process either generating or revising such distinctions. Kant and his successors pursue the first option, claiming there are processes of reason, available to all rational beings whatever their social and physical environments, yielding conclusions about the status of actions. Social contract accounts of the traditional kinds develop the second, suggesting that ethical distinctions among actions derive from the deliberations people would make under ideal conditions. The third is developed in the next chapter, and, since the present focus is on rival perspectives, further consideration of it is postponed till then.

Both realists and constructivists have several ways of developing the thesis that ethics is governed by external constraints. All the alternatives so far reviewed adopt the framework of truth as correspondence. One rival approach deserves a mention. Instead of considering the *structure* of truth, how the truth of statements arises (what makes truths true), one may adopt a *functional* account, seeking to understand what we *aim at* in various areas of inquiry.[21] For descriptive statements about the physical world, from common sense to refined science, a correspondence account seems to deliver both structure and function. By contrast, applying a correspondence theory of truth to mathematics appears problematic, precisely because it requires us to suppose the existence of a realm of abstract objects, whose properties are ever-more precisely and more completely described by the mathematical statements accepted, and because the ways in which mathematical statements come to be accepted make it utterly mysterious how mathematicians are gaining access to that realm. The great figures in the history of the subject, those to whom we attribute the most significant advances, respond to earlier problems by introducing new notation, often inspired by attempts to carry out symbolic manipulations on a broader scale. They seem to be expanding the lan-

---

21. The distinction is drawn in Michael Dummett's important essay "Truth" (reprinted in his *Truth and Other Enigmas* [Cambridge, MA: Harvard University Press, 1978], 1–24). Dummett does not use the terms I employ to mark the distinction.

guage, thereby introducing new games for mathematicians to play. Perhaps a superior account of mathematics would abandon the enterprise of saying what mathematics is *about*, and concentrate on characterizing the kinds of statements at which mathematical practice *aims*. It would offer a functionalist account of truth.

We might approach ethical practice in similar fashion, seeking the functions that ethical prescriptions and other parts of ethical codes are to serve. In the next chapter, ethical progress and ethical truth are considered in this light. As we shall see, the result is very different from the usual ways of viewing ethical progress as conformity to external constraints.

We now have an array of possibilities for developing a response to the prima facie concerns of §29. Is any of them adequate?

§31. The Sources of the Troubles

We can profitably begin by countering one attractive gambit. Section 29 disclosed some important historical actors—Wollstonecraft and Woolman—as discovering important *factual* truths. They use their discoveries to argue, against the background ethical practice of their contemporaries, for the favored modifications (education for girls, freedom for slaves). Perhaps, then, novelties occurring within the ethical project are always of a special kind: ethical progress consists in recognizing true factual statements—statements, for example, about what women do under particular circumstances or about the attitudes of those treated as disposable property; these statements are integrated with previously accepted ethical statements to yield new consequences. Appearances to the contrary, there are no fundamental ethical innovations. No statements involving ethical vocabulary are ever added, subtracted, or modified, without making inferences from premises consisting either of previously recognized ethical truths or of new factual truths; ethical concepts are never introduced or refined. (These formulations rely on everyday understanding of the distinction between ethical and factual statements. Many writers have recognized "mixed terms" ("thick ethical concepts") involving both a factual and an ethical component: concepts such as *cruel*. In my usage, a concept with any ethical component is ethical.

Debates about borderline cases are irrelevant for present purposes. For the issues about ethical truth and progress turn on examples where the distinction between the ethical and the factual is quite uncontroversial: it is a fact that educated women do certain things; "Slavery is wrong" is an ethical statement.)

If the denial of fundamental innovations were sustained, realists could dismiss a large part of the challenge outright, for there would be a uniform account of progressive ethical transitions. Once the ethical project was begun, each generation inherited a body of established ethical beliefs, indeed ethical truths, from which people drew false consequences because the most general and basic ethical truths were conjoined with incorrect factual statements. Wollstonecraft discerned the truth about women's behavior under conditions of education, and Woolman recognized the psychological effects (on slaves and slave owners alike) of chattel slavery. They made ethical progress by replacing falsehood with truth.

Realists would still have to explain how the entire project was started, how, in the dim mists of the Paleolithic, small human groups arrived at all the fundamental concepts and principles required to yield, with different factual premises, the divergent conclusions adopted by rival traditions at distinct epochs. Although nobody should deny the significance of new factual knowledge in the progress of our ethical practices, it is hard to view the proposed story as explaining the entire evolution of ethics—can we really suppose the origins were ethically rich enough to allow the explanation envisaged? In denying ethical novelty, the account assumes a collection of elaborate ethical principles was not only available for Wollstonecraft and Woolman to work with, but present at the very beginning. The history of Part I took the initial institution of socially embedded normative guidance, undertaken in response to the fragile and tense hominid society it transcended, to be relatively crude and limited—to comprise principles about sharing and the like. To suppose ideas about roles, expanded altruism, and the good life were already part of the Pleistocene package strains credulity. Furthermore, at the dawn of written history, societies made important transitions not easily assimilated to the story: the refined conception of the individual figuring in the reformulation of the lex talionis, the reshaping of the concept of courage, the acceptance of the policy of forgiving your enemies.

Realists must allow occasions on which people persuade their contemporaries to accept ethical statements not yet adopted by those around them (even though most efforts at persuasion are rejected). When such episodes culminate in ethical progress, the thesis of external constraints takes the resultant practice to accord more closely with those constraints. Articulating this in terms of acquiring truth, it is assumed that true ethical statements replace previously held falsehoods (or, perhaps, prior lack of commitment). On the more ambitious version of the appeal to external constraints, the people who initiate the progressive transition apprehend these truths. Twenty thousand years ago, some people saw it was good to extend some of the protections accorded to their bands to the neighbors; three thousand or so years ago, people apprehended that it is wrong to kill the daughter of a man who has murdered another man's daughter; after Wollstonecraft, Britons accepted the value of education for women; after Woolman, Americans gradually rejected slavery; and, in the last decades, many have learned that the combination of sexes in loving relationships has no ethical significance. Here are statements of the form "*E* is *F*," where *E* is some event, state of affairs, or pattern of conduct and *F* is some basic ethical property (goodness, for example) that historical agents are supposed, somehow, to have grasped.

Suppose the basic ethical properties are simple and irreducible. Just how did those who made the changes apprehend them? We do not know exactly when the extension of ethical rules to the neighbors occurred, but we can easily imagine how it might have happened. Members of a band have ascertained facts about the resources commanded by those across the river and envisaged opportunities for exchange. Sitting around the campfire, they agree to make gestures of restraint and nonhostility to instigate mutually profitable interactions. The practice succeeds and is eventually accompanied by explicit declaration that, on the pertinent occasions, peaceful conduct is required. Speculative though it is, this sociopsychological account is more plausible than the suggestion that, one day, one band member enjoys some experience of the rightness of a pattern of behavior the group has not yet tried and that he communicates his experience in some way, enabling them to share his recognition.

We know a lot more about Woolman's journey to his judgment against slavery—and people of my generation can examine their own changing views about the ethical character of homosexual relationships. Woolman's dedicated efforts to persuade others to share his convictions about slavery offer all sorts of considerations that integrate factual information with prevalent ethical precepts, but his writings never indicate how readers might put themselves into a position to apprehend the simple and irreducible wrongness of owning other people. If he had some episode of apprehension, he hid the character of his revelation. Unless other people are different from me and those contemporaries I know well, shifting ideas about same-sex relationships are products of wider knowledge about varieties of human love and of extensive conversations with people who live in different ways. Factual knowledge and social exchange are prominent in this important ethical advance, but I find no place for apprehension that same-sex love has a simple and irreducible property with which the conventional wisdom of the 1950s failed to credit it.

Versions of realism holding ethical properties to be complex and reducible look more promising. Consider the suggestion that one important constituent of rightness is acting to diminish suffering and that another is promoting unsatisfied desires. Realists adopting this suggestion can view ethical innovators as appreciating facts about suffering and the frustration of human aspirations—Woolman recognizes, in perfectly straightforward ways, that slaves are subjected to all kinds of pain, and Wollstonecraft understands that her own desires, as well as the wishes of women she knows, are thwarted. Perhaps those who imagine peaceful exchanges with neighboring bands foresee extended possibilities of meeting the needs of the groups. How do they use their perceptions to justify judgments about the wrongness of slavery, the rightness of educating girls, or the goodness of extending the protections applying within the local group? Neither acting to relieve suffering nor satisfying others' desires is sufficient for doing what is right. The ethical codes within which the innovators operate deny that the sufferings of slaves are always wrongly inflicted, view women's desires for education as problematic and misguided, see the wishes of the neighbors across the river as irrelevant to goodness. Behind these conventional judgments stand other views about what is gained, achieved in the way of goodness or rightness, by allowing the sufferings, quashing the desires, or treating the neighbors with hostility. Woolman, Wollstonecraft, and my hypothetical advocate of trade question those views, effectively because they do not find anything strong enough to counterbalance the perceived pain relief and desire satisfaction.

Do the pioneers or their interlocutors have any articulated account of the good and the right, clearly held in view, an account showing the exact circumstances under which desire satisfaction and relieving pain contribute to goodness or rightness? Apparently not. Woolman and Wollstonecraft are aware of contentions about valuable goals achieved by permitting the sufferings of slavery and not responding to female demands for education—and they actively and specifically oppose those contentions. They express their own assessment of what is gained and what is lost, but they do not operate with any independent understanding of goodness and rightness beyond those available in the formulations of their opponents. They are *catalysts of a renewed social exchange*, a conversation of the sort that initiated the ethical project, and *the properties of goodness or rightness can be seen as fixed through such exchanges.* It is even harder to imagine how the early champion of neighborly exchange could appeal, either in his own thinking or in his efforts to persuade, to some reduction of goodness or rightness sanctioning his proposed modification of the code.

The point can be illuminated by focusing on another example. The modification of the lex talionis relieves one form of pain but substitutes another of the same kind: in either version, someone's life will be truncated. Perhaps that fact is itself ethically problematic, but the substitution of perpetrator for daughter surely looks like ethical progress. We know nothing about the people who proposed and argued for the change. Yet, on the realist account, there must be some connection between facts about suffering and mortality, and ethical properties, that differentiates the cases in which daughter and perpetrator suffer and die. If the innovators and those whom they convince are to apprehend the rightness of making the transition, they have to recognize that connection. How? We can dimly recognize the *form* of the explanation realists want, but we have no idea about how to give it substance. By contrast, it is easy to suppose that the ancient societies in which the transition occurred were engaged in frequent debate, that many voices participated, and that eventually continuing the old practice became *socially problematic. Conversation within* a social group replaces the nebulous *contact with* some external standard.

The realism just considered supposes some externally fixed connection between the sorts of things ethical innovators apprehend—things the examples reveal them as recognizing in perfectly straightforward ways—and fundamental ethical properties. The challenge is to say what the connection is, to explain how it could be apprehended, and to support the hypothesis of actual (albeit dim) apprehension. I now turn to what seems the most promising realist position.

According to that position, ethical properties are conceived by analogy with colors, and colors identified with dispositional properties of objects: redness, for example, is a disposition to cause us to enter certain neuropsychological states (triggered by the impinging on our retinas of light of particular wavelengths).[22] Goodness and badness, rightness and wrongness, apply to actions (say) in virtue of the tendency of those actions to generate reactive emotions, feelings of approbation and repugnance, for example. A great advantage of this account is its license of justified ethical assessment without any complicated cognition (backed by some unspecified process). You see the torturing of the cat, you feel the repugnance, and your reaction both prompts and justifies your ethical judgment.

The reactive emotions that figure in this proposal are not merely affective states available to be triggered across a wide range of social environments. If they were, it would be impossible to account for the phenomena we are attempting to understand, cases of progressive ethical change (where in *similar* environments people react quite differently to the same events). Instead, the reactive emotions—approbation, gratitude, resentment, repugnance—involve cognitive and volitional states con-

---

22. Working out the details of any such account of color properties is itself a large task, but I shall simply suppose that it can be done. Many contemporary thinkers have been attracted to the thesis that ethical properties of things, states of affairs, actions, and patterns of behavior can be treated analogously. The version offered by John McDowell ("Values and Secondary Qualities," in his *Mind, Value, and Reality* [Cambridge, MA: Harvard University Press, 1998]) is, in my judgment, the best, precisely because it is sensitive to the thought that the responses in the individual are shaped by the culture in which he or she grows.

nected, in ways nobody yet knows how to specify (§4), with affective states. What reactive emotions are available to an individual, the extent to which a person is sensitive to such emotions, and the entities triggering the emotions are all subject to social shaping (§4).

It is important to appreciate this triple dependence on the environmental conditions, including the social environments, in which we find ourselves. Features of the *ambient* environment at the time at which we encounter some occurrence affect whether we have an emotional reaction to it, and what form that reaction takes. Features of the *developmental* environment, the surroundings in which we learn and grow, shape the ways in which our emotional reactions are directed. Even more fundamentally, that developmental environment affects our emotional repertoire. For, although there may be some physiological responses relatively invariant across regimes of socialization, the emotions pertinent to ethical assessment are more complex than any such affective reactions, having socially shaped cognitive components. No realist approach to ethical properties that ignores these three modes of environmental influence can be adequate, for the simple reason that certain types of environments can produce reactions strikingly at variance with one another. Realists need a distinction between types of environments, proposing that particular reactions of people who have been socialized in *normal* developmental environments, and who find themselves in *normal* ambient environments, signal the goodness (say) of the states of affairs to which they respond.

A full version of this form of realism has to suppose an external standard fixing some environments as the pertinent class in which specified reactive emotions suffice for specified ethical properties—the fact that in some situations (death camps, the near-starvation conditions of the Ik) people fail to respond to acts of cruelty does not affect the ethical properties of those acts. Let us concede that the specification can be provided, and the realist can even explain why the "normal" environments are privileged.

Consider, now, the ethical innovator. At last we appear to have an articulated, and convincing, account of how that innovator responds to the external constraints on ethics. An ordinary experience produces a strong reactive emotion, say, a violent feeling of repugnance against some form of behavior tolerated by the surrounding community. The innovator is prompted to a judgment of the wrongness of the conduct, and the reactive emotion provides support for the assessment.

Yet surely we—and the innovator—should hesitate. Reflective people know they can feel different, quite complex, emotions toward many types of events and actions; they know that their background moods, their idiosyncratic affections, and any number of other contextual features can shape their feelings. Faced with the indifference of those around them, they should reasonably wonder about their own reactions. Are they expressing personal idiosyncrasies? Is something awry with the environment in which they find themselves (is it "abnormal" in whatever sense the realist has provided for that term)? Is the emotion what they take it to be, ethical repugnance rather than some variety of nonethical disgust, or even oversensitive distaste? There is nothing "inner" to which they can point to settle these doubts. Hence, they must either have some more explicit knowledge enabling them to convince themselves that the sentiment is a genuinely ethical one, or they must enter into dialogue with their fellows, attempting to explain just how and why they are moved as they are.

Woolman was a tireless campaigner for his abolitionist views, a man whose "unease" only deepened as he debated with people who disagreed with him. His initial response, however, was appropriately modest—he completed the bill of sale. Despite his unhappiness with the transaction, he surely worried that he was being "over-nice." Who was he, after all, to question practices other godly men and women, the Friends among whom he lived, accepted without demur? Only the persistence of the feelings, together with the extensive confrontation with many alternative points of view, intensified his conviction that slavery was wrong. If we view him as justified, it is because he comes to recognize his emotional reaction is not transitory—it cannot be displaced by the most severe attempts to revise some of the associated cognitions and volitions. *It is the social exchange, not any awareness of an external standard endorsing his feelings, that gives Woolman whatever ethical insight he has.*

There are gaps between unrelieved suffering and wrongness, between thwarted desires and wrongness, and between negative emotions and wrongness. The commitment to external constraints requires those gaps to be filled with a standard, obtaining independently of the individual and of society, which justifies the move from perception of suffering, or of frustrated desire, or the feeling of the emotion, to a judgment of wrongness. Innovators *appropriately* worry whether their willingness to make this move is idiosyncratic. If they are justifiably to reach the conclusion, on the realist account, they must have some apprehension of the standard. Not only do latter-day analysts struggle to explain just what the standard is, but there is not a shred of evidence of historical recognition of it. As a result, realism is fundamentally flawed.

So, too, are the idealizing forms of constructivism, and for the same reasons. Consider the individualist proposal attributing capacities for moral reasoning independent of social environment (in some views, absolutely independent of all experience). Let us (charitably) suppose the processes giving rise to the distinctions the constructivist claims to draw are thoroughly and precisely characterized. Not only are we told that "wrong" applies to actions when the principles on which they rest cannot be universalized, or that the word marks out actions that would be rejected by people deliberating under ideal conditions, but we are given a complete and precise account of universalization or of the ideal conditions.[23] Constructivists take the external constraints to which ethics is to conform to consist in the privileged status of these procedures. When ethical practice is properly pursued, people apply ethical predicates in ways matching those sanctioned by the procedures. In the approach currently under consideration, historical actors are assumed to apprehend the accord between their innovations and the deliverances of the privileged processes.

23. The examples I offer here recapitulate the most famous versions of constructivist approaches, the first evident in Kant, *Groundwork of the Metaphysics of Morals* (Mary Gregor, trans.) (Cambridge: Cambridge University Press, 1988), and the second attributable to Rawls, *A Theory of Justice* (Cambridge, MA: Harvard University Press, 1971), and his successors [most obviously T. M. Scanlon, *What We Owe to Each Other* (Cambridge, MA: Harvard University Press, 1998)]. I do not think any of these constructivists succeeds in providing a completely clear and precise account of the processes envisaged—Kant's critics, for example, have rightly complained that his appeals to "contradictions" and his implicit restrictions on the types of principles of action (maxims) permitted are vague and loose. The challenge I am presenting, however, to integrate the account of ethical truth with the evolution of the ethical project can be more accommodating and can concede to constructivists successes so far not achieved.

Given the complexity of the special procedures, constructivists probably do not want to contend that innovators recognize all the details, that they can formulate the correct account of the ethical vocabulary and see clearly how it applies in the case at hand. At best, they have inklings of the right version of constructivism and can apply some rough approximation to the tests it supplies—they cannot universalize in the strict sense of the theory, but they can ask, "What if everyone acted like that?" Not even this much can be traced in the historical record. We might attribute some inchoate universalizing thought to Wollstonecraft or to Woolman, but it does not figure in the works they write to persuade others. The absence is comprehensible, for relatively imprecise thoughts about universalization would have little force in the debates engaging them, debates centered on allegedly important distinctions among kinds of people (men and women, Africans and people of European descent). Much more precise constructivist ideas, not inklings, would be required to make any headway, and Wollstonecraft and Woolman do not have those tools. For earlier transitions, attributions of incomplete apprehension of the constructivist's preferred standard are less plausible, even ludicrous. The unknown modifiers of the lex talionis and the hypothetical pioneers who proposed extending the code to allow for trade almost certainly lack the concepts appeals to universalization require.

Like realists, constructivists face a more fundamental difficulty. Imagine some hypothetical innovator, as insightful as you please, who tries to follow your favorite constructivist procedure for applying ethical vocabulary. On the basis of the attempt, she announces her proposed change in ethical practice. It is clear to her, from the beginning, that others, including those counted as authoritative within her community, disagree with her. What should she make of that fact? One obvious self-diagnosis is that, for all her efforts, she has failed to carry out the procedure properly. To the extent she *cannot articulate* the grounds of her judgment, discussing them with those around her who are reputed to be wisest in ethical judgment, she should have doubts about her own competence. If she could explain and convince, laying out the procedure she has followed in detail, explaining its privilege, and challenging others to conclude differently, she might have justified confidence in what she has done. No historical actor shows even the slightest signs of these abilities.

Constructivists often suppose individual "reason" can generate justified applications of ethical predicates, without attending to potential worries that what has been done fails to get "reason" quite right (perhaps even bungles badly). In circumstances of innovation, those worries are especially apt. How could they properly be suppressed? The question fails to arise because constructivists often attribute a special self-certifying property to "reason" (it is a priori, luminous, or whatever). With respect to mathematics, where people often, but by no means always, reach agreement, that ascription might enjoy some small initial plausibility—although even there it is, I believe, profoundly misguided.[24] In the context of ethics, where disagreements are rife, it has no plausibility, and in situations where someone is challenging features of the prevalent code it should surely be discarded.

To sum up the discussion so far: neither realist nor standard constructivist approaches to ethical truth can provide a satisfactory account of progressive transitions in the evolution of ethics that will attribute to innovators some apprehension of external constraints. Before we take up the less ambitious idea of viewing historical actors as "sleepwalkers," whose fortuitous conformity to the constraints is appreciated by their more enlightened successors, it is worth exploring another position distinguished above. This position supposes ethical statements are, properly speaking, neither true nor false. To attribute wrongness to an action is not to ascribe some property to it, but to express an attitude. Some philosophers, disconcerted by the "strangeness" of the idea of preexistent divisions among acts and states of affairs whose contours ethical truths limn, have claimed ethical statements lack truth-values, taking them instead to express the emotional reactions of those making them, coupled with an injunction to share the emotional reactions.[25]

24. I defend this unpopular assessment in *The Nature of Mathematical Knowledge* (New York: Oxford University Press, 1983), chaps. 1–4.

25. The very simplest versions emerge in the early writings of A. J. Ayer and Charles Stevenson, although there are previous sources of inspiration in Hume. Both Ayer and Stevenson articulate the position with increasing sophistication, and there are intricate debates about the adequacy of the proposed semantic treatment for all contexts. A well-developed version of the view is provided by Allan Gibbard, *Wise Choices, Apt Feelings* (Cambridge, MA: Harvard University Press, 1990).

As characterized so far, noncognitivism (to use the standard name) provides no account of ethical progress. Reactive emotions of different sorts could arise in different societies, being directed toward different actions and states of affairs, without there being any standard against which the rival types of emotions, or the deeds triggering them, could be assessed. Some versions of noncognitivism might abandon any search for ethical progress. In exploring the challenge for the thesis of external constraints, however, we should focus on positions seeking a distinction among transitions. Those views take the ethical project to have made progress when specific types of emotions became available to us (we learned to feel shame, guilt, or resentment) and when such emotions were generated by particular types of situations. The notion of ethical truth is abandoned, but constraints remain. Some capacities for emotion and some emotional reactions are *apt*; others are not.[26]

Any version of noncognitivism hoping to go in this direction, and to avoid viewing ethical evolution as a matter of mere change, has further work to do. It must specify what is meant by taking emotional responses to acts and states of affairs to be improved; it needs a conception of "emotional progress."[27] However any such conception is developed, if it is to buttress the idea of historical actors apprehending external constraints—presumably by feeling things they can recognize as "more apt" than those of their fellows—it will face problems exactly parallel to those discerned above. Appeals to emotional reactions work no better in the context of noncognitivism than they did when ethical properties were seen as dispositions to induce particular kinds of emotional responses.

The challenge defeats the thesis that ethics is a form of inquiry in which *innovators* recognize and respond to external constraints. Retreat, then, to viewing the attainment of ethical novelty as sleepwalking. Human beings, individually and collectively, stumble along, sometimes responding to the difficulties of their social lives, sometimes feeling

26. Here I echo the language used by Gibbard, whose version of noncognitivism is the best I know.

27. Many noncognitivist writings accept the relativist conclusion (or are uninterested in resisting it). There are hints of a more progressive view in Gibbard's *Wise Choices*, but, for all its attention to many details of noncognitivism, I have not been able to draw from that work any clear account of ethical progress.

confined by the ethical codes they inherit, and consequently modifying those codes. Some innovations replace ethical falsehood with ethical truth (or less apt emotional responses with reactions that are more apt). Although the processes through which such novelties become incorporated in ethical practice are various and complex, typically having little to do with any apprehension of ethical divisions among actions and among states of affairs, there are occasions of progress, as communities align their codes with external constraints.

How could anyone know? Some who come later, after the changes, must be able to pronounce on what has occurred. Looking back, the enlightened recognize ethical progress, perceiving new ethical truth incorporated into an ethical code. How do they make their judgments? If their judgments are formed like those of the historical predecessors, there is no reason to believe they are anything other than further ventures in sleepwalking, providing no new knowledge. To avoid skepticism, the defender of external constraints must show how the latecomers do better than their predecessors, who, it is admitted, had no knowledge of novel ethical truths. Analysts like us are in a better position to judge.

Many philosophers hold a parallel view of mathematics. Committed to foundationalism, they propose basic principles, axioms, somehow apprehended, from which mathematics proceeds by deductive arguments, codified in proofs. One problem with foundationalism stems from the elusiveness of the ways in which the basic truths are justifiably apprehended: most authors content themselves with a phrase or a label.[28] A second difficulty lies in the schism made between the "sleepwalkers" of history and the enlightened folk of the present. Since the basic axioms come late in the history of mathematics—the route to them is full of erroneous formulations—the great mathematicians of the past did not apprehend them. Euclid, Descartes, Newton, Euler, Gauss, Weierstrass, and company formulated mathematical truths but did not know them; only in the twentieth century did mathematical knowledge become possible. Pragmatic naturalism rejects the odd combination of elusive

---

28. Most popular is "intuition," a term Frege rightly saw as overused. It stands to Kant's credit that he attempted to give the notion of intuition some clear content. See Kitcher, *Nature of Mathematical Knowledge*, chap. 3.

epistemology with a peculiarly self-confident attitude toward the present and the recent past.

Mathematics, however, has the advantage of consensus. Scholars who reflect on how mathematical knowledge is attained offer divergent accounts but concur in supposing a body of knowledge, shared among mathematicians. Ethical theory resembles mathematics in lacking any convincing proposal for how anyone, past or present, has access to any external constraints. At the same time, ethical practice lacks the power to achieve consensus that supports the attribution of mathematical knowledge.

If, in the mathematical case, we should regard the practitioners of the present as akin to the mathematicians who preceded them—rejecting a division into "sleepwalkers" and "enlightened"—stronger reasons favor a comparable assimilation with respect to ethics. What exactly has been learned about the external constraints on ethical inquiry, constraints the unwitting innovators could not apprehend? Ethical theorists often make confident assertions about fundamental constraints. Moore declares it to be evident there are only two types of intrinsic goods, human relationships and beautiful things. Kant locates the moral law within us, identifying it with Reason, which generates the Categorical Imperative. These radically dissimilar suggestions would be defended along similar lines. There are supposedly fundamental processes giving rise to the conclusions defended, intuition of the good (Moore), and deliverances of a priori reason (Kant). If you worry about nebulous processes, incompletely characterized, supposedly giving rise to knowledge of basic mathematical axioms, in a domain where people largely agree, you should be profoundly skeptical about equally mysterious invocations generating assertions of incompatible claims.

Was Moore a novel type of *Homo ethicus*, equipped for the first time with an ethical analog of the microscope enabling him to discern previously unrecognizable constraints? Psychosocial explanation appears more plausible. Moore's judgments about beautiful things and human relationships arise from a long psychological journey. The development of his ideas began in the Victorian socialization of his childhood, winding through his school experiences and interactions at Cambridge, the social and sexual entanglements of the elite university community of which he was a prominent member, the chafing of the late Victorian ethos and

the various revolts against it. Detailed biographical research would explain Moore's forthright dicta about value—and the attractions his views
held for his contemporaries.

Equally for Kant. His brilliant account of ethics in terms of Pure Practical Reason can be understood as a reaction to the moral law as commanded from on high. Attracted by ideals of autonomy and repelled by
the thought of subjection to a divine commander, Kant solves his conflict with a perennially popular account of ethics by placing the lawgiver
within, elaborating his revisionary theory in a framework of powerful—
but purely hypothetical—faculties. Kant can be assimilated to the great
innovators in the history of the ethical project. He responded to the state
of ethical practice he found in the Prussia of his day in an especially ingenious way, but there is no reason to think he transcends the predicament of his fellow pioneers.

Theorizing about the ethical project has been hampered by assuming
there must be some *authority* in ethics, some point of view from which
truth can be reliably discerned. Philosophers have cast themselves as
enlightened replacements for the religious teachers who previously pretended to insight. But why credit any individual participant in the project
with such special authority? Ethics may simply be something we work
out together (Chapters 6 and 9). If latecomers improve on previous efforts,
their success might be based, not in occupying some privileged epistemic vantage point, but on study of the history of the project as it has led
up to the practices of their day.[29]

We can now see why attacks on the idea of the divine commander fail
to dislodge the idea (§27). The alternatives seem so mysterious. Ironically,
supposing ethical truths are claims a deity endorses and reveals does

---

29. Dewey is optimistic that ethics can find its own analog of scientific method [see, for
example, chaps. 10 and 11 of Dewey, *The Quest for Certainty*, vol. 4 of *John Dewey: The
Later Works* (Carbondale: Southern Illinois University Press, 1984)]. I share the hope that
a properly informed understanding of the ethical project and its evolution might lead our
successors to pursue it more sure-footedly. That will depend, however, on more systematic
studies of changes in ethical practice (whose absence I lamented—§28). It would be foolish
to suppose the historical cases inadequately sketched in Part II can do anything more than
reorient discussions of ethics so as to prepare eventually for synthetic understanding and
improved ethical practice.

better at integrating the thesis of external constraints with the evolution of the ethical project than any sophisticated secular rival. Adherents of divine-command theories suppose that much, if not all, of the correct ethical system has been disclosed at some point in the past, and even that many apparent innovators receive special messages from on high. They can resist the mere-change view.

Unfortunately, these theories collapse under Plato's objection (§27) and, more fundamentally, because of the demise of the principal character. Is there any remaining option for resisting the evolution of ethics as mere change?

CHAPTER 6

Possibilities of Progress

§32. The Centrality of Ethical Progress

It is tempting to think resisting the mere-change view *requires* making sense of the concept of ethical truth. Confronted with the deviant practices of others, including people who came before us, we want to distinguish our ethical principles from theirs: so we must claim ours are true, theirs false. Hence the troubles elaborated in the last chapter appear to leave no options: even though the mere-change view conflicts with spontaneous judgments about historical episodes (Chapter 4), we have to acquiesce in it.

Truth is readily seen as prior to other notions used to explain the objectivity of our practices, concepts like progress, justification, and knowledge. To make progress is to accumulate truth, to be justified is to proceed in ways reliably generating true beliefs, to know is to have a true belief generated by a reliable process.[1] In this way of relating the

---

1. Here I follow the reliabilism pioneered within epistemology by Alvin Goldman, *Epistemology and Cognition* (Cambridge, MA: Harvard University Press, 1987). There are well-known problems of saying exactly what kinds of reliability are at stake in securing justification or knowledge; these are lucidly presented by Robert Brandom, *Articulating Reasons* (Cambridge, MA: Harvard University Press, 2000, chap. 3); solving these problems

concepts, an adequate response to the mere-change view has to start with ethical truth—and we are returned to the array of options of the last chapter. We could escape by permuting some elements in the picture. Here is a proposal: ethical progress is prior to ethical truth, and truth is what you get by making progressive steps (truth is attained in the limit of progressive transitions; truth "happens to an idea").[2] Pragmatic naturalism retains a notion of ethical truth for expository purposes, but it starts from the concept of ethical progress.

The demand for some idea of objectivity—for talking of ethical truth, ethical progress, or ethical knowledge—arises when we consider alternatives to our own ethical codes. Some of these are codes adopted in the past, others are codes different people follow in the present, yet others are codes we envisage ourselves coming to follow. In all instances, the practical question concerns the possibility of amendment. The mere-change view finds no basis for choice: there is no sense in which substituting one code for another would be objectively better or worse. An extreme version of objectivism would declare that, for any two genuinely distinct codes, either one is objectively better than the other, or there is a third code, constructible from elements of the two rivals, objectively better than both. Pluralism, as I conceive it, proposes there are (1) some pairs of codes for which one is objectively better than the other; (2) some pairs of codes for which neither is objectively better than the other, but there is a third code constructible from elements of both, objectively better than either; and (3) some pairs of codes for which neither (1) nor (2) obtains. *To salvage the notion of "objectively better than" that occurs in these claims and counterclaims, we do not need any concept of ethical truth. It is enough to recognize which kinds of changes would be progressive or regressive.* Opposition to the mere-change view can be formulated by

requires pragmatic judgments, dependent on the values endorsed. Those problems will not be addressed here, since the task is to understand how appeals to truth and justification in ethics might be reconceived in terms of the notion of ethical progress.

2. The parenthetical characterizations connect the permuted picture with the pragmatist tradition, the first with Peirce, C.S. Peirce "The Fixation of Belief" in Nathan Houser and Christian Kloesel, eds., *The Essential Peirce*, vol. 1 (Bloomington: University of Indiana Press, 2009) 109–123 the second with James, William James *The Meaning of Truth* (Cambridge, MA: Harvard University Press, 1978), 169.

saying there are pairs of codes such that the transition in one direction is progressive (and, in the opposite direction, correspondingly regressive).

Looking at the codes of the past, we are concerned to evaluate them because of the possibility of going back. Fundamental to our assessment is the question whether a particular transition, one reversing the actual course of history, would be progressive. Faced with the codes of actual rival people, we are interested in possibilities of shifting to their commitments, or of constructing from the elements of their current code and the elements of ours a new amalgam. We ask whether the change to their code, or to the amalgam, would be progressive. Similarly for codes we envisage. What concerns us is the possibility of a progressive transition.

In fact, it is *better* to approach these decisions by starting with the notion of progress rather than that of truth. Thinking in terms of truth narrows the focus. For truth applies to statements, so we are led to conceive the decision as one about descriptive counterparts of rules of the alternative code. There are other components of ethical codes—concepts, exemplars, habits, emotions, modes of inducing compliance—and improvements to our own practice could occur in each of these respects. A rival code whose rules agree with ours might do substantially better at preventing relevant forms of blindness. Thinking in terms of progress responds more directly to the practical choices we face.

Attending to contexts of comparison, in which issues of progress arise, poses an obvious question. Is it right to take the overall progressiveness of an ethical transition as fundamental, or should we think instead of changes as progressive in particular respects? Historical changes often seem to involve losses as well as gains. Rejecting the glorification of the Homeric hero in favor of social solidarity impresses us overall as progressive, but we may want to concede something to Nietzsche's nostalgia for lost creativity, daring, and freedom. On balance we can talk of a progressive transition, but that is because we can weigh the relative importance of the aspects with respect to which progress occurs. So, it might be suggested, the fundamental notion is that of a transition progressive *in certain respects*: judgments of overall progress are made by adding the weights of the progressive respects and subtracting the weights of the regressive respects. Sometimes the sum can be done, and we can talk confidently of a progressive (or a regressive) transition, but there will be occasions on which no common scale for comparison can be found, when no determinate weights can be assigned, and no overall judgment can be reached.

We should reject this universal atomism. Although it may sometimes be a useful strategy to decompose an overall judgment of progressiveness by identifying advantages and disadvantages of an ethical change (and showing the former outweigh the latter), the thought that *all* overall judgments are built up out of atomic judgments involving no commitment to weighing gains and losses is suspect. With respect to any change whatsoever, are there really atoms, possessing only advantages or disadvantages? Human behavior is so multifaceted that even if we focus on one aspect of the change—the bonding together of the citizens of the *polis*, or the increased freedom felt by women who see new doors open for them—are these facets uncontroversially positive in all respects? Perhaps they bring relatively modest problems in their train, or at least increased risks of small disadvantages (the anxieties of newfound freedom, comrades' blindness to outsiders), slight, to be sure, in contrast to the large positive gains. If so, further decomposition is needed to find the genuine atoms, whose weights will figure in the eventual sum. However far we dissect, new instances of the same phenomenon are likely to arise. Overall assessment of progressiveness is essential to ethical judgment. The basic concept is that of a progressive transition (period). Yet we can allow for dissection of transitions to proceed where, and to the extent, it is useful—for there will be cases of overall judgment in which a consideration of respects, without commitment to the idea that they are "atomic," clarifies the basis for judgment.

So far, I have tried to show how appraisals of progressiveness could do the work often achieved by appeal to truth. Issues relating to justification have been ignored. That is because the need for objectivity is felt primarily in contexts of live choice, when we wonder whether an envisaged transition would be worth making. On the picture of truth as fundamental we ask whether we would be gaining new ethical truth and eliminating old ethical falsehood; in my replacement view, we are concerned with the progressiveness of the transition. If and when we need a notion of ethical justification, it is easily found: people are justified when their decisions are generated by processes likely to yield progressive changes.

Reliability in the production of ethical truth gives way to reliability in the genesis of progressive transitions.[3]

§33. Generalizations from History

Part I portrayed the evolution of ethics as driven by forces of selection. Darwinian considerations figured in the prehistory of ethics (the emergence of the preconditions for normative guidance and experiments of living), possibly retaining a role in subsequent stages of cultural competition (as when experiments lead groups who practice them to wither and go extinct); they were clearly present indirectly when ethical codes became attractive to other groups because of biological benefits they appeared to confer (greater survivorship of young members, for example). Forces of cultural selection, dependent on the attractiveness of particular ethical ideas and thus answering to human desires, desires possibly independent of Darwinian advantages (§16), have also shaped the evolution of the ethical project. Friends of ethical truth find no reason to think forces like these are likely to generate true ethical beliefs; by the same token, nobody should suppose them conducive to ethical progress. That means, however, only that progressive transitions are not to be identified with those promoting Darwinian or cultural success, a point simply recapitulating the unsteadiness of ethical progress—as well as the failure of crude evolutionary reductionism.

What could ethical progress be, if not accumulation of ethical truth? An obvious (empiricist) strategy for finding an answer is to start with the motivating examples of Chapter 4 and examine what, if anything, they have in common.[4]

Suppose we were to have a comprehensive vision of the progressive transitions occurring in the evolution of ethics, from the beginnings of socially embedded normative guidance to the present. Merely listing all

3. Issues about justification and knowledge, and how these notions apply to historical actors, will occupy us briefly at the end of this chapter.

4. This is the empiricist strategy pursued by Hume in his second *Enquiry* [*Enquiry Concerning the Principles of Morals* (Indianapolis, IN: Hackett, 1986)], a strategy in accordance with his understanding of Newtonian method. (On this topic, I have been helped by Matthew Jones.)

these transitions and then declaring any ethical change progressive just in case it occurs on the list would plainly be unsatisfactory. That would fail to allow for future progressive transitions. More important, the skeptical challenge (§28) would arise again, in the pointed form of asking why anyone should care about ethical progress, if it is just a matter of making one of the officially designated transitions. We want an account of ethical progress isolating common features of the favored transitions, features revealing why we might be concerned to make ethical changes of this particular kind.

Chapters 2 and 3 offer only an infinitesimal fragment of the envisaged list. Yet their constituent examples are often in line with the most popular idea about ethical progress: insofar as they have attended to ethical progress, historians and philosophers have singled out a particular kind of movement as constitutive of advances. They have talked of "the circle expanding" or "the expanding circle."[5]

The proposal is motivated by a striking feature of some progressive transitions in the history (and prehistory) of ethical practices. In the relatively recent examples of the abolition of slavery and the recognition of women as equal participants in public life, individuals who had previously not been brought within the purview and protections of ethical precepts routinely applied to others (white men) became recognized as *full people*, as proper subjects for the application of principles hitherto applied on a restricted scale. Slave owners were committed to precepts forbidding certain kinds of behavior toward full people: full people were not to be permanently separated from their families; they were not to be bought and sold. Before the ethical change, black men and women did not count as full people; after it they did, and old proscriptions now applied to them too.

Earlier examples reinforce the idea. A prominent Gospel message denies that boundaries of aid and compassion embrace only the local group. Even in prehistory, interaction with other groups, temporary fusion of

5. The former phrase is used by W. E. H. Lecky in his *History of European Morals from Augustus to Charlemagne* (New York: Braziller, 1955); the latter is the title of Peter Singer's book-length attempt to identify ethical progress, *The Expanding Circle* (New York: Farrar, Straus, Giroux, 1981).

smaller bands into larger units, negotiation, and trade become possible when some rules were extended to cover outsiders.

Reflections of this kind inspire a straightforward theory about ethical progress. Ethical codes begin by increasing the scope of altruism within a group (enlarging the set of contexts across which members are prepared to respond altruistically; §5); they progress by further expansions of scope and by expansions of range (the set of individuals toward whom altruistic responses are made; §5). Progressive transitions occur exactly when the modified code contains precepts enjoining altruism of wider scope or greater range than the code it replaces.[6]

There are difficulties. Not all our paradigms of progressive ethical change fall under the envisaged rubric. Consider the modification of the lex talionis. Here there is a shift from inflicting suffering on individuals specially related to the perpetrator to punishing the person who did the deed. The transition does not begin with a class of people initially protected by an ethical precept and another class of people not so protected, and transfer some of the latter individuals to the protected class. Initially, anyone is vulnerable to harm, provided he or she stands in a particular relationship to the crime—being the son or daughter of someone who killed the son or daughter of another person; after the transition, anyone is vulnerable to harm if he or she stands in a *different* particular relation to the crime (being the doer of the deed). No circle is expanded; one circle is replaced by another.

Sometimes, what changes is not the group attitude toward a particular individual or class of individuals, but the group attitude toward the desires those individuals have. Consider the withering of vice (§26). With some strain, you could say there are two classes of people, those with preferences solely for sexual activity with members of the opposite sex and those who sometimes want to engage in intercourse with a member of the same sex; before the change, there is toleration (an altruistic response?) of the desires of the former class, but toleration is not extended to similar desires of the latter class; after the transition, toleration extends to both.

---

6. As I read Singer, he is attracted by a theory of this sort (although possibly not one so extremely rudimentary); at an early stage of my own thinking about ethical progress, I was too.

Quite apart from the issue of whether talk of altruism is really illuminating (or appropriate) here, the crucial step consists not in gathering persons under a protective umbrella from which they were previously excluded, but in recognizing their desires as worthy of expression. Inclusion and exclusion apply to *desires*, not people.[7]

Attention to some changes that must have occurred in prehistory reinforces these conclusions. During the Paleolithic, our human ancestors modified their ethical codes to require individuals to use their talents for the benefit of the group, they introduced roles and role-specific prescriptions, they added requirements for development of talent, they came to appreciate the higher forms of altruism that play a role in human relationships, and, in doing all these things, they developed a richer conception of human life and of the human individual.[8] Among the members of the earliest societies to have left written records are at least some who have a conception of what it is to live well, far more elaborate than any present in the groups that began the ethical project: for well-born Babylonians and Egyptians, a life satisfying the basic desires felt by their distant ancestors would not be good enough.[9] To be sure, aspects of their understanding of what is good constitute no advance at all (their eager acquisition of gaudy luxuries); others are simply regressive (desires to be

7. Defenders of the "expanding-circle" proposal might reply that they allow for extension of *scope* as well as of *range*. Yet the example is importantly different from central instances in which an ethical advance consists of increasing the collection of contexts in which people are willing to help one another (for example, by commanding the response, even when the costs for agents would previously have inhibited addressing the other's plight). Failure of response arises not from the burdens of altruism but from the character of the desire. Precisely because of what the other person wants—sexual relations with another man or another woman—there is condemnation, assault, punishment, even murder. Progress does not consist in expanding any circle but in recognizing facts that normalize desires.

8. In citing these changes I do not presuppose my "how possibly" explanations of them, but simply *that* these transitions occurred. That is a simple consequence of the difference between the initial stages of normative guidance and the practices present at the dawn of history.

9. Whether it would be enough for the distant ancestors themselves is a question to be considered. One might well attribute to them desires for cooperation and for making contributions to joint projects. If so, the *seeds* of a richer notion of living well are already present.

viewed as superior). Important parts of their conceptions, however—the wish for relationships marked by higher-order altruism, the wish for a life of service to the community—mark genuine ethical progress. In these respects, what has evolved is an advanced appreciation of human possibilities.

The experiments of living of our prehistoric ancestors generated an important mode of progress that fits the "expanding circle" account, in extending their prescriptions beyond the small bands of the first human beings, as well as increasing the scope of (behavioral) altruism. Besides that, they improved techniques of securing compliance (in large measure by inventing the unseen enforcer; §17). They also changed radically the framework in which altruism, and its failures, could occur, by enlarging the repertoire of human desires. Part of the story of ethical progress must consist in understanding how *acquiring* new desires, not merely *satisfying* them, counts as progressive.

My formulations contain important modifying words: *some* people in Near Eastern societies had more elaborate views of the good human life, and *some* desires in the enlarged repertoire represent real advances. These words signal the fact that the lineages extending from the first small bands of ethicists to the far larger and more complex cities of Egypt and Mesopotamia made ethical losses as well as ethical gains. One obvious loss consists in replacing an initially egalitarian society with something highly pyramidal. That shift renders many members of the society invisible to others—and dramatically expands the class of altruism failures. For many of those acquiring the richer conception of the good life, the invisible people no longer figure as potential targets of altruistic response. Another loss, interconnected with the development of larger, hierarchical societies, is the shift from a situation in which all are involved in socially embedded normative guidance to one in which rules are set by particular figures (typically representatives of the unseen enforcer). These are the principal loci, at which it becomes attractive to talk of ethical progress *in particular respects*, not overall progress. The significance of a mixture of gains and losses will be apparent in later chapters.

There are other modes of ethical progress besides "expanding the circle." The next sections offer a more inclusive account.

§34. Problems, Functions, and Progress

One way to break the stranglehold of the idea that progress consists in accumulation of truth is to consider an area in which advances are understood differently. Technology serves as a paradigm. Our world is full of instruments, machines, and devices that improve on previous efforts. The chair in which I sit, the light illuminating my desk, and the computer on which I type these words are all refinements of similar things I used a decade ago, and spectacular advances on things my ancestors employed to similar ends.

Progress with respect to these artifacts, and in the domain of technology generally, is readily understood as functional refinement. We start with a function to be fulfilled and an initial device that does the job. From first success descends a sequence of improvements, things performing the task more reliably, more quickly, more cheaply, and with less demands on the user. Decreasing the error rates is important because those who use the devices usually want the job done correctly on a series of occasions. Speed, cost, and ease of use are worth having because we have other jobs to do, and these compete for our time, our resources, and our energy. More reliable, faster, cheaper, and less effortful performances are better, contributing to functional refinement and to progress.[10]

Behind the original device stands a person and a problem, or a person with a problem—usually a class of people with one or more problems. The people want to solve the problem, and that wish grounds the function of the successful device. A particular thing, introduced into a context, has the function F in that context if and only if that thing is present because someone wanted something to do F and the thing was introduced into that context to satisfy the wish.[11] Functional refinement consists in

10. In many cases, people have to make decisions about what factors are most important to them, trading reliability against cost, or speed against ease of use. Here, as in the ethical case, we can recognize progress in particular respects. Sometimes it is possible to talk of overall progress, but sometimes not.

11. Here I adapt ideas from a seminal discussion by Larry Wright ("Functions," *Philosophical Review* 82 [1973]: 139–68). There are long and intricate discussions of how to elaborate Wright's "etiological" approach to functions, but details are not pertinent to present purposes. The version of the account I favor, and that underlies my general claims about

satisfying the wish more reliably or more completely, and doing so in ways that generate fewer problems for potential users (that is, allowing satisfaction of their other desires to proceed more freely).

If a device is complex, its function frequently generates functions for its parts. One function of my desk chair is to enable me to sit at my desk for long periods without back pain; the slide letting the surface against which I lean move up and down has the function of enabling me to direct the support where I most need it. In general, functions of parts are determined in virtue of their causal contributions to the function of the whole.[12]

Attribution of functions is straightforward where there are clearsighted potential users who can express their desires and identify their problems. On other occasions, we can talk of functions even though cognitive beings with desires and needs are not involved. Biologists and physicians routinely discuss the functions of organs, bodily systems, cells, and molecules. They identify causal processes contributing to the survival and reproduction of the organisms in which the organic parts are found. An animal has teeth of a certain shape, and we recognize them as having the function of grinding tough woody matter into digestible form. That is to say, teeth of this shape have a causal power to break down the exterior walls of the plants in the animal's environment, modifying this material so it can serve, given the background properties of the animal's digestive system, as nutrition, and hence contribute to the organism's persistence. Although there is no cognitive being whose wishes are fulfilled by having animals survive—no benevolent creative deity, no Mother Nature—there is an analog of the problem background figuring in the technological case. Here the problem background is constituted by the Darwinian forces shaping the characteristics of life and its history: at the basis of these forces lies the general problem of reproducing, from which results, at a slightly less general level, the problems of

---

functions here, is given in my essay "Function and Design," *Midwest Studies in Philosophy* 18 (1993): 379–397.

12. This is an important insight of Robert Cummins ("Functional Analysis," *Journal of Philosophy* 72 [1975]: 741–65). I amend Cummins's original account by requiring that the causal contribution relate to some function of the whole system (and thus there must be a prior notion of overall function). For defense, see my "Function and Design."

surviving, being fertile, and finding mates (for sexually reproducing organisms).

Functional refinement comes with modifications of organic parts so they are able to discharge the function more thoroughly, more reliably, more quickly, or with less strain on other bodily systems. Teeth making plant casings more digestible, less likely to leave indigestible lumps, demanding less chewing time, or diminishing the effort required to chew are refined versions of teeth that came before. As in the original case of technological devices, some improvements are directly aimed at the job to be done, while others respond to the ancillary conditions of user or organism. Speed and cheapness are good because people have other things to do with their time and money; speed and diminished effort are worth having because organisms have other demands on their time and energy.

Physiologists routinely identify the functions of molecules without knowing very much about the details of the evolutionary processes that led to the presence of the molecules in the cells carrying them. They can neglect history because they see the *problem background* for the organisms in question and can pick out the problem-solving contributions the molecules make. The problem background results from the most general Darwinian pressures (the need to reproduce, the need to survive long enough, the need to be fertile) in light of the organism's prior constitution. Animals require food; so, if they are already committed to living— locked in by their anatomy and physiology—in a particular environment where only a certain food source is available, they need capacities to process that food; if the digestive system is hard to modify, they will need ways of transforming the raw material into something digestible, given the present constitution of the stomach; hence the problem of breaking down the casings, solved by the teeth; from this come attributions of functions to molecules deployed in building teeth with the required properties. Reconstructing that problem structure, with its cascade of evermore specific difficulties to be overcome, is compatible with any number of historical scenarios, most of which physiologists cannot formulate and among which they do not have to choose.[13]

---

13. For more detail about the differences between this approach to functions and that taken by etiological theories, see my "Function and Design."

Between clear-headed recognition of problem structures and the biological cases, in which no cognitive subject who sees the problems and designs the solutions is present, stand intermediate cases. On occasion, people recognize difficulties—they know not all is well—even though they cannot frame the troubles exactly. You feel twinges of discomfort and sometimes pain when you perform particular motions; your doctor formulates the problem precisely and prescribes a supportive device or a program of exercises. In examples like this, we can distinguish the different perspectives of the problem facer and the problem solver, the former encountering the problem background and the latter recognizing its structure. From one pole, at which problem facers know everything a problem solver might grasp, there is a sequence of cases in which problem facers are increasingly incapable of seeing what is wrong, leading to the opposite pole, biological examples in which problem facers lack any cognitive grasp or are not even cognitive subjects at all.

Given this general view of functions and functional refinement, I propose that socially embedded normative guidance is a social technology responding to the problem background confronting our first full human ancestors. None of them had a clear understanding of that problem background. Moved by a sense of the fragilities and tensions of their social life, they first guided their behavior by regularities to help them avoid trouble and later discussed with one another rules to govern conduct, to be applied in increasingly explicit systems of punishment. Crucially, the problems arise not for a single individual, but for the social group. Each member of this band is committed to a particular environment, in this case, a social one: they have to hang together (§§9, 14). Each of them feels the difficulties the circumstances of their shared life impose, the frequent tensions, the long episodes of peacemaking. *The problems are felt by all.* Ethical codes serve the function of solving the original difficulties, dimly understood by these ancestors. Initially, they offer only partial amelioration. Ethical progress consists in functional refinement, first aimed at solving the original problems more thoroughly, more reliably, and with less costly effort (substitutes for hours of peacemaking; §10). In the course of progress, however, the problem background itself changes, generating new functions for ethics to serve, and hence new modes of functional refinement.

So far the schema of an account of ethical progress: one needing to be filled in by specifying the original function ethics is to serve, by explaining the ways in which functional refinement takes place, articulating the idea of generation of new functions, and understanding those further functions. Pragmatic naturalism aims to analyze the situation in ways the original participants could not, to bring into focus the dimly perceived problems our ancestors felt. The historical narrative of Part I provides the materials for doing that. The remainder of this section explores the question of original function.

The tensions and fragilities of hominid (and chimpanzee) social life arise from the limited altruism of the participants. Altruism failures lead to conflict, to pain inflicted, to rough discipline, and lengthy peacemaking. To the extent altruism failures can be avoided, life goes more smoothly, with increased opportunities for cooperation and, consequently, greater mutual benefits. Group members satisfy more of their desires and protest less. The first ethicists did not recognize themselves as responding to the problem of altruism failure—they simply wanted relief from social instability. All of them felt fear and dislike of the episodes of conflict; all wanted their interactions to go more smoothly. From the analyst's point of view, the members of these small societies were aware only of the symptoms of their predicament. The underlying disease was the prevalence of altruism failures. Remedying altruism failure is the original function of ethics.

To specify the function more exactly, we must examine cases of failure to respond to the desires of others. Suppose A fails to respond to the desire he attributes to B; his solitary desire is unmodified. A potential rule would direct him to act on the basis of a desire incorporating, at some level of intensity, B's preference. Imagine, for concreteness, the rule prescribes golden-rule altruism (§5) for this context. Will this rule ameliorate the difficulties our early humans experience? That depends on whether promoting B's desire in this context clears up B's dissatisfaction, *without simultaneously generating trouble for other members of the group*. If B's desire is to do something that would block the satisfaction of the desires felt by B^*, the rule may simply substitute one altruism failure (toward B^*) for another (toward B). Perhaps there is a cascade of effects on the desires of other group members. The bump in the rug may be shifted and even made larger through the effort.

In discussing the early phases of the ethical project (§19), I introduced the concept of *endorsing* others' desires. Desires are endorsable just in case there are possible environments in which they could be satisfied for all our fellows. So, for example, to remain close to the predicament of our early human ancestors, the desire to have adequate food is endorsable, whereas the desire to monopolize reproduction is not. With respect to many desires, it makes sense to think of the extent to which, in the current environment, the desire goes unsatisfied. Two members of the group may each be hungry, even though the food available to one is less than that available to the other. A desire will be said to be more *urgent* just in case the shortfall in satisfying it is greater. One way of analyzing the problem background faced by our ancestors is to see the function of ethics as allowing a smoother, more peaceful, and more cooperative social life, through remedying altruism failures, and, more specifically, through clearing up those altruism failures involving the most urgent endorsable desires on the part of the potential beneficiaries.[14]

Although early ventures in socially embedded normative guidance— particularly the most successful ones—may have framed principles according with this conception of the function of ethics, that was not our ancestors' perspective. They surely began with paradigms: desires for food, for shelter, for protection against attacks. In responding to the most urgent of these desires, they remedied a class of altruism failures and reaped the benefits of decreased conflict. Retrospectively, we later analysts can view attention to cases of altruism failure where the desires are endorsable and urgent as a preliminary way to specify the problem background, but it cannot be the final story. The original function of ethics is to remedy *those altruism failures provoking social conflict*, and the problematic class is only partially specified by deploying notions of endorsability and urgency.

Consider another aspect of the lives of our early human ancestors, aggressive behavior inhibiting another group member's attempts to satisfy desires. B attacks B^*, while A looks on. As things stand, two altruism

---

14. To conceive the problem background in this way is to introduce familiar ideas from ethical theories. The emphasis on endorsability has obvious connections to universalizing principles and can be viewed as echoing a Kantian idea; the focus on the most urgent desires has kinship with Rawls's different principle; see also T. M. Scanlon, "Preference and Urgency," *Journal of Philosophy* 82 [1985]: 655–69.

failures have occurred. One way for *A* to "remedy" the failures of altruism would be for her to form a desire that responds positively to *B*'s manifest wish—to join the attack on *B**. Often that would be to intensify social conflict. Some forms of positive responses to the desires of others are *contaminated* because what the other wants to do is to initiate interference with the expression of the desires of a third party. Societies with rules supporting such interference are likely to fan the flames of social conflict; societies directing altruism away from contaminated desires, and toward the wishes of the targets of the interference, will probably diminish the tensions. They will develop a system of punishment directing retribution toward those who initiate aggression.

So we can indicate another class of altruism failures to be remedied, those in which the desires of victims to be protected go unsupported. As with the earlier effort at more exact characterization in terms of urgent, endorsable desires, this too is incomplete. Societies that fasten on actions initiating aggression, take altruistic responses to the desires those actions express to be contaminated, and institute rules for remedying altruism failures toward the targets of aggression make a large first step toward solving the problem of social conflict. There will, however, be occasions for further refinement. Excluding contaminated desires is not the whole story.

The large hope of ethical systems is to formulate an overarching account of fundamental ethical properties (goodness, rightness) or a collection of basic principles sufficient to adjudicate all cases. Those who share the hope will believe that the function of ethics can be specified exactly by latecomers in the evolution of our ethical practices. To the extent they accept the framework in which I have posed the issue, they will seek an exact representation of the problem background and a precise characterization of the altruism failures to be remedied. I view the situation differently. We latecomers know *more* about the problem background than the pioneers who developed ethical systems partially responsive to it: we can see how rules directing altruistic responses toward urgent endorsable desires and away from contaminated desires are likely to diminish social tensions and conflicts. Codes incorporating those rules have taken a first step, but they leave room for functional refinement. Extrapolating from whatever progress can be discerned in the past, we envisage an indefinite sequence of adjustments dealing more thoroughly with the original problem. *Neither we nor our descendants are likely to achieve a complete solution, one that will correspond to a complete and exact characterization of the altruism failures to be remedied—or, correspondingly, to a complete system of ethical principles.* At its most progressive, the evolution of ethics is (at least as far as its original function goes) a series of responses to the most powerful sources of residual social conflict.

Pragmatic naturalism makes the following claims about the original function of ethics. (1) The problem background consists in social instability and conflict caused by altruism failures. (2) The original function of ethics is to promote social harmony through the remedying of altruism failures. (3) Our ethically pioneering ancestors had only a dim appreciation of the problem background, responding to the difficulties and discomforts of a tense and fragile social life. (4) We know more about the problem background than they did and can offer partial and incomplete diagnoses of the types of altruism failures to be remedied; we can understand the success of certain kinds of ethical systems in terms of these partial characterizations. (5) Even with respect to the original function, the project of refining the codes we have continues.

I anticipate a family of obvious objections to this approach to the original function of ethics, and the rest of this section is devoted to responding to them. The common strategy of all the worries is to specify ways of discharging the function, ways that would elicit from many—perhaps almost all—people the judgment that they are ethically repugnant or, at best, ethically indifferent. The simplest and crudest version remarks that greater social harmony can be produced by brutal means. We can envisage a society in which a powerful dictator physically prevents conflict from ever occurring. As it stands, this fantasy does not expose a real difficulty with the approach, since, although it may be conceded that the envisaged society is unsatisfactory, it does not fulfill the function identified in thesis 2, the remedying of altruism failures. To use an obvious analogy, it palliates the symptoms without attending to the underlying cause, as if the doctor were to offer to cure your aching knee by amputating your leg. In situations where conflict would previously have arisen, nothing is done to modify the desires of the participants;

instead, they are simply prevented from expressing their—altruistically failing—desires.

Let the dictator work differently. He lays down rules compelling people to take the desires of others into account in their own conduct. How does he achieve that? He makes the penalties for not responding to the needs of others truly severe. If the modification of desire occurs across the board—everyone responds to the perceived wishes of those with whom he or she is in contact—it is not obvious that the dictator's creation of the system fails as an ethically progressive step. Behavioral altruism replaces previously selfish conduct. The resultant system is a variant on the sorts of successful ethical practices envisaged in the early period of experimentation (§§11, 17): a system of rules produces behavioral altruism in all members of the society, and following these rules diminishes social tension; the main difference comes in the way that compliance is achieved; the unseen enforcer is replaced by a looming secular presence. Further progress would be made if a more intricate mix of emotions—awe, respect, solidarity, and genuine sympathy with others—superseded fear.

To generate trouble the dictator must proceed at cross-purposes with the kinds of altruism failures we later analysts see as the ones to be remedied. Perhaps he does that by instituting rules we take to be unfair: some members of the society are designated as servants, others as masters; subject to the usual severe threats, the servants must respond to the wishes (even the whims) of the masters. Assume the wishes of servants go unsatisfied. If the class of servants is large, only continued repressive actions will keep their frustrations in check: social conflict erupts in new ways. Perhaps it is tightly restrained. Yet that requires coercive action, and, once again, the dictator has failed to fulfill the function of remedying altruism failures. Indeed, his procedures constitute a drastic modification of human life as lived by the original participants in the ethical project, for he has suppressed some of the conditions under which the admittedly limited human capacities for psychological altruism can be expressed.

If the class of servants is small, however, the dictator has not remedied altruism failures so much as created a vast class of them, to wit the failures of masters to respond to the desires of servants. Suppose, on the other hand, the wishes of servants are satisfied, and this is done through rules encouraging the masters to respond to those wishes. Now the distinction between masters and servants lacks the significance it appeared to possess, and, again, we have a recognizable system of ethics with an odd means of ensuring compliance.

Consider yet another scenario. Social harmony is achieved because, in the cases that would have generated conflict, some parties just do not want whatever it was that caused a quarrel. If we focus the idea by imagining the original experiments of living, the previously contested items will be basic necessities, and the fantasy consists in supposing that some members of the group do not want the resources they need to keep them going. The other members respond to their abstemious wishes, and they wither and die. In this instance, however, although a certain class of altruism failures may be remedied, an extreme type of altruism failure remains—for those who profit from the resources their unfortunate comrades forego fail to take any steps in the direction of paternalism. As the plight of the ascetics becomes ever more dire, identification with their actual wishes, rather than with the wishes they would have had if they clearly perceived the consequences for themselves, ceases to be a form of altruism.

Imagine a last variation on the theme. Social tension within the group is decreased by expelling some of the members.[15] There are two main possibilities. Perhaps those expelled are the primary causes of social trouble: they sometimes benefit from the altruism of other group members but consistently fail to make altruistic responses themselves. Ending the practice of interacting with them genuinely would remedy the altruism failures by which the band has been beset and so would fulfill what has been identified as the original function. If these are the circumstances, however, the account of the original function is not challenged, since there is no basis for describing the envisaged expulsion as regressive—it is of a piece with prevalent societal practices of sequestering troublemakers. On the other hand, if those driven out are only marginally more likely than their fellows to lapse from psychological altruism, excluding

15. I am grateful to Christian Nimtz and Boris Hennig for forceful presentation of this possibility.

them from further social interaction will achieve correspondingly little in remedying altruism failures. In this case, expulsion is not ethically progressive, but we no longer have an instance in which the original function is fulfilled.

Under the original conditions of the ethical project, diminishing the size of the group was, where even possible, a costly option: recall that the coalition game culminates with bands of roughly equal strength partitioning the environment (§9). Further, the members of these bands are bound to one another by dispositions to (limited) psychological altruism. Fragmentation of the group is both socially dangerous and psychologically difficult. Appreciating this point enables us to develop an important perspective on the purported counterexamples to the account of original function in terms of the remedying of altruism failure. Not only do the supposedly problematic scenarios fail to undermine the account, for the reasons given—when they are scrutinized, either they fail to meet the stipulated condition (remedying altruism failure) or they involve genuinely progressive transitions—but they would not be *available* in the contexts in which our ancestors began the ethical project. Those ancestors could not have used alternative methods of attending to the *symptoms* of altruism failure, the social tensions.

Dictators cannot institute practices remedying altruism failures unless those dictators are so wise and benign they deserve less pejorative labels. What dictators can apparently do is to relieve social tension by coercive measures. At the dawn of the ethical project, however, dictatorship was not a realistic possibility. Dominance of a chimpanzee group is hard enough (§10), but for language-using, tool-wielding animals with less sexual dimorphism, repressive rule would be much harder. Failure to accommodate the wishes of each band member runs serious risks of fragmentation, and a reversion to the conditions of the coalition game—made more difficult because of the psychologically altruistic dispositions thwarted in the group's splintering. The scenarios that might subvert my account of original function not only fail as contrary instances but also are unrealistic as ways in which, in the early phases of the ethical project, social tensions could be ameliorated.

The evolution of the ethical project envisaged (§§19–20) created possibilities for hierarchical societies, within which dictatorial impositions become real options. These societies can address social tensions in the brutal ways the scenarios propose. When they do so, they cannot be conceived as fulfilling the original function of ethics and thus making ethical progress—for, to recapitulate, they do not remedy altruism failures. Nevertheless, these societies do raise an important question for the approach to ethical progress I am developing. *Does the original function of ethics continue to bind those who come later?* The question brings home the important fact that there are two sides to questions about ethical progress, only one of which has as yet been considered. The facet under study, here, as well as in Chapter 5, concerns whether one can make sense of some notion of ethical progress integrable with the narrative of Part I. Even given success in that venture, however, it would not follow that progress, understood in the preferred way, would have any force on later participants in the ethical project. Is "progress" of this sort something we should aspire to make? Do the functions our ancestors attempted to fulfill matter to us?

These important issues will occupy us in Chapters 7–9. So far, they are merely registered—but their turn will come. For the moment, it is enough to note the circumstances in which the problem background originally arose, the situations of our first human ancestors with their uneasy social lives; these are very different from those in which many latter-day participants in the ethical project have lived (during the past ten thousand years). The project began in small, egalitarian societies, in which people with limited tendencies to psychological altruism lived together. Feeling the tensions of their social lives, they had no successful options except to address the (unrecognized) cause—and ethics was born with the function of remedying altruism failures. That original function is refined and gives rise to further functions, in ways we shall now explore.

§35. Modes of Refinement

Because ethical codes are multidimensional, not all forms of ethical progress consist in amending rules so the functions of ethical practice are better discharged. There can be improvements both in socializing the young and in integrating the precepts with an effective system of punishment. I consider three modes of ethical refinement, beginning with advances in techniques of socialization.

The idea of an unseen enforcer serves as a prominent example of functional refinement. In the earliest experiments with socially embedded normative guidance, the agreed-upon prescriptions are connected with the system of punishment in an obvious way: those recognized as breaking the rules will be disciplined. At this early stage, there is little besides fear of retribution to motivate compliance, and hence, when people think recognition of their conduct is highly unlikely or impossible, they may follow egoistic desires. Altruism failures addressed by the rules persist even after those rules have been publicly accepted; sometimes other members of the group know a breach of the ethical code has occurred, even though the perpetrator cannot be identified. Mutual suspicion produces social conflict. Adding more rules does not help, for existing rules are directed at the troublesome behavior. The group needs a device for making the rules more effective when people are deliberating what to do.

The unseen enforcer is just the right sort of device, for, once the idea is accepted, deliberators never escape recognition. Using the existing mechanism for bringing group members into line, the fear of bad consequences from deeds the code prohibits, the thought of perpetual surveillance shrinks the class of altruism failures and diminishes social conflict. So far, so good. On other grounds, however, this step is problematic. From the perspective of assessing the *cognitive* state of the group, it is not progressive. More importantly, the unseen enforcer sets the stage for later abuses. When the system of normative guidance has become more elaborate, when instituting changes to the system is no longer a matter of a small band deliberating together and trying to arrive at prescriptions to which all can subscribe, the task of specifying the rules can be left to a small class of privileged people. Conceived as the enforcer's representatives, these social leaders easily inscribe their own dislikes, wishes, and fears as expressions of the divine will. In the course of subsequent history, that can render some people's desires invisible, prohibiting their expression with such severity that they are only rarely manifest to others who do not share them, and thus fostering the common judgment that they are abnormal. Because of its power to secure greater compliance, the idea of the enforcer counts as a major progressive step *along one dimension*, even though it creates obstacles to progress of other types. This is an important general point: advances in the social technology that is ethics can shape the way in which further ethical progress becomes possible or impossible.

Introducing myths about unseen powers is not the only way in which our ancestors have improved the techniques of promoting compliance. As previously noted (§§17, 21), the thought of the enforcer can serve as a conduit to association of a far richer set of emotions with the ethical code. Initially, rules are followed out of fear. Later, once the belief that the commandments express the divine will is entrenched, fear of the unseen power can become awe, respect, even love and gratitude, so, still later, following the code appears an act of reverent obedience, a gesture of respect and love. The imagined wrathful countenance of a punitive deity gives way to joy in the prospect of conforming action to shared ethical rules. People may even describe themselves as acting out of respect for the moral law, a law they suppose the deity to have inscribed on their own hearts.[16] In my account, these steps count as ethically progressive, not because they eventually take us to some wonderful place—the privileged "ethical point of view"—but because the multiplication of potential motivations increases the reliability of compliance to the ethical code. They are refinements of the original function of promoting social harmony through eliminating altruism failures.

Exactly similar points apply to the extension of human sympathy. Socially embedded normative guidance first replaces altruism failures with behavioral altruism, but, as remarked (§§11, 21), continued interactions

16. Kantians will probably not approve of this way of viewing the stance they identify with a special ethical point of view. They are likely to suppose a decisive break between the crude and inadequate ethical systems relying on divine commands and the supposedly a priori unfolding of the moral law. It is, however, completely opaque to me how one could arrive at correct ethical principles a priori, even if they were (in what sense?) "within," and it is worth considering whether Kant's ethical innovations can be—and should be—viewed as the recombination of elements from earlier traditions: the idea of divine command approaches that ethical principles are prescriptions not grounded in good consequences for human beings, the suggestion of natural law theories (e.g., Aquinas) that the law is written within our hearts, a long-standing philosophical celebration of reason and a corresponding suspicion of emotion.

with people, each responding to the other's desires, can generate sympathetic emotions. You begin the regular work of helping because that is prescribed, but, as you continue, and as others react to your help, the responses occur from different causes. This, too, is a progressive shift—and again not because it attains a special "ethical point of view" in which mutual sympathy holds sway, but because an increase in the potential sources of compliance makes fulfilling the original function proceed more reliably.

People with enlarged sympathies, with respect for the ethical code and for those who have made conspicuous sacrifices to follow it, are more likely to comply than those deficient in one of these respects. For them, failure to respond to the socially approved desires of others opposes several distinct psychological dispositions. They are not perfect, and, on occasion, all the mechanisms for leading them to behavioral altruism fail. Failure occurs more readily as societies grow, as the ethical project introduces more distinctions and divisions, assigns status differentially, and makes some members of the large society invisible to others. Even as some developments in the project buttress old mechanisms and introduce new ones, they simultaneously make psychological altruism of certain types more difficult—sympathies expand, but only as far as the limits of a group with which the higher-order altruist identifies him- or herself. Transitions progressive in some respects are regressive in others and even block off further ethical advances. In changed social circumstances, inherited modes of socialization encourage some forms of altruism failure.

A second mode of ethical progress consists in integrating the code with the system of punishment. Normative guidance emerges in a context in which a rough system of punishment is already present (§12), and it builds on psychological capacities for avoiding behavior that provokes punishment. In part, the introduction of ethics substitutes prevention for punishment. It does so by combining specification of commands with the threat of punishment if they are broken. The systems of punishment found in chimpanzees inflict harm on individuals who have done certain things. An ethical code must approach such raw reactions to behavior more self-consciously. Implicit in any endorsement of punishment is a suspension of altruistic response: at least some of the desires of the individual who has performed the rule-breaking action do not count; they are not to be accommodated in dealing with him. The perpetrator wants to avoid all the harm others in the group intend to visit upon him, but this is not an occasion for responding to his desire. Integrating the ethical code with a system of punishment requires some things resembling altruism failures *not* to be remedied, because, as we would ordinarily say, the rule breaker has made some of his desires forfeit.

The ethical code connects with the system of punishment through three questions: When? Who? How? The first issue is addressed by appealing to the prescriptions of the code: retribution is to occur just when one of these rules is broken. The second is the translation of the earlier raw reaction into the social practice of normative guidance: we look to the individual(s) who did the action that broke the rule. For the third, the ethical code must itself pronounce, declaring which kinds of harms are to be inflicted. It must decide what accommodations with the desires of the perpetrator are to be made. As noted, his desire to avoid all harm is not to be the basis of an altruistic response; in carrying out the punishment, ignoring that desire is not an altruism failure to be remedied. That does not mean, however, that the code must be indifferent to everything the rule breaker wants.

The original function of the code is to remedy altruism failures causing social discord, and that is achieved if punishment and the threat of punishment decrease the frequency with which failures occur. Setting some of the perpetrator's desires outside the scope of altruistic responses expresses a commitment to this function: if those desires are ignored, he and others will be less likely to engage in future rule-breaking behavior.[17] In many situations of rule breaking, other members of the group are compelled to choose the target of their altruistic response, forced to decide whether to align their desires with one person or another. The apparatus of punishment makes this decision for them by distinguishing between perpetrator and victim. In ignoring the desires of the perpetrator and attending to the desires of the victim, one prominent—and readily

17. Here my functional approach to ethical progress aligns itself with views of punishment that are forward-looking, that view punishment as a means to prevent future harm. The raw reaction of a desire for revenge is transmuted into a piece of social technology.

understandable—thought is to focus on the exact way in which the rule breaker has overridden the victim's desires and to make exactly the adjustment that will restore the status quo. Hence arise punishment rules demanding precise reciprocity: as in the lex talionis.

The attraction of such rules is evident from the early Mesopotamian law codes, and it is highly probable that previous generations of ethical experimenters had used similar prescriptions. Yet, as we saw (§23), the lex talionis was elaborated in a peculiar—and disturbing—way, through the literal idea of compensating for human loss through the death of a "corresponding" person. Someone who has murdered the child of another is to be punished by having one of his or her own children (of the same sex as the victim) put to death; wives of rapists are to be raped. Behind this law lies an apparent dyadic calculus of violated desire: the code focuses on only two members of the society, perpetrator and victim; each desires enduring relationships with other individuals; the perpetrator has violated one of these desires, and the victim is compensated through violation of the perpetrator's corresponding desire. One altruism failure is canceled by its precise counterpart.

Societies made progress when they rejected this appalling law in favor of one harming the perpetrator him- or herself. That can be viewed as a refinement of the function of the ethical code. Central to the ethically mediated practice of punishment is, as we have seen, delineating some desires of the perpetrator as outside the scope of altruistic response. Yet punishments striking at the perpetrator "through" his or her relatives—in the prominent instances, his children and wife—introduce a vast new class of altruism failures, violations of the wishes of those who are going to be the immediate targets of retributive action. Precisely because *individuals*, not groups or clans, are the bearers of desires, efforts to remedy altruism failures are better directed by considering the wishes of all the members, not simply those of the family, the clan, or some other collective (or even some single person taken to "represent" this collective), and the change in focus counts as a progressive step. Acceptance of the ethical code and its enforcement through punishment requires that members of the society commit themselves to noninterference in the inflicting of harm on perpetrators, but they have made no similar commitment to allowing suffering to be imposed upon others. The articulation of the lex talionis using the "corresponding relative" as a vehicle for retribution is likely to inspire protests and reactions from those who sympathize with the people targeted. In its early form, the law introduces altruism failures instead of remedying them and, in doing so, destroys rather than promotes social harmony. The modified version confining discipline to the perpetrator can thus be seen as a refinement of the original function of ethics.

Another important aspect of ethically mediated punishment deserves consideration. The initial stages of ethical experimentation had to operate with a conception of rule breakers, and it is overwhelmingly probable that they introduced the idea in the most straightforward fashion, thinking of perpetrators as people whose behavior violated rules. The function of punishment is to decrease the future occasions on which similar altruism failures occur, and realizing that function depends on deliberators' ability to recognize in advance when a potential action is at odds with the code. As prescriptions and the concepts involved in them become more complex, it is apparent that there are episodes in which people unwittingly break rules. Inflicting harm on those who tried to honor the code is likely to sap the motivation to comply, since dedicated efforts still lead to trouble. Acknowledging human limitations, incomplete knowledge, and life histories yielding a distorted perspective makes progress in fulfilling the original function. Breaking the rules is forgiven when violators convince others they were doing their best to comply.

The extent to which an ethical code should take account of our limitations of knowledge and the environmentally caused distortion of our sympathies is something on which the code itself must pronounce. The strategy of conceding something to human frailty is a progressive step in the evolution of ethics. Our understanding of how to make such concessions remains incomplete. Too rigid an insistence on the fact of rule violation proliferates altruism failures by weakening the motivation to try; too easygoing an approach multiplies altruism failures by undermining the rules themselves. The challenge, for our predecessors and for us, is to operate in the space between these extremes and to sharpen the concept of an excusable offense. Because of the variety of ways in which human effort can fall short, our progress remains incomplete—we continue to work out the extent to which we excuse human imperfection.

For the last of my three modes of functional refinement, return to the attractive, but oversimple, account of ethical progress in terms of "the expanding circle" (§33). It is relatively easy to understand how many examples of ethical progress are functional refinements achieved by widening the scope of existing ethical precepts. A Gospel message exhorts us to love the downtrodden and despised among us. Incorporating the prescription in our ethical code identifies members of our society who have previously been outside the scope of altruism for many other people—we are directed to remedy a class of altruism failures, allowed by the rules previously articulated (at least under the common understanding), and doing so enhances social harmony. Imagine we come to view slaves as "full people," or recognize women's desires for participation in public life, or understand how some people can find love with members of the same sex.[18] In all these instances, prior altruism failures are to be remedied: we now try to act in ways that take into account the desires of those whose wishes have been overridden, neglected, condemned, or simply not seen; the result is a diminution of social conflict.

An especially important instance of "the circle expanding" came in the lives of our Paleolithic ancestors when they extended certain of their ethical principles to cover outsiders in neighboring bands. According to my "how possibly" explanation (§19), the impetus to partial recognition of people outside the local band came from the perceived possibility of trade. If so, the result would have been something that was neither a full society, of the kind already constituted in the small groups of cooperating human beings, nor a confrontation of unremittingly hostile bands. We can think of it as a limited society, introduced to promote a particular purpose (exchange of surplus goods). Because of the desirability of such exchange, a restricted analog of social harmony was also desired in this limited society, to be achieved through the extension of some parts of the ethical code. The problem background is not that of decreasing social conflict within an already existing society, but of creating a limited society that requires decreased conflict at its founding. To obtain the level

18. With respect to this last example, as §32 noted, the altruism failure consists of an attitude toward certain types of desires (desires for fulfillment of same-sex love) that the progressive transition modifies.

of social harmony enabling this limited society to emerge, classes of previously occurring altruism failures (in aggressive interactions among neighboring bands) had to be remedied. Expanding the circle does just that.

Major instances of ethical progress can thus be seen in terms of functional refinement, where the function in question is that identified as the original function of ethics (§34). Some of the difficulties recognized in §33 are now resolved. Not all, however. Progressive changes of a particular type—those offering new ideas of the human being and of human life remain problematic. The next task is to show how the functional approach to ethical progress can be extended to cope with these cases.

§36. Functional Generation

In the evolution of ethics there is an original function that grounds the concept of ethical progress, but there are also secondary functions, generated from the ways in which that original function is discharged. The possibility is evident from the domains serving as analogs: in Darwinian evolution and in the development of technology, responses to older problems generate new problems and new functions.

Consider our hominid ancestors. To simplify considerably: they moved from forest to savannah, taking on an increasingly upright posture; for reasons still much debated, they also grew bigger brains. The two trends combined to cause a new difficulty. Females, whose pelvic structures are constrained by the requirements of walking upright, give birth to young with larger heads. Adaptive responses to old problems generate a new problem of giving birth. The hominid/human solution is to truncate the pregnancy so the baby's head does not grow too big. That generates a further problem, one of nurturing an infant that cannot do much for itself. The original functions of upright gait and big brains (whatever exactly they were) generate the function of curtailing pregnancy, and the latter function generates the function of extensive parental care.

When new biological functions emerge, much is held constant throughout the process. Organisms continue to be subjected to the basic pressure of natural selection, competing for reproductive success. Because of solutions achieved by their ancestors, they are locked into particular structures and modes of behavior. Their inheritance frames the new problems generated for them by selection. Successively solving problems, they make *local progress*. Whether they also make *global progress* is another matter. Is there any Darwinian basis for comparing the primates in the trees and the hominids on the ground?

Or consider the development of systems of mechanized transportation. The initial problem background is one in which people feel spatially confined: they want ways of going more speedily over longer distances and of reaching previously inaccessible places. Mechanized vehicles fulfill these functions, but, once they have been provided, new problems must be solved. Appropriate channels (roads, railways, canals) have to be built and maintained. Traffic must be coordinated, signals designed, regulations for responding to them enacted and enforced. All sorts of auxiliary systems arise, as responses to the problems generated along the way—driving schools, traffic courts, fuel-distribution networks, insurance requirements, inspection stations, the familiar paraphernalia of the car-infested world. Local progress is easy to discern. Early automobiles made more destinations accessible. They generated new problems of road building. Technology supplied new surface materials: a progressive step. Standing back from the individual transitions, however, questions of progress become harder. Did the automobile improve the world? Was there *global* progress?

Does a notion of functional generation apply in the ethical case? If so, does it complicate judgments of global progressiveness? I shall argue for affirmative answers. Begin with an example figuring in the transition from the first ventures in normative guidance to the ethical approaches present at the dawn of history: the emergence of a conception of a good life involving some type of individual self-development (§§20–21). The initial problem is to remedy the altruism failures causing social conflict. In the earliest phases, our ancestors focus on the basic desires of group members—desires for food, for shelter, for protection. Some desires are endorsed, others seen as contaminated. Preliminary development of these notions is a first ethical step—desires for food, shelter, and protection are endorsed; desires to engage in unprovoked violence are taken to be contaminated; simple rules express these judgments. The society now attempts to find ways of enabling all its members to satisfy the endorsable desires.

That is the starting point for a process leading, a very large number of generations later, to societies with a far richer conception of living well. My "how possibly" explanation links the two by starting with divisions of labor (§19). Dividing the labor helps satisfy endorsable desires; specification of roles and assigning people to roles on the basis of talent refine the social technology. In the process, new desires emerge: people find some roles more attractive than others. The conception of the good life expands, as members of the society view living well not just as meeting the needs identified at the beginning—it is no longer enough to have a full belly and a warm, safe place to sleep.[19] Additionally, living well involves attaining roles one wants and having the chance to choose one's roles.

The original problem remains constant throughout this process. Ethics must continue to promote social harmony through remedying altruism failure. Now it must do so on an expanded field of desires. If the means to satisfy the endorsable desires present at earlier stages have become available, it is in principle possible to solve the older problems (although, in practice, that may not be done because of violations of the code). The residual problem is to address social tensions caused by altruism failures with respect to desires that the (in principle) successful solution has generated. Ethical principles are also required to respond to conflicts within the individual's expanded repertoire of desires, and in this sphere, prescriptions for character development emerge.

This is a story about the generation of functions, where the notion of generation is that discerned in the biological and technological examples.[20] To make the structure explicit, consider the following sequence:

19. So simple a menu of basic desires was probably never sufficient to the idea of the good life, even at the beginning. But this elementary formulation suggests the appropriate contrast.

20. The account is a *story*—a "how possibly" explanation. It was originally provided to show how ethical evolution could have led by gradual steps from the original form of the ethical project to the complex practices recognizable in the ancient world. It is used here to illustrate the notion of functional generation with respect to ethics, and also to show how a challenging example—the emergence of richer notions of the good life—can be seen as proceeding by locally progressive steps. Neither of these purposes requires my story to be an account of actual history.

1. An initial problem (P_o) requires the existence of something to fulfill an initial function (F_o).
2. Something (E_o) is introduced to fulfill F_o.
3. For E_o successfully to fulfill F_o, a new problem (P_1) must be solved.
4. P_1 requires something to fulfill a new function F_1.

In cases that exemplify this sequence, I shall say that P_o *generates* P_1, and that F_o *generates* F_1. A *function* for ethics is anything resulting from the original function (the remedying of altruism failures) by a finite number of steps of functional generation, that is, from a finite number of iterations of the sequence 1–4.[21] The narrative of the last paragraph identifies responding to the endorsed (elementary) desires of all group members as a function of ethics, introducing and elaborating the division of labor as a function of ethics, specifying roles as a function of ethics, and ultimately endorsing an expanded set of desires as a function of ethics. (Those identifications would be preserved, even if my "how possibly" story is not historically accurate, given the minimal assumption that the steps leading to the richer conception of the good life involved group responses to antecedent problems.)

People need not always be aware of their problems. Proliferation of mechanized vehicles might give rise to a period during which traffic was unregulated. Some of the population, people unaffected by the accidents befalling others, might see no difficulty in the status quo. The same can occur with the ethical project. The evolution of ethical practice can give rise to codes whose shortcomings and burdens are felt by only a few. When that occurs, the first task of would-be reformers is to make the problem apparent to all members of the society. That may occur through verbal presentations—as with Woolman and Wollstonecraft—or through the exerting of pressure from people whose voices have not previously been heard. In the earliest phases of the ethical project, however, when groups are small and everyone can make his or her particular perspective known, the problems generated are visible. Stratified societies create

---

21. At risk of being unnecessarily explicit: this specification counts the original function as a function, since it is obtainable from itself by zero steps of functional generation.

conditions under which it is harder to recognize when the original function of ethics is not being fulfilled.

The significance of this point will occupy us later (§§38, 46). The immediate task is to understand the difficulties with global progress. It will help to have a notion of *functional conflict*. *External* functional conflict occurs when there are two different domains of human practice such that the optimal available ways of discharging some of the functions of one of the practices are incompatible with the optimal available ways of discharging some of the functions of the other. *Internal* functional conflict occurs when there are two subsets of the functions of a practice such that the optimal available ways of discharging the functions in one subset are incompatible with the optimal available ways of discharging the functions in the other. It is easy to understand how there can be external functional conflicts between economic practices and ethical practices, or between ethical practices and sexual strategies, but my interest will be in internal functional conflict within ethics.

The story of the emergence of a richer conception of the good life reveals conflict of this sort. In the transition to hierarchical societies with pronounced division of labor and of status, fulfillment of the generated functions (supplying enough to satisfy the previously endorsed desires of all) is obtained at the cost of compromising the satisfaction of the original function. Similarly, in the invocation of the unseen enforcer, there is functional conflict between the original function of remedying altruism failure and improving compliance to the rules of the code. The societies who emerge from these changes have the choice of whether to take steps to improve the fulfillment of the original function at cost to the solutions they have achieved with respect to the generated problems. (That is not to say they will recognize this choice unless it is forced upon them by some militant reformer or through the collective pressure exerted by some large group.)

Armed with notions of functional generation and functional conflict, we can now take up questions about local and global progress. To what extent are global comparisons available in appraising the evolution of the ethical project?

§37. Local and Global Progress

In the evolutionary and technological cases certain judgments about progress are easily made: we compare the first organisms who invade a new environment with their later descendants and view the latter as having made progress through the refinement of organs or forms of behavior that were initially able to provide only clumsy solutions to the problems the environment poses. Similarly, later surface materials for roads provide a smoother ride and thus constitute a progressive step. Judgments of local progress are facilitated because it is possible to focus on a single function, or concordant set of functions, and examine how completely (thoroughly, speedily, etc.) they are discharged. Assessments of global progress are difficult because multiple functions come into play. With respect to technology, we recognize external conflicts between practices: our ancestors who lacked mechanized transport clearly had less functional forms of transportation, but in other domains of human practice—their communal interactions with one another and their responses to the physical environment, perhaps—they may have been better able to fulfill particular functions.[22] So, too, in the ethical case.

Some examples of ethical change, like the easy cases for the evolution of life and of technology, focus on single functions or on concordant sets of functions. They can be treated as progressive by concentrating on the appropriate function and applying the ideas of previous sections. A more ambitious account would attempt to extend talk of ethical progress to comparisons of temporally distant forms of ethical practice, with functions that conflict. Can an account of this sort be given—and, if not, can one manage without it?

Imagine two societies. The *Primitive* live under conditions in which it is hard to satisfy the most basic desires: food is in short supply, shelter is hard to find, danger constantly looms. Their ethical codes respond to the altruism failures that arise under these conditions and are tightly focused on responses to the basic desires. The *Civilized*, by contrast, have

---

22. Whether this is so requires deeper scrutiny of their lives than I can offer here. I do not endorse the conclusion that they were better at discharging some functions, but simply recognize this as a question.

developed means of acquiring resources sufficient to satisfy all the group members' basic desires. They have done this by introducing divisions of labor and distinguishing social roles. Their ethical code enjoins sharing resources so everyone's basic desires are met, and let us suppose, optimistically, that the Civilized live up to their code in this respect. Because they know an efficient division of labor is needed if they are to provide for the basic desires of all, they have introduced rules, assigning roles on the basis of perceived talent and enjoining all members of their society to develop their talents and to contribute to the social effort. Each of these members has wishes about the form of his or her social contribution, and, because the roles are differentially attractive, many of these wishes are frustrated. Amending the system, however, allowing rotation among roles, would cause shortages of basic resources.

Can we talk of progress across the Primitive/Civilized divide? The Civilized view themselves as having made progress, talking of their improved conception of the individual and of the good life—they make witty remarks about disgruntled sages and contented pigs—but that only means that, having a wider repertoire of desires, they cannot be happy with a way of life that satisfies only some of those desires. Progress, as we saw earlier, cannot be grounded in these subjective criteria (§28). We who come late in the sequence of "experiments of living" may be appalled at the thought of going back to life as the Primitives lead it. We may think of ourselves as unable to do that, incapable of simply wiping away the kinds of wants and aspirations our social environment has equipped us with; we may imagine descendants of ourselves for whom those desires and aspirations did not arise as truncated and bereft. So far, however, there is no basis for claiming that what has occurred between our Primitive ancestors and ourselves is progress, rather than simply a historical development resulting from people's ingenious attempts to satisfy their basic desires.

To support any claim of global progress, it is necessary to resolve a functional conflict. The Civilized have solved the *original* problem of altruism failures, they have satisfied the endorsable *basic* wishes of all members of their society, and they have expanded the repertoire of desires. This latter expansion has also introduced occasions—even long periods of time—when, for many people, many of the newer desires go unsatisfied. No response to these frustrated desires is permitted. That seems to constitute regression through the creation of a large class of altruism failures (failures provoking social tensions).

One might classify the unsatisfied desires as not endorsable: members of Civilized society would not want a situation in which the participatory wishes of all were satisfied (say, through rotation of roles). That, however, is simply to announce a preference for fulfilling some functions rather than others. Given that participatory desires are as central to the psychological lives of the Civilized as basic desires are to the lives of the Primitives, there is pressure to endorse them. If those desires were endorsed, the Civilized would have made progress over the Primitive by fulfilling some of the *generated* functions of the ethical project but have compromised the *original* function. Can this count as an overall advance?

The concept of progress is central because judgments about progress figure in our evaluations of past transitions and our deliberations about what further modifications to make (§32). While the conception of ethical progress so far elaborated cannot pronounce on questions about some large historical or prehistorical trends ("Was the shift from Primitive life to Civilized life progressive?"), that does not leave us uncertain about how to go on from where we are. Faced with the question "Should we modify our way of life to become Primitive again?," we cannot answer by arguing that the reversal would be regressive (or that it would be progressive!). But, from the standpoint of pragmatic naturalism, the question is idle. Whether the historical processes that have led us to where we are embodied a fundamental mistake or not, we simply cannot go back. We are equipped with wants that cannot be wished away: there is no genuine possibility of arranging a new Primitive future for ourselves. The lack of a concept of global progress does not *make* a difference in the situations where a notion of ethical progress is most needed—and that is enough for a pragmatist not to be moved to settle the question.

The contemporary human situation is hardly that of the Civilized— they are already better than we are at responding to altruism failures. Yet it is not hard to envisage how they might make further progress, in ways responding to the function they currently slight. They could introduce new conceptions of social roles, so people who adapt to forms of social contribution they did not initially prefer are especially highly valued, or so what is hailed as most important is energetic service to the society rather than the form it takes. Even without a principled stand on the functional conflict involved in comparing them with the Primitive, a notion of local progress suffices for (their) practical purposes.

In striving to improve our practices, we cannot foresee everything. A messianic[23] perspective evaluating their evolution in the indefinitely long run is impossible for us. Yet issues of functional conflict cannot always be avoided. As we shall see (§§39, 43) functional conflict lies at the core of the most important challenge to my version of ethical naturalism and is the source of the normative questions addressed in Part III (Chapters 8 and 9). Although the account of ethical progress so far presented answers to many of the uses for which a notion of ethical objectivity is required, it will eventually need further articulation on the basis of a proposal about method in ethical practice.

§38. Ethical Truth Revisited

My conception of ethical progress does not rely on any prior notion of ethical truth. Sections 30 and 32 already recognized the possibility of approaching truth *constructively*: true statements are those arrived at by following particular procedures. The familiar versions of constructivism are ahistorical. They presuppose a set of privileged procedures individuals can follow in whatever epoch they find themselves. Kant, for example, offers us a supposedly a priori method and understands the truths of ethics to be those generated by following that method.[24]

23. I borrow the term from Isaac Levi (*The Enterprise of Knowledge* [Cambridge, MA: The MIT Press, 1982]), who deploys it to make a kindred point about the importance of local ("myopic") standards.

24. The method is specified by the first formulation of the Categorical Imperative, in which potential maxims are to be scrutinized to see if they can be willed without contradiction. Ethical truths are just statements that record the results of the process—thus, if "Make a promise not intending to keep it" fails the test, then "Promise breaking is impermissible" constitutes an ethical truth. I shall not try to work out the intricacies of this theory, since its procedure seems so indeterminate. A more plausible constructivist approach can be derived from Rawls, who can be read as viewing ethical truths as expressing the

Pragmatic naturalism articulates the constructive approach differently, attending to the historical processes out of which statements come to be accepted and following the approach to truth suggested by Peirce and James.[25]

Descriptive counterparts of ethical rules count as true just in case those rules would be adopted in ethical codes as the result of progressive transitions and would be retained through an indefinite sequence of further progressive transitions. There is no prior conception of ethical truth, so that people make ethical progress when they discover (or stumble on) independently constituted ethical truths. Progress is the prior notion, and descriptive counterparts of rules *come to count as true* in virtue of the fact that they enter and remain in ethical codes that unfold in a progressive sequence ("truth happens to an idea"). Derivatively, Tarskian machinery of correspondence truth allows the extensions of ethical predicates ("wrong," "good," and so forth) to be fixed to make the counterparts of rules accepted under indefinitely proceeding progressive transitions true (§30), but there are no prior independent properties to which those who formulate and preserve the rules respond.

Given this understanding of ethical truth, can we make claims about the truth or falsehood of ethical statements? We are never at the end of the ethical project, never at some hypothetical limit of any progressive sequence of ethical practices. Yet we confidently assert some, relatively imprecise and vague, ethical statements, declaring that honesty is typically right and that murdering people who have done no harm is typically wrong.[26] How can we be confident these statements will endure as our ethical practices progress?

---

decisions arrived at in the original position. There are well-known difficulties in specifying which version of the original position is privileged (and why).

25. See Peirce, "The Fixation of Belief," and James, *Pragmatism* (Cambridge, MA: Harvard University Press, 1978). In connection with ordinary statements about the physical world, as well as scientific statements, I regard James's development of the thought as compatible with a correspondence theory [see my "Scientific Realism: The Truth in Pragmatism" forthcoming in Wenceslao Gonzalez (ed), *Scientific Realism and Democratic Society: The Philosophy of Philip Kitcher*, (Rodopi, Poznan Studies in the Philosophy of Science) and "The Road Not Taken," to appear in German in a volume edited by Marcus Willaschek, celebrating the centennial of William James' *Pragmatism* (Suhrkamp].

26. As with many generalizations in other areas, it is hard to state these with any great precision. This has sometimes inspired authors to think of ethical precepts as analogous to

We cannot. There is no good reason to think the concepts we use to formulate such generalizations will be endorsed by our successors. They may introduce quite different vocabulary for organizing their ethical practices, and appreciation of that possibility should incline us to be tentative and vague in generalizing. The confidence we have is that something very like what our imprecise statements express will be found in later progressive extensions of our own practice, and our assurance rests on two related thoughts. First, despite the variation we find in the world's ethical experiments, some themes are discovered again and again, not easily captured in any exact formulation, but universally present. Traditions lacking them would make progress by adopting them. Second, as we think about possibilities of jettisoning our own versions of the rules expressing these themes, we find it hard to conceive how any society that abandoned them, that lacked anything like our prohibitions against murder and lying, would have an ethical system able to reduce social tension through remedying altruism failures. We make a well-based conjecture about which rules are "destined" to endure, using our own, quite possibly crude and inapt, language to gesture toward a theme that will be present in the "last words of the last man."

Can we be sure? The recognition of functional conflict internal to ethical practice prompts a genuine skeptical question. The considerations of the last paragraph turned critically on supposing that the original function of ethics must continue to be fulfilled, that it cannot be overridden by other functions—at least, not to the extent of tolerating the sweeping altruism failures expressed in widespread lying and indifference to human life. To rely on the continued commitment to the original function of ethics in so minimal a form appears no less reasonable than supposing that something akin to a predictively successful scientific generalization is likely to figure in future scientific practice. The point is buttressed further (Chapter 9), by insisting on the priority of the original function of ethics.

"Aristotelian categoricals"—generalizations about murder or stealing are akin to generalizations about the properties of types of plants and animals; see, for example, Philippa Foot, *Natural Goodness* (Oxford: Oxford University Press, 2001), and Michael Thompson, *Life and Action* (Cambridge, MA: Harvard University Press, 2009). I think the connection with biology is a fruitful one, often spoiled by elaborating the biology as if Aristotle had written the last word on the subject.

There is a core set of imprecise precepts we can hail as ethical truths. There are also areas in which the convergence of progressive traditions is genuinely in doubt. Rival traditions offer different elaborations of the ethical project, alternative cultural lineages, and societies belonging to those lineages may by no means find the same problems salient. As they progressively modify their ethical codes to meet distinct challenges, their ethical concepts and the rules, roles, and institutions they introduce and develop may be quite various. Perhaps the solutions achieved are readily combinable, so that, faced with a diversity of codes, there is a way of assembling ingredients from them to produce something progressive with respect to all. But perhaps not. We can imagine two different ethical traditions proceeding indefinitely, making a series of progressive transitions, without its ever being possible to integrate their differing accomplishments.

We saw this already in the contrast between the Primitive and the Civilized. They take different stances with respect to a functional conflict. We can envisage the two societies meeting, and each recognizing the functions the other has learned to fulfill most completely. Inspired by the comparison, they might continue to evolve their ethical practices—just as the Civilized respond to the problem of altruism failures, so the Primitive absorb some of the devices for increasing the supply of basic resources, perhaps even introducing versions of those roles that generate the richer conception of human life. So, as time goes by, more and more roles come to be held worthy among the Civilized, and the society becomes more egalitarian; cooperation increases and altruism expands, but no practice in the lineage that descends from the Civilized ever frees itself from competition, status differences, and inequality. By contrast, the cultural offspring of the Primitive emphasize, from the beginning, the maintenance and intensification of cooperation on equal terms; the result is a certain inefficiency; surpluses are much smaller, roles are not so finely divided; with time, there is some proliferation of different forms of social contribution, but it is always constrained by an emphasis on cooperation and the preservation of equality; the descendants of the Primitive are offered fewer options about how to live, but they live on terms of greater equality than their Civilized cousins.

The societies in these lineages do not overlook the functions the other fulfills better. They judge it good to have what the alternative tradition achieves. In addressing functional conflict, however, they set priorities. Discharging one function comes first—one value takes precedence. During the course of time, they may come closer, even though, however long their unfolding of the ethical project continues, significant differences between them remain. The lineages have exhibited different varieties of ethical progress, and this points to a real incommensurability of practices.

This is the possibility of pluralism (§32).[27] Where that possibility is realized there will be no determinate ethical truth. The central thrust of this chapter has been that we can make enough sense of ethical progress, prior to any conception of truth, to combat the thought that the evolution of ethics is a history of mere changes. Once ethical progress is understood, the concept allows for a constructive development of ethical truth, one useful in enabling us to characterize inexact core statements we take to be shared among ethically progressive traditions. Notions of truth and falsehood do not always apply in the ethical domain, for the core of ethical truth is surrounded by a periphery of pluralism.

§39. Residual Concerns

The inquiry of this chapter began with a pair of difficulties: (1) resisting the mere-change view requires making sense of the concept of ethical progress; (2) if progress is the accumulation of (prior) truth, then the concept does not fit the history of Part I. The solution is to revise the concept of progress so it is prior to the notion of truth. The previous sections attempt to carry out this revision in some detail.

Important concerns remain. In closing, I consider three. First, it has not been shown how to connect ethical progress with the psychological processes moving the great innovators, despite the prominence of the demand for a similar connection in the arguments of the previous chapter. Second, grounding the concept of progress in the problems experienced

27. The pluralism envisaged here is close to that defended by Isaiah Berlin. One important feature of it is the recognition by each of the rival traditions of the values taken as fundamental by the other—each feels the tug of what the other does so well. (I am grateful to Chris Peacocke for emphasizing to me this crucial facet of Berlin's pluralism, and thereby prompting me to make my acceptance of it explicit.)

by the ethical pioneers threatens to make the account vulnerable to the arguments against subjectivism.[28] Third, recognizing that something has (or, worse, had) a function lacks normative force: if the original function of the ethical project was to remedy altruism failures, why should that matter to us, who come so late in it?[29]

To address the first concern, we need a notion of justification of the kind sketched in §32: an innovator justifiably makes a revisionary proposal if the underlying processes are likely to lead to ethical progress. Empathetic understanding, especially when sensitive to (endorsable, uncontaminated) desires that are typically suppressed can foster societal discussion to remedy existing altruism failures. People who call attention to the frustrated desires, beginning a broader deliberation about them, are likely to move the ethical project toward refinement of its original function. Wollstonecraft could not frame the issue in this (analyst's) way, but her sensitivity to women's suppressed aspirations would be expected to refine the original function of ethics. Woolman is a more ambiguous figure. His account of his route to rejecting slavery reveals the role his own anxieties about personal salvation played. Insofar as these were dominant, we cannot view the psychological processes that generated his proposed reform as reliable. Nevertheless, some considerations underlying his hesitation about writing the bill of sale—concerns for the pains and abuses visited upon slaves, worries that official interest in their well-being (their salvation) is a sham—indicate the operation of an empathetic capacity. To the extent he resembles Wollstonecraft, we can view him, too, as a justified agent of ethical change.

The figure who emerges from the Gospels shows many inconsistent tendencies. Yet, if you focus on some passages (and ignore others), Jesus appears as the most striking example of empathetic identification in recorded history. The sweeping exhortations to forgive, to cherish and nurture those whom society neglects or despises, are inadequately described by noting that they point out potential altruism failures on an

28. I am indebted to Sharon Street for a forceful presentation of this worry.

29. Several people have expressed this concern, offering it in different (sharp) formulations. I am indebted to Mario Branhorst, Laura Franklin-Hall, Kirsten Meyer, and an anonymous reader.

extraordinary scale. Because of the scope of his sensitivity to others, you might credit Jesus with a capacity for empathetic understanding whose exercise reliably generates progressive proposals for reform. That would be a misdescription. Jesus does not make *proposals* for criticism, debate, endorsement from others with different points of view. He presents commandments. He is portrayed as an ethical authority who recognizes his role. In my view of the ethical project, there are *no* "moral teachers"—it would be wrong to hail him as one. There are justified *reformers who make proposals*, and we can count him as one of the greatest—even if that does not fit his own self-image or the image of his followers.[30]

The second worry stems from the connection made between functions and their problem backgrounds. Something is only a problem, it seems, if it is felt as interfering with the satisfaction of desires; hence the account of progress offered here is based on subjective considerations of the sorts found wanting in §28. The partial justice of this charge is expressed in some of my formulations, for example, in characterizations of people as wanting relief from the tensions of chimp/hominid social life. With *equal* justice, however, the problems can be regarded as *objective* features of the social situation to which, in our evolutionary past, our ancestors were led. Desires for relief are in no way idiosyncratic—they would be felt by virtually all members of our species. Indeed, the discomforts of chimp/hominid social life would affect human beings and hominids alike (and possibly chimps as well).

The subjective/objective dichotomy is too blunt to apply to the concepts of function and problem background that lie at the basis of my account of progress. Forcing the dichotomy into the discussion is bound to mislead. Dewey saw the point clearly, remarking that "moral conceptions and processes grow naturally out of the very conditions of human life."[31]

---

30. Few people in the history of the ethical project have offered progressive proposals, and few have had the insight into possibilities of altruism failure achieved by the three figures whose status I have (briefly, unsympathetically, and coldly) evaluated. One of the wise sayings attributed to the most wide-ranging of these reformers is that judgment is dangerous. Those whose contributions to the ethical project are infinitesimal in comparison (as mine are) should acknowledge that.

31. John Dewey and James Tufts, *Ethics*, 2nd ed. (New York: Henry Holt, 1932). In the original, the sentence is italicized.

The approach to ethical progress developed here articulates that thought.

The third concern cuts deepest, and it will require extension of the account given in this chapter (including the adoption of a substantive normative stance). As the worry clearly recognizes, people are not bound to use a device to do the job for which it was originally introduced. Why, then, should the "original function" of ethics have any purchase on us? Why should we suppose "progressive" transitions—those contributing to the original function—are to be valued? This chapter has recognized functional generation and functional conflict. An extreme case of the latter occurs when earlier functions are simply dropped in favor of the functions they have generated. Why would the ethical project be immune from that?

The next three chapters build resources for answering these questions (and others besides). We start with some metaethical issues often seen as devastating for any form of naturalism.

CHAPTER 7

Naturalistic Fallacies?

§40. Hume's Challenge

Naturalists, so the story goes, inevitably commit "the naturalistic fallacy." There is, however, no such thing as *the* naturalistic fallacy. While critics agree that naturalistic ventures are inevitably embroiled in some error to which they attach this name, they offer very different diagnoses of what mistake is made.[1] This chapter considers several distinct objections, attempting to show that, whatever their merits against other naturalistic accounts, they hold no terrors for pragmatic naturalism. Yet as we shall discover (§§43, 56), substantive ethical work is needed to absolve pragmatic naturalism completely.

The most venerable challenge derives, ironically enough, from one of the greatest naturalistically inclined philosophers. In a famous passage, Hume writes:

---

1. This is very clearly argued by Richard Joyce in *The Evolution of Morality* (Cambridge, MA: The MIT Press, 2006). Some of the following discussions present points akin to those made in this informed and well-argued book, but, since my pragmatic naturalism diverges from Joyce's approach in important respects, the formulations are sometimes different.

> In every system of morality, which I have hitherto met with, I have
> always remarked, that the author proceeds for some time in the ordi-
> nary way of reasoning, and establishes the being of a God, or makes
> observations concerning human affairs; when of a sudden I am sur-
> prized to find, that instead of the usual copulations of propositions,
> is, and is not, I meet with no proposition that is not connected with
> an ought, or an ought not. This change is imperceptible; but is, how-
> ever, of the last consequence. For as this ought, or ought not, ex-
> presses some new relation or affirmation, it is necessary that it should
> be observed and explained; and at the same time that a reason should
> be given, for what seems altogether inconceivable, how this new re-
> lation can be a deduction from others, which are entirely different
> from it.[2]

Hume is commonly (and with good reason) interpreted as making a claim
about inference, one formulated as the denial that you can derive "ought
statements" from "is statements." This is taken to threaten naturalistic
approaches to ethics because these are allegedly committed to claiming
that normative conclusions can be based upon factual premises.

It is not easy to state Hume's supposed claim about the impossibility
of certain kinds of warranted inference in a form in which it is both precise and true. Plainly some statements in which ethical vocabulary occurs are logical consequences of others in which no such vocabulary
occurs—for there are logical truths involving ethical terms, and logical
truths are derivable from any set of premises (including the empty set).
You can evade this trickery by requiring the ethical vocabulary to occur "essentially," but cases can be devised in which, although the criteria for essential occurrence are met, a conclusion involving ethical vocabulary is derivable from purely factual premises.[3] An example:
premise: some Jews (those who won the Iron Cross in World War I) have

2. David Hume, *Treatise of Human Nature,* 3.1.1. (Oxford: Oxford University Press,
1978).

3. The notion of "essential occurrence" is developed by writers on logic from Bolzano
on and is especially clearly worked out by W. V. Quine (*Philosophy of Logic* [Englewood
Cliffs, NJ: Prentice Hall, 1970], 80–81). Counterexamples to the thesis that statements essentially involving ethical vocabulary cannot be derived from factual statements were of-

served Germany with honor; conclusion: if we ought to respect and protect those who have served Germany with honor, there are some Jews we ought to respect and protect.[4]

Moreover, some ethical statements have factual presuppositions, and, in consequence, statements involving ethical vocabulary ("essentially") can be derived from the denial of these propositions. Thus, if "X ought to do A" implies "X can do A," then "It is not the case that X ought to do A" would follow from the factual premise "X cannot do A."[5] More generally, if E is a statement involving ethical vocabulary ("essentially"), and E implies some factual statement F, the inference of not-E from not-F will be logically valid.

It would be sophistry to take these points to address Hume's challenge. Whatever the difficulty involved in specifying just which types of inferences are unwarranted, something is uncontroversially correct about the Humean worry. It is no victory for naturalism to contend that some negative ethical statements can be derived from factual premises, for that fails to touch the fundamental question of whether a naturalistic approach to ethics can justify conclusions offering positive guidance for our actions. Hume was reflecting on the "systems of morality" offered by books he had read, and he observed that the kinds of conclusions they wanted to reach about proper conduct received only a problematic justification from the factual premises invoked. The problems are well illustrated by Darwinian ventures in ethics that postdate Hume's challenge. When late-nineteenth-century social Darwinists appealed to the struggle for existence to justify allowing, or intensifying, competition in human societies, their inferences were unjustified in exactly the way Hume diagnosed. A century later, when human sociobiologists and fellow travelers proposed that prohibitions against incest can be derived

---

fered by Arthur Prior ("The Autonomy of Ethics," *Australasian Journal of Philosophy* 38 [1960]: 197–206).

4. Troubled by this inference, Nazi leaders hit on the idea of sending Jews who had won the Iron Cross to Theresienstadt, not as a way station to Auschwitz but as a terminal destination.

5. The awkward formulation here is important—it would be wrong to suppose that "X ought not to do A" follows from the factual premise. The negation has to be "external," applying to the whole sentence that ascribes the obligation.

from the alleged fact that inbreeding is evolutionarily deleterious, they were repeating the mistake.[6]

For the moment, I shall take naturalism to be committed to the thesis that *in some sense* ethical conclusions are susceptible of justification.[7] Hume's challenge applies to a simple, historically prominent, way of elaborating this commitment. If you think there are external constraints on ethics, constraints whose sources human beings are struggling to fathom, and also think we have straightforward epistemic access to (and only to) factual truths about the world, the justificatory problem will arise in a way that recalls Hume's objection. For you will suppose that people can justifiably arrive at a set of factual claims and that justification of ethical statements must be through inferences from these claims. In our own personal development, or at some point in the history of the cultural lineage in which we stand, the only justified beliefs are factual; later we have justified ethical beliefs; these must be based on inferences from the earlier corpus of facts.

In envisaging some "ethics-free" state, this formulation overlooks the possibility elaborated in Part I. Pragmatic naturalism's favored history views human beings as always having been committed to ethical claims: they have governed their lives together according to agreed-upon rules and have believed the descriptive counterparts of those rules (when they have accepted an injunction to do some deed, they have also believed the deed to be right, or good, or virtuous). On occasion, modifications of ethical practice have been effected through inferences, but the clearest cases

6. For social Darwinism, see the classic study by Richard Hofstadter (*Social Darwinism in American Thought* [Boston: Beacon, 1955]). The most prominent versions of the sociobiological inferences come in E. O. Wilson, *On Human Nature* (Cambridge, MA: Harvard University Press, 1978), and in Michael Ruse and E. O. Wilson, "Moral Philosophy as Applied Science," *Philosophy* 61 (1986): 173–92. For extended critique, see the final chapter of my *Vaulting Ambition: Sociobiology and the Quest for Human Nature* (Cambridge, MA: The MIT Press, 1985) and "Four Ways of 'Biologicizing' Ethics," in *Conceptual Issues in Evolutionary Biology*, ed. Elliott Sober (Cambridge, MA: Bradford Books, The MIT Press, 1994).

7. As the discussion of method in substantive ethics, to be presented in Chapter 9, will reveal, the sense in which ethical proposals are justified is rather special and involves a collective procedure. For the moment, the complications of this method are ignored, and I shall operate with a blanket notion of justification that figures in naturalism's commitment.

from history reveal the premises as partly normative and partly factual. Wollstonecraft's premises are (1) women ought to be capable of wifely and maternal behavior (in line with the prevalent ethical code), and (2) educated women are more likely to have these capacities. She infers that women ought to be educated (§24). Her inferences are not touched by Hume's challenge.

Merely making this observation is no more satisfactory than appealing to the fact that the inferences Hume criticizes are hard to circumscribe. To say that "from the beginning" human beings (*human* beings) have been committed to normative judgments would only release naturalism from Humean suspicions if all subsequent inferences could be assimilated to the pattern discerned in some of my examples (for example, those that figure in the reasoning of Wollstonecraft and Woolman). Unfortunately, if you suppose all modifications of ethical practice involve logical inferences from previously adopted ethical judgments and factual statements (typically newly discovered), ethical innovation will be viewed as an illusion. Clear-eyed champions of the Humean challenge would wonder how, if there were never any novelty in ethics, it was possible for human ethical practice to achieve, at the beginning, a strong enough collection of normative premises, and to do so in a justified way (§31). If pragmatic naturalism is to scotch concerns about its reliance on illicit inferences, more will have to be said.

One thing that can, and should, be said immediately is that not all cogent inference is deductive. From Hume himself on, no sophisticated devotee of the challenge has believed that the inferences leading from factual premises to normative conclusions have to be cast in a form revealing them as deductively valid. Naturalists can avail themselves of nondeductive modes of inference. Pragmatic naturalism can plausibly suppose that some modifications of ethical practice proceed through a search for reflective equilibrium: given the normative judgments currently accepted, one looks for principles subsuming them and perhaps jettisons particular normative judgments that do not accord with candidate subsumptive principles.[8] Recognizing this possibility (and others akin

8. Rawls's conception of reflective equilibrium, apparent in his "Outline of a Decision Procedure for Ethics" Chapter 1 of John Rawls *Collected Papers* (Sam Freeman, ed.)

to it) enables pragmatic naturalism to account for some instances of ethical novelty, and thus to avoid attributing to the first pioneers possession of the full complement of ethical resources. Unless it could be shown, however, how all later transitions can be reconstructed by appeal to reflective equilibrium, pragmatic naturalism would still face the task of explaining the remainder—and, in any event, it would have to show how the enterprise began in a justified way.

Given these preliminary reflections, the Humean challenge can be tackled head-on. The worry concerns the justification of the elements of ethical practice. Naturalists must elaborate an account of the justified beginning and growth of the ethical project, showing it to be free of the sorts of inferences Hume questioned, the types of inferences appearing, for example, in social Darwinism and human sociobiology. Without the narrative of Part I, and the conception of ethical progress presented in Chapter 6, any serious response would be impossible—which is why consideration of those worries has been postponed.

The framework developed takes ethical progress, not ethical truth, to be the fundamental notion. So, instead of posing the challenge as a question about whether inferences from factual statements to normative statements would be likely to yield correct conclusions from true premises, the issue has to be reformulated. Is it possible to understand how our ancestors made progressive transitions, and did so on the basis of processes (observations, emotional responses, modes of reasoning) likely to promote progressive transitions?[9] An affirmative answer would remove the sting from the challenge. For it would demonstrate how any inferences made accord with the fundamental criteria for good inference, and thus are exempt from the mysteries Hume rightly queried.

(Cambridge, MA: Harvard University Press, 1999), and fully explicit in *A Theory of Justice* (Cambridge, MA: Harvard University Press, 1971), is well suited to a dynamic understanding of ethical justification. Interestingly, Rawls's source for the appeal to reflective equilibrium, Nelson Goodman, deployed the idea as part of a naturalistic solution to his own "new riddle of induction" (see the final chapter of *Fact, Fiction, and Forecast* [Indianapolis, IN: Bobbs-Merrill, 1956]). I shall discuss the credentials of appeals to reflective equilibrium at greater length in Chapter 9.

9. Here I draw on my reconceptualization of a reliabilist approach to justification. See §§32, 39.

Divide the problem into parts. Start with the pioneers, those who began ethical practice. They belonged to a tense society, in which altruism failures constantly produced social conflict. If their newly introduced rules address the problem underlying the original function of ethics (they fulfill the function of remedying altruism failures), they take a progressive step. Despite my emphasis on the sporadic character of ethical progress (episodes of "sleepwalking" might occur with significant frequency), the codes introduced by groups successful in cultural competition discharged this function. It is implausible that their success was entirely accidental. The first ethical deliberators surely perceived clearly some sources of trouble—failures to share, unprovoked aggression, and so forth. Almost certainly, the rules they formulated to address their problems were imperfect, but they were an advance over the unregulated state in which they suffered social tension. Imagine, for example, early ventures in regulating alliances and mating, rough-and-ready delineations of "the elementary structures of kinship," adjusting the conduct of group members so it became more frequently behaviorally altruistic and less likely to provoke trouble. We cannot suppose that deliberators who recommended progressive transitions were entirely clear about what they were doing—they were not in the position of later analysts who can identify functions and show how they would be promoted by proposed changes—but the successes of these early ethicists did not emerge from blind guesswork either. In the beginning, the successful pioneers made ethical progress through processes (diagnosis of prevalent social problems, joint deliberation) likely to generate progressive transitions.

Their successors expanded the conceptual framework to introduce ideas (of good, of right, of virtue—or of what it is to be "one of us") enabling them to express descriptive counterparts of the rules previously adopted. They have a rule enjoining or forbidding a class of actions, and, within the expanded framework, they declare that actions in this class are right or are wrong (what members of the group do, or avoid). If adopting the rule was based on deliberation likely to generate progressive transitions, promulgating the rule is justified—and so derivatively is acceptance of its descriptive counterpart.

Turn now to the subsequent modifications. As emphasized (§§36–37), the evolution of ethics brings new functions beyond the original one.

Hence, at later stages the deliberations, arguments, and observations prompting new transitions should be evaluated by seeing how well they serve the overall set of functions ethical practice is to serve. In the simplest cases, the processes through which transitions are effected refine some functions without compromising others and can be vindicated as of types likely to do so. Thus, the extension of some rules, already protecting group members, to cover trade with others, produces satisfaction of endorsable desires for group members, as well as diminishing conflict with neighbors; the extension can result from perception of likely advances in fulfilling these functions, so that processes bringing about the transition are well adapted to generate ethical progress. As with the original case, we can recognize the justification of the modification—and derivatively of the descriptive counterparts of the extended rules.

More difficult are cases in which one function is served at cost to others, or a modification of ethical practice threatens future regressions. Consider, for example, the introduction of the unseen enforcer as a device for promoting compliance. Insofar as the modification increases the frequency with which members of the group obey the prescriptions, this is a progressive step, but it paves the way for damage to the project of joint deliberation (some group members have direct access to the divine will—§35). In such instances there is a propensity both for progress and for regress. The transition cannot be forthrightly classified as justified. In general, when there are gains and losses with respect to different functions, three possibilities arise: (1) because the balance is significantly greater on one side (the gains are much larger than the losses), the modification is overall progressive (or regressive); (2) although there is no overall verdict, the modification can be partitioned, and some newly introduced elements make progress, while the rest are regressive; (3) the situation is so thoroughly mixed that neither an overall judgment nor a recognition of progressive and regressive aspects is possible. Instances of the first two types can be assimilated to instances in which there is no functional conflict, by recognizing that the change in practice is justified as a whole (case 1), or by separating out justified parts of the change (case 2). With respect to transitions of type 3, the change is not justified.

Let us differentiate standards of progressiveness and of justification. So far, in responding to Hume, relatively weak, *short-term criteria* of progress and of justification have been used. We can, however, mimic the distinctions between local and global progress (§37). *Weak* conceptions of progress and justification can be formulated explicitly:

> (WP) A change is progressive just in case it leads to greater fulfillment of those functions of ethical practice that have emerged at the pertinent stage.
>
> (WJ) A change is justified just in case it is generated through processes likely to lead to progressive changes (as specified by WP).

These criteria make allowances for the circumstances in which those who modify their ethical practices find themselves, focusing on the functions ethics serves for them and on the immediate effects of the transitions. It is possible to demand more.

> (SP) A change is progressive just in case it introduces elements that will be preserved in any indefinitely extended sequence of WP-progressive changes.
>
> (SJ) A change is justified just in case it is generated through processes likely to lead to SP-progressive changes.

Parts of the history of ethics proceeded according to WJ-justified transitions. Although we cannot be sure our ancestors always saw how the rules they proposed addressed the functions ethical practice served for them, in some instances the difficulties they faced would have been obvious (lessening conflict within the group, dealing with hostile neighbors) and their deliberations would have been overwhelmingly likely to follow reliable procedures for dealing with those difficulties: we can expect them to recognize how rules of alliance would work, how offers to trade peaceably with neighbors would satisfy desires endorsed by their existing rules. The standards for weak justification were probably met quite frequently in the progressive transitions our ancestors made.

Can a stronger conclusion be defended? Perhaps. Consider our predicament, if, as pragmatic naturalists, we analyze what has occurred. Honesty exemplifies the "vague core themes" that serve as the best candidates for ethical truth (§38; see also §46). Analysts can offer lines of reasoning for thinking some elements of our current practice—like rules enjoining honesty—would be preserved in any indefinitely proceeding WP-progressive sequence of transitions. The injunction to truthfulness remedies a class of altruism failures that will continue to be important to us as long as human beings communicate with one another. By identifying permanent human needs, analysts can conclude that some elements in ethical practice will probably figure in any progressive development of what we now have. When they do that, their judgments are SJ-justified.[10]

Because of the complications just noted, it is worth stating explicitly how Hume's challenge has been met. Hume worries that any naturalist account of ethics is committed to forms of inference incapable of justifying the ethical conclusions obtained. Pragmatic naturalism understands notions of ethical truth and justification in terms of the fundamental notion of progress, conceived as functional fulfillment and refinement. Introducing ethical novelties, whether at the beginning of ethical practice or in subsequent modification, is justified when those who make the change do so by following processes likely to lead to better functional fulfillment. Any inferences they make are thus evaluated by appeal to fundamental criteria, which either may make allowances for their historical circumstances (WJ) or may consider the indefinite evolution of the ethical project (SJ). Many of the processes through which historical actors made the changes they did are likely to satisfy the less demanding criterion, and later analysts can meet the more exacting standard (SJ) for some (perhaps all) core themes in contemporary ethics.

The key to this response is quite simple. Once ethics is viewed as a social technology, directed at particular functions, recognizable facts about how those functions can better be served can be adduced in inferences justifying ethical novelties. The mystery that worried Hume disappears.

---

10. In making the particular claim about honesty, I offer no judgment about whether the line of reasoning that counts as SJ-justified would have been available to our predecessors. It is possible (but not obviously correct) that SJ-justification requires explicit articulation of a pragmatic naturalist perspective. If so, it does not detract from the WJ-justification I take to be present throughout the history of ethics.

§41. Authority Undermined?

Replying to Hume in this way may exacerbate other concerns. If ethics is a human construction, devised to discharge particular functions, can it have the authority usually attributed to it? The question expresses another version of the charge that naturalism commits a fallacy. The first task is to understand what the challenge might mean.

Pragmatic naturalism offers a simple explanation of why people *feel* the authority of ethical prescriptions. Each of us is born into a society that inculcates a body of lore, and, because each of us has a capacity for normative guidance, more or less cleverly stimulated by processes of ethical education, we find ourselves, as children and as adults, feeling a tug to align our will in a particular direction. Accounting for the authority of ethics might proceed not by adducing some metaphysical claim about the status of ethics, but by pointing to sociological and psychological facts. A sense of what is required of us ("the internal sanction of duty") stems from a cluster of capacities and dispositions with which we have been equipped ("Conscience," or a "mass of feeling which must be broken through in order to do what violates our standard of right").[11] Perhaps parts of this subjective basis derive from some innate feeling, while others are "implanted."[12]

---

11. The quoted phrases are from Mill, who favors explaining the "sanction of ethics" in sociopsychological terms. See Mill, *Works* (Toronto: University of Toronto Press, 1970) 10:229, 228.

12. Mill is inclined to the latter explanation: see *Works* 10:230. In accordance with the discussion of §14, I recommend framing the questions in terms of interactions between genotype and environment, and remaining open about the precise contributions of each.

Mill offers an eloquent formulation of the picture he views as most plausible. Our cooperative social existence arouses this feeling, and it is given direction by the community around us:

> The smallest germs of the feeling are laid hold of and nourished by the contagion of sympathy and the influences of education; and a complete web of corroborative association is woven round it, by the powerful agency of the external sanctions. (*Works* 10:232)

As I read him, he tries to address the question of the authority of ethics (the "sanction of morality") by appealing to something like the account of normative guidance and of conscience offered in §§11–13.

Does a sociopsychological approach "evacuate [ethics] of all authority"?[13] The concern is clear. Explaining *felt* authority is no problem for naturalism (specifically for pragmatic naturalism), for the (philosophical) history it offers attends to the psychological processes underlying our obedience to the ethical precepts we learn. That is not enough: *felt* authority is different from *real* authority. Naturalists shift the question and fail to provide any explanation for why ethical prescriptions have real authority over us. Instead of recognizing the authority of ethics over our social arrangements, naturalists focus on the authority of our social arrangements over individual conduct and judgments.[14]

What exactly does the objector understand by "the authority of ethics"? At first glance, it would appear that the authority of anything has to be mediated by psychosocial processes, for how could one's will be forced in a particular direction unless there were some such causal rela-

13. The quoted phrase is from Dewey, who extends Mill's sociopsychological approach. He confronts what is supposed to be a great difficulty:

> It is said that to derive moral standards from social customs is to evacuate the latter of all authority. Morals, it is said, imply the subordination of fact to ideal considerations, while the view presented makes morals secondary to bare fact, which is equal to depriving them of dignity and jurisdiction. (*Human Nature and Conduct*, 79)

Dewey claims that the worry rests on a "false separation," the supposition that cultural practices are merely "accidental by-products" (ibid.). Here he recapitulates the point made in his summary of his ethical perspective, that "moral conceptions and processes grow out of the very conditions of human life" (John Dewey and James Tufts, *Ethics*, 2nd ed. [New York: Henry Holt, 1932], 343). This prepares the way for his answer to the challenge. After arguing that, strictly speaking, no view of the "origin and sanction" of moral obligation can provide what some people seek for the authority of ethics, he suggests that, in "an empirical sense" there is a simple answer: "The authority is that of life" [*Human Nature and Conduct*, (Amherst NY: Prometheus Books, 2002), 80–81; see also 98, 232, 326]. My approach to the objection that naturalism loses the authority of ethics will endeavor to articulate what I take Dewey to have had in mind.

14. Some of Mill's formulations make him vulnerable to the charge that he has changed the topic in this way. Yet that accusation is a little unfair, since the discussion in *Utilitarianism,* in Mill *On Liberty and other Essays* [Oxford: Oxford University Press (The World's Classics), 1998], does make some efforts to show how the conditions of community life will orient our subjective feelings (conscience) in particular directions—specifically in leading us to view our own preference and perspective as one among many. It is easy to see, however, how my emphasis on processes of socialization provokes the complaint.

tionship? Lurking in the background, however, is the view of ethics as a form of inquiry subject to external constraints. Ethics is seen as answerable to something outside (or beyond?) us, as individuals and as groups, something to which our conduct and our social practices ought to conform. Even if we honor the authority we feel, complying with the directives laid down for us in the society in which we live, that society may not be aligned with the external constraints: mistakes may have been made.

An analogy with other parts of human culture helps. Consider the formation of belief, in everyday deliberations and in scientific investigations. As individuals, or collectively, our inquiries can go astray, failing to measure up against an external standard: facts, one might say, are authoritative over beliefs. This slogan is easily overinterpreted. Consider the following readings:

(a) For any true statement p one ought to believe p.

(b) For any true statement p one ought not believe any statement q that is incompatible with p.

Plausible as they may initially appear, both a and b are false.

There are at least three sources of trouble. First, there are far too many true statements for finite animals like ourselves to believe all of them. Our world could be described in any of some nondenumerable infinity of potential languages, and within each of these languages there are infinitely many statements describing the course of events within some small spatial region (a room, say) during some short temporal period (for example, an hour). Only a truly tiny minority of such statements is worth bothering about, only an infinitesimal fraction of them has any significance for any human being—where the standards of significance are set by human needs, interests, and projects.[15] We can rightly criticize people for investigating the wrong things (or posing the wrong questions). Even true answers to irrelevant questions do not merit belief, contradicting a.

---

15. For discussions of this theme, see chaps. 5 and 6 of my *Science, Truth, and Democracy* (New York: Oxford University Press, 2001).

Second, it is sometimes impossible for us to achieve our purposes by trying to discover and work with the exact truth. Much of our practical use of "knowledge" is grounded in accepting simplified versions of the world, approximations to the truth rather than the genuine article. When the phenomena are too complex for us to render accurately, and when an accurate rendering would only interfere with our aspirations to predict, control, and intervene, we are better served by believing something that is not, strictly speaking, true. On such occasions, both a and b are false: we ought not believe the exact truth and we are entitled to believe statements incompatible with the exact truth.

Third are considerations about the limits of honesty (§46). Seriously ill patients are sometimes better off believing false statements. Not only are they not obliged to believe what is true about their condition, but they can permissibly believe things incompatible with the true description. Once again, both a and b fail.

The common trouble lies in human purposes ill served by true belief. Proponents of the "authority of facts" take attaining truth as an ultimate standard. It is not. More fundamental is the role of inquiry and belief in advancing human concerns, and the cases briefly surveyed show how, occasionally, promoting such concerns can come apart from acquiring true belief. When it does so, the concerns set the more basic standard and thus release people from any obligation to believe the truth. Of course, much of the time, even almost all the time, our human purposes are best served by discovering and believing what is true—that is why a and b are plausible and why we easily talk of the "authority of facts." From a naturalistic perspective, however, one should probe more deeply, exposing the purposes having true beliefs typically serves, and viewing those purposes as constituting the fundamental standard from which authority arises.[16]

The analogy prepares for a similar reconnection in understanding the "authority of ethics." We have reviewed the idea of external constraints that "prescribe" to us in a fashion similar to that in which the world is supposed to "prescribe" to our beliefs (§§29–31). Critics might take the

16. This is to overcome the "separation" of which Dewey was suspicious, and to reconnect the standards of inquiry with human life.

pragmatic naturalist mistake to lie in abandoning any external standard. Or they might believe pragmatic naturalism cannot find a counterpart for the "constraints on practical reason" by which we are bound.[17]

As with Hume's challenge, the key is to recognize the effect of substituting progress for truth as the fundamental notion. The objection can be cast as follows. When naturalists explain the authority of ethics, they point to contingent ways in which societies formulate rules for their members and the contingent processes through which they socialize people into subordinating their behavior to such rules. The real authority of ethics consists in the fact that certain prescriptions are binding on our conduct, whether or not they figure in an ethical code and whether or not ethical practice includes effective ways of inducing conformity to its precepts. If naturalism is to succeed, it must differentiate felt authority from the genuine article and show why particular rules are authoritative.

By appealing to the functions of ethical practice, pragmatic naturalism can directly answer the first part of this demand. There is no difficulty in showing how people who follow the ethical code of their community, in cases where the critic would view the code as diverging from ethical truth, are mistaken. Consider the following possibilities:

(i) Someone brought up in a society conforms to its ethical maxims, maxims that would be replaced in a progressive modification of the ethical practice.

(ii) Someone brought up in a society conforms to ethical maxims that would be incorporated in a progressive modification of the ethical practice and does so through processes likely to promote progressive alterations of practice.

(iii) Someone brought up in a society conforms to ethical maxims that would be incorporated in a progressive modification of the ethical practice but does so through processes not likely to promote progressive alterations of practice.

17. Kantians will incline to the latter formulation, resisting any suggestion of an independent ethical reality to which our judgments should conform. Their approach to objectivity emphasizes constructions taken to have a priori validity, and they will argue that pragmatic pluralism can supply only contingent—second-rate—constraints.

Possibility i focuses on subjects who conform to what they have been taught, even though progressive change would replace the precepts on which they rely—here there is a clear distinction between "felt authority" and a standard to which the subjects (and their society) are not living up. Possibility ii, by contrast, reveals those who comply with the divergent standard and do so in a justified fashion—those whom we might see as appreciating the "real authority" of ethics and consequently as overriding the felt authority current in their community. That type of reformer is to be distinguished from the characters who emerge in possibility iii, who base their conduct on a progressive shift, even though they have no justification for taking it to be so; like the subjects of possibility i, they exhibit shortcomings, and it would be appropriate to view them as failing to appreciate the "real authority" of ethics.[18]

Pragmatic naturalism can *represent* a difference between people who submit to local authority in an unsatisfactory way and those who overcome the shortcomings of their peers. Moreover, it can *explain* that difference in terms paralleling the critics' preferred accounts. For the critics suggest that inadequate conformity to the prevailing code is separated from appreciation of real authority, in virtue of the fact that the appreciators acknowledge external constraints—ethical truth or the commands of practical reason. Pragmatic naturalism sees these formulations as misguided ways of trying to resist the mere-change view and substitutes its preferred account: the appreciators make progressive modifications (better fulfilling ethical functions), and they do so through processes disposed to yield progressive changes. On the basis of the parallel, pragmatic naturalism proposes that the demand for this sort of progressive change constitutes the "real authority" of ethics.

Although this rebuts the criticism in its original form, it may not satisfy critics. For they may declare that the naturalistic account introduces

18. The notions of progress and justification that figure here can be construed according to either of the conceptions that figured in the last section, as weak or strong. The strong concepts provide a close analog of the split between conformity to local authority and the recognition of ethical truth, on which critics want to insist, but the weak concepts mark another important distinction, one accommodating our fallibility and the historical contexts in which we find ourselves.

a contingent element, one failing to capture the *absolute* authority of ethics. So long as the functions of ethics, the original function of remedying altruism failure and the derivative functions emerging through the historical ways of fulfilling that original function, are endorsed, it is possible to recognize the authority of ethics. Those functions could be rejected in a way in which it is not possible to reject the realm of ethical fact or the demands of practical reason. The analogy with belief formation exposes the critics' error. The thesis that "facts have authority over belief" must be understood, in the end, as a claim about the functions our beliefs serve: strictly speaking, *a* and *b* are false, and their import can be properly recognized only by seeing the acquisition of true beliefs as typically important to the functions of belief formation. In similar fashion, if the "absolute authority" of practical reason were anything more than a dogmatic pronouncement, it would rest on showing the role of practical reason in discharging human functions. *The accounts proposed by pragmatic naturalism and by the critics are on a par.* If any of them can deliver an explanation of the "real, absolute authority" of ethics, pragmatic naturalism can do so. If something more is wanted, all are deficient—and it will be reasonable to judge that the sort of authority demanded is a myth.

In portraying naturalism as introducing a "merely contingent" authority for ethics, one depending on endorsement of functions that might be rejected, the criticism does make a "false separation"—for it fails to appreciate the connection between the problems behind those functions and the circumstances of human life (§39). The demands leading to ethics are not arbitrary or conventional; they grow out of human needs. To what higher—or less contingent—standard could one appeal to ground the authority of ethics?

§42. Troublesome Characters

Surely a major issue remains. Although parallels are provided, and explanations of the "authority" of ethics offered, traditional (nonnaturalistic) conceptions of ethics may seem to do something naturalistic surrogates cannot. They can answer the troublesome characters—skeptics, egoists, nihilists—who crop up here and there in the history of ethical discussions.[19] In relatively polite versions, these troublemakers have already surfaced in earlier sections, for example, in considering a subjective criterion for ethical progress (§28). They conceded that particular transitions might be labeled as "progressive" but saw this as merely providing a sense of false contentment. They level a basic challenge: once ethics has been seen as the evolving enterprise pragmatic naturalism describes, why should anyone conform his or her behavior to its deliverances?

So far, the issue has been framed as *comparative*. Allegedly, there are questions pragmatic naturalism cannot answer, but some other (nonnaturalistic) conception of ethics can. Naturalistic proposals are inarticulate, or at least unconvincing, in addressing skeptical challenges to which nonnaturalism can provide fluent responses. My main aim is to show that anything rivals can do, pragmatic naturalism can do just as well (if not better).[20]

Start by disambiguating the skeptical question. The challenger wants to know why he should conform to "the deliverances" of ethics, given their emergence from a historical process. What exactly are these "deliverances"? Skeptics often present their demands as questions to the authority of *current* codes, but an evolutionary account will not suppose all the rules included in such codes are to be followed. We can distinguish:

(a) rules present in current codes
(b) rules that would be sustained in an indefinite sequence of progressive modifications from current codes
(c) rules we are justified in thinking of as likely to be sustained in an indefinite sequence of progressive modifications from current codes

19. I shall not attempt any fine distinction between skepticism and nihilism: my focus will be on a range of questions that might be covered with either label. Egoists will be considered as skeptics/nihilists who view ethics as a constraint on their personal (solitary—§3) desires.

20. As we shall see, discussion of the comparative issue leads to a deeper challenge, one already identified at the end of the last chapter. I am indebted to an anonymous reader who recommended being explicit about this.

Even this basic differentiation allows various ways of reading the skeptical query.

Nonnaturalistic approaches to ethics clearly suppose that the skeptic is raising a question about some *true* ethical precept. Nobody thinks a skeptic who inquires why he or she should conform to the injunction to lie whenever one feels like it deserves any more than correction about what ethics commands. So a very straightforward reading of the challenge would adopt *b* as indicating the precepts with respect to which skepticism is to be addressed. That formulation, however, might seem to underestimate human fallibility. We know we do not have any ultrareliable access to the rules that would be adopted in any indefinite sequence of progressive modifications, and so, perhaps, we should take ourselves as reasonably conforming to the rules our best justifications single out as likely to have the status assigned in *b*—and take skeptical questions to be directed against *c*. Or we might go further, viewing our individual selves as unlikely to improve on the wisdom of the tradition in which we stand, and modestly disclaiming our own abilities to make judgments about future progressive modifications. Consequently, we might take the rules adopted in our current code as the best available candidates for *b*, supposing the skeptical challenge focuses on *a*. Even formulating the skeptical worry turns out, ironically, to involve a judgment about the allowances we *ought* to make for our acknowledged fallibility.

As before, it is not enough to point to inexact or problematic formulations of an objection. However you circumscribe the queried rules, the challenge can be posed with respect to examples. Consider any likely candidate, and imagine the demand directed at it: given that injunctions to honesty (say) emerge as stable elements in all progressive traditions, why should that provide grounds for compliance? Skeptics want to know why a sequence of progressive transitions, conceived as addressing the functions of the ethical project, should deliver rules to be obeyed, even when the rules in question seem overwhelmingly likely to be maintained in all future progressive modifications of current practice.

What conditions should an answer to the skeptical question meet? One thought is that, to be satisfactory, a response should *silence* the skeptic. Given the response, nothing can be said to continue the skeptical challenge. Plainly, that sets the standards of success very high, and, in line with viewing the issue as comparative, we should begin by asking if the silencing criterion can be satisfied given *any* understanding of ethics.

Some well-known conceptions clearly fail. Consider the reductive naturalism often attributed to utilitarianism, according to which one defines the good as that which promotes the greatest total happiness. A familiar claim about any reductive definition of an ethical concept—the good or the right, say—is that it leaves a question open: given the definition, you can always ask, "Why should I act in a way that conforms to the definition?"[21] The naturalistic reduction of the good thus fails to address the skeptical challenge, precisely because it does not silence the skeptic.

Do things go better on alternative pictures? Consider some approaches found in or inspired by Kant. Ethics expresses the requirements of pure practical reason: to deny the moral law reason generates is to fall into a mode of irrationality, in which one contradicts oneself. Or it might be said: ethics consists in a set of principles ideally rational agents would agree to under ideal circumstances, so failing to abide by its precepts is to violate conditions of rationality. Tough-minded skeptics would hardly be brought to silence by either of these dicta or by any plausible emendations of them. The skeptic speaks: "You can call the procedures you use to generate the rules you favor 'pure practical reason,' if you like, and suppose those who don't go along with them are involved in some sort of contradiction, but the mere label doesn't frighten me, and the effects you envisage don't appear particularly dreadful. If I reject these rules, I am hardly doing something similar to asserting a statement and its negation—and even if I were, it's not obvious anything I care about would be compromised by doing so. The history of science contains episodes in which people have worked quite well with internally inconsistent ideas (think of the Bohr model of the atom[22]), and I have no reason to think my 'practical

21. The "open-question" argument, in just this form, is usually attributed to Moore. But a close reading of what Moore actually says reveals a much more complex and rather different line of reasoning. For good discussion, see Joyce, *The Evolution of Morality* (Cambridge, MA: The MIT Press, 2006). Whatever the merits of the interpretation of Moore, the familiar objection is an important one.

22. As has frequently been observed, the conception of discrete orbits is at odds with the principles of electromagnetism. Another interesting example is the concept of the rigid body in relativistic dynamics.

irrationality' will pose difficulties for me. Nor am I much moved by the thought of rules hypothetically ideal people (supposedly better than me) in some fictional situation would agree to. Why should I be bound by what they would decide? There is no argument for thinking the purposes I care about would be ill served by flouting any such precepts."

Orthodox Kantians or contractarians may think this alleged reply is incoherent, tacitly self-undermining, or something like that. Yet however hard they struggle to reveal the skeptic's irrationality, they do not render him speechless—he can cheerfully wave away the gloomy descriptions of his state by pointing out how well he is able to pursue the things that matter to him. Silencing is hard to do.

Maybe, though, the skeptic's answers have exposed something wrong with him: he has failed to meet ideal conditions of rationality. If that is so, the criterion for a successful reply to the skeptic is modified. He does not have to be silenced; one must merely have an account of why his responses are problematic. Given this understanding, however, pragmatic naturalism can do just as well as the allegedly superior nonnaturalistic approaches. Where Kantians and contractarians see failures of ideal rationality, pragmatic naturalism diagnoses an inability to appreciate how central the ethical project is to human life. Pragmatic naturalism will begin answering the challenge by pointing out that the achievement of normative guidance was central to the origin and development of fully human society, that ethics served an important original function, and that progressive shifts in ethical practice consist in fulfilling that function, and those generated from it, more effectively. Once again, the skeptic speaks: "Why should I be bound by the rules emerging from 'progressive transitions' in this 'project'? No matter how they have helped discharge the functions you identify, I can manage perfectly well by flouting them. There is no compelling motivation for me to continue in any ethical tradition." Pragmatic naturalists will want to insist that desires the skeptic wants to satisfy by breaking the rules have been made possible only by the project he rejects, that he fails to understand how the origin and evolution of ethical practice have framed his life, giving him the options he wants to pursue. That insistence will not silence the skeptic—it will be no more effective than appeals to practical reason or charges of irrationality. Yet it fares no worse as a diagnosis of the skeptic's mistake: he wants to reject something that has made his envisaged way of life and his preferred choices possible.

This judgment can be elaborated further and defended by considering some familiar characters from the history of philosophical ethics. Thrasymachus challenges Socrates, suggesting that ethical principles (more exactly, the claims of justice) are put forward to advance the interests of those in power.[23] He views ethical practice as a device employed to keep weaker members of the society in line and refuses to go along. Pragmatic naturalism credits Thrasymachus with an insight and faults him for an error. The insight: in many ethical traditions, rules are introduced or emended by a powerful minority, supposedly especially good at recognizing the will of the unseen enforcer, and the precepts of the social practice can be arbitrary and oppressive (§§35, 26). The error is to overlook the fact that pragmatic naturalism diagnoses the oppressive injunctions as stemming from regressive transitions. *Functional* ethical practice is not a tool for asserting the will of the strong and mighty, but rather grounded in attempts to take into account the desires of all members of a society. The original function of remedying altruism failures acknowledges the wishes and aspirations of all. Thrasymachus can be enlisted as an ally and invited to continue in the evolving project of ethics by responding to places at which it is *dys*functional. He is wrong in his characterization of the project as a whole.

A different challenge comes from a later figure, the "sensible knave" who pops up in the final paragraphs of Hume's *Enquiry Concerning the Principles of Morals*.[24] Knave does not offer any general theory of what ethical practice is and how it has gone wrong—he simply wants his own plans not to be constrained by it. He has egoistic (solitary) desires contravening some of the ethical principles in force, and, where he can get away with it, he would prefer to satisfy those desires, instead of conforming

23. See Plato *Republic* Book 1. Socrates replies with a convoluted and unconvincing sequence of questions that eventually reduce Thrasymachus to silence. Later, Glaucon and Adeimantus present the challenge in a milder form, eliciting a far more sophisticated and interesting account of justice. Whether an adept real-life counterpart of Thrasymachus would really be rendered speechless by either version of the response is highly dubious.

24. David Hume, *Enquiry Concerning the Principles of Morals* (Indianapolis, IN: Hackett, 1983), 81.

to the principles. Knave has no interest in persuading others the rules need reform—in many instances, he is delighted others are prepared to follow them and happy to be a "free rider." What can be said to bring him into line?

Pragmatic naturalism can start with the points Hume makes, in his quick, almost casual, attempt to deal with the problem. [25] We can inform knave that he will be disturbed and worried, that he will forfeit that tranquility of mind good conscience bestows. We can try to show him he will not actually achieve the things he wants if he pursues his knavish tricks. In many instances, these responses will be ineffective. Knave correctly appreciates that the chances of detection are tiny, that he is the sort of person who can sleep serenely after his knavery is done, and that his knavery does not interfere with his other goals. Nor does it help to tell him he is exhibiting practical irrationality—for he will describe the ends he has in view and point out how they are attainable even if he is "practically irrational." The specific point pragmatic naturalism will make focuses on his rejection of the very practice that has made his knavish aspirations possible.

What exactly does knave want to do? If his aim is to satisfy certain basic desires—the kinds of desires he shares with hominid ancestors who preceded, and human ancestors who began, the ethical project— he deserves help and sympathy. If the only way for him to satisfy those desires is to break the prevalent precepts, something has gone badly wrong, either with the rules themselves or with the conduct of others. A knave in desperate straits (Jean Valjean) is no knave at all. Hume's character is not in this plight: he simply wants to increase his wealth, his power, or his position in ways forbidden by ethical rules, even rules embodying those core themes that are the best candidates for ethical truths. Pragmatic naturalism points out that the aspirations knave harbors are possible only in virtue of the ethical project that began with the acquisition of socially embedded normative guidance, and dependent on just the rules knave proposes to disregard. If our ancestors had not instituted those rules and found sufficiently effective ways of making their fellows comply with them, the options he wishes to pursue and the goals he hopes

25. Ibid., 81–82.

to attain would not be available to him. Knave replies that this is all past history. He is grateful so many of his predecessors went along with the codes in force and modified them in the progressive ways that have made contemporary society possible—although he hints that there have probably been a few of his sort before, people who have quietly made use of the obedient docility of their fellows. Now that ethical practice has made a rich and complex society possible, he can take advantage of that fact and pursue his egoistic goals.

Knave cannot be silenced. Yet pragmatic naturalism can offer a diagnosis of what he is doing. He wants to take advantage of the products of social evolution without acknowledging the functions that have made those products possible. He wants to operate within a society without feeling that sympathy to others, that altruistic response to their desires, whose evolution has formed that society. Being human, we tell him, consists in participating in this project through which altruism failures are remedied and further ethical functions generated and fulfilled. He shrugs his shoulders, unmoved by this rhetoric. Yet the diagnosis seems no worse than that offered by the major rival approaches to ethics. Indeed, one way of elaborating the notion of practical rationality might suppose that the knavish incoherence consists in making an exception of oneself while simultaneously relying on the ethical practice that sustains human cooperation, a reading that would erase the differences between the naturalistic and nonnaturalistic alternatives.[26]

The third and last troublemaker is a Nietzschean persona, the "free spirit." Unlike Thrasymachus, free spirit does not want to provide a general characterization to convince everyone to view ethics as oppression. Unlike knave, free spirit is not concerned only with himself and with his ability to use institutions presupposing the ethical project to advance his own solitary ends. Free spirit writes for his peers. *They* are oppressed, confined, by ethical practice; how matters stand with the rest, with the herd, is of no concern to him.

---

26. Pragmatic naturalism does point out a kind of "contradiction" in knave's attitudes—and Kantian-inspired approaches might identify this as the crucial failure of practical rationality. If so, the accounts would be not only on a par, but also almost identical. The residual difference would lie in the ability of pragmatic naturalism to point to the failure of sympathy knave exhibits.

The charge leveled by free spirit is so far ambiguous. Free spirit might be rejecting the ethical codes actually developed (or developed in a particular tradition), seeing these as failing to fulfill important human functions, and, on this basis, demanding a "revaluation of values."[27] Viewed in this way, free spirit is a reformer, a participant within the evolution of ethics, not someone who rejects the rules it would deliver if it were to proceed progressively. He calls attention to particular places—possibly quite fundamental places—at which he takes the historical development of ethics to have been regressive. Assuming we can reach agreement with him about whether the pertinent modifications refine the functions of ethics, there will be no difficulty.[28] For, if the transitions turn out to be progressive after all, he will acknowledge the rules. If they do not, we shall not insist on his acknowledgment of the rules actually generated, and he will honor the prescriptions resulting from progressive replacements of the dysfunctional precepts.

Free spirit may want more, however. He asks why he should care about the specific recommendations emerging from those transitions pragmatic naturalism counts as progressive. His question is best met with another: what alternative does he have in mind? To conceive of the historical evolution of ethical practice, taken as a whole, as oppressive is vacuous, unless one can do more than wave vaguely in the direction of unarticulated possibilities. The more limited version of free spirit, who is willing to acknowledge the functions of ethics and who clamors for

27. I think most of Nietzsche's many-sided polemic against "morality" can be interpreted in this way. So understood, he turns out to be an interesting and insightful *ally* of the pragmatic naturalist project. The derogatory remarks about "English" genealogies, and the attacks on the value of altruism, [see the early sections of *Genealogy of Morals,* Friedrich Nietzsche *On the Genealogy of Morality* (Cambridge: Cambridge University Press, 2007)], are fundamentally attacks on the thesis that a historical understanding of the emergence of contemporary ethical ideas must vindicate those ideas. Instead, Nietzsche wants to use history in the interests of reform—with the aim, one might say, of advancing the ethical project. That is entirely in accord with pragmatic naturalism, which is receptive to the thought that increased historical understanding might expose regressive transitions and open up new possibilities for us. I am grateful to Jessica Berry, whose insightful comments and questions in response to a lecture I gave at Emory University have led me to view Nietzsche as less of a threat and more of a fellow traveler.

28. The next section will begin to scrutinize this assumption.

reform so those functions will be more effectively fulfilled, is readily comprehensible, for he is committed to continuing the progressive evolution of the ethical project. To reject that project entirely is to take a further step, and we need to be told what free spirit aspires to.

A life cut off from any society, perhaps? That, as Nietzsche sees clearly, would be dangerous and brutish.[29] But what forms of alternative community or society are available, once the ethical project is renounced? We do know something of social life outside the tradition of ethical practice, for a life of this sort is the lot of our evolutionary cousins, the chimpanzees. Given the psychological dispositions free spirit has acquired, this cannot be a serious possibility for him. Can he offer another? If not, our response to the challenge is, in the end, a simple one: operating within the evolving practice of ethics is a central part of our humanity. Until we are given some description of an alternative—or until the Übermensch actually arrives—our choices are confined to the human, the ethically guided, life and the social state of the chimpanzees, a state transcended by our first human ancestors.

Three famous troublemakers are enough. They have been taken seriously, as if the task were to find arguments to bring them into line. A moment's reflection reveals this to be a distortion of the actual task. Were we faced with actual figures for whom resistance to ethical precepts was a live option, it would be foolish to rely on philosophical debate alone. We know too well that ethical injunctions, even those most stable and central to the main ethical codes, are often ignored or flouted, and only the most otherworldly of philosophers would think sufficiently well-crafted arguments would bring potential offenders into compliance. In societies with a high incidence of failure to obey their ethical precepts, it is usually appropriate to recognize deficiencies in techniques of socialization and to seek to improve them. Yet some people are beyond the reach of even the most effective methods for inculcating ethical attitudes. Some real-life knavery results from inadequate education; some requires sterner measures.[30]

29. Nietzsche, *Genealogy of Morals*, 2:9.
30. This point is well made by Philippa Foot, *Natural Goodness* (Oxford UK: Oxford University Press, 2001).

Understanding that the task is not to quell the sociopath with some brilliant philosophical formula helps us see more clearly what the troublesome characters really represent. We should envisage the skeptical challenge as posed by ordinary people, whose socialization is reasonably effective and who feel the tug of ethical commands. They have a tendency to conform but also want to know why they should be glad to have this disposition. The troublemakers are devices for giving substance to this worry, personae with which the skeptical questioner can vicariously identify but for whom following through on the identification would be psychologically disastrous. The skeptical question is not a demand to be talked into complying (typically, disobeying the rules is not a live option), but a request for reassurance. The questioner needs to feel at home with his or her ethical propensities.

We can now see why certain approaches to ethics, particularly the nonnaturalistic Kantian and contractarian varieties, seem appropriate replies to skeptical challenges. They provide reassurance by delineating an ideal of rational thought and behavior, more or less thoroughly articulated, so people who already feel the ethical tug can identify a mistake deviants would be making. These philosophical replies cannot (to repeat) silence deviants or bring sociopaths to heel. But they succeed at a more modest task.

So too does pragmatic naturalism. To the extent people who wonder whether they should be glad to have ethical dispositions can be satisfied with explanations invoking practical rationality, they should be (at least) equally content with the pragmatic naturalist account. For that account places ethical practice at the center of our humanity, viewing ever-more refined attention to altruism failures, ever-increased recognition of the wants of others, as preconditions of the kinds of lives we live and the kinds of societies we have. Although one may challenge parts of the ethical practices we have inherited, there is no escaping the ethical project. The only social alternative we know is that of our hominid ancestors and our chimpanzee contemporaries, an alternative from which the introduction of ethics originally liberated us. That reply should be reassurance enough.

In a justly famous image, Otto Neurath specified our epistemological predicament, comparing us to sailors who must constantly rebuild the vessel on which they sail.[31] Pragmatic naturalism takes a similarly anti-foundationalist approach to ethics, denying any serious human alternative to the ethical project. Skepticism is, in the end, nothing more than an invitation to jump into unknown, and potentially dangerous, waters.

§43. Settling Disputes

One final version of the objection needs to be confronted. In discussing the challenge posed by free spirit, I supposed there would be no threat, provided free spirit was prepared to participate in the ethical project. Under those conditions, I envisaged him questioning the progressiveness of certain transitions and assumed agreement on this issue would be followed either by his conformity to the rules (if the episodes scrutinized were endorsed as progressive) or by vindication of his objection (if the episodes were found to be regressive). By what right, however, can one suppose that *internal* disputes, those arising among parties equally committed to the ethical project, can be resolved? Champions of rival approaches to ethics can argue that their favored accounts provide ways of settling disputes—they offer fundamental principles or methods for arriving at such principles—and so far pragmatic naturalism provides no counterpart.

This last concern mixes elements in two previous complaints about naturalism. In §40, pragmatic naturalism was absolved from any *general* commitment to fallacious, allegedly justificatory, inferences. The last section addressed skeptical challenges by distinguishing the rejection of the ethical project in toto from attempts to reform ethical practice. To decide the validity of these local challenges, however, requires methods of resolving questions about progressiveness, and these must embody something more specific than the general possibility recognized in the response to Hume (§40).

---

31. Otto Neurath, "Protokolsätze," *Erkenntnis* 3 (1932–33): 206; translated and reprinted in *Logical Positivism*, ed. A. J. Ayer (New York: Free Press, 1959), 199–208, 201. Quine made the passage famous by using it as the epigraph for his influential book, *Word and Object* (Cambridge, MA: The MIT Press, 1960).

Functional conflict is a source of potential trouble. Different participants in the ethical project may proceed from different views about the
functions to be discharged, and this may lead them to quite radically
divergent judgments about what precepts are to be endorsed and what
actions allowed. Pragmatic naturalism allows for pluralism (§38), but
one might wonder if it has ways of resisting truly *rampant* pluralism.
Could the troublesome characters of the last section moderate their
stances to commit themselves to the ethical project but recommend pursuing it in deviant and threatening ways? The possibility generalizes the
threat perceived at the end of the last chapter, where I envisaged overriding, or abandoning, the original function of ethics.

If pragmatic naturalism is to serve as a framework for substantive ethical discussion, it will have to discharge the basic task of showing how
particular disputes can be resolved, how particular claims can be justified. More exactly, it will need to explain exactly what sorts of justification are available, to actors or to analysts, when live issues about the
continuation of the ethical project arise. Hume's original challenge was
met by identifying a criterion for the cogency of ethical inference. The
latest concern asks for the specification of inferences that meet this criterion. The specification has not yet been given. Hence, as so far elaborated, pragmatic naturalism is inferior to classical accounts of ethics that
appeal to overarching general principles in the resolution of ethical debate. It does not follow that the deficiency is irremediable. To infer that no
method *can* be given would be a (counternaturalistic) fallacy. The next
two chapters take up this task.

To summarize: without some work in substantive ethics, a normative
stance, the metaethical picture of pragmatic naturalism remains incomplete. For the moment, however, without knowing the extent to which it
is possible to elaborate a cogent theory of ethical method, we can offer a
preliminary reply to the final worry. Proposed reforms of ethical practice are adjudicated by modes of argument tending to promote ethical
progress. If a challenger denies our conclusions, we make explicit the inferences we have made. If the inferences are questioned, we show how
they accord with general rules. If the challenger disputes the rules, we
demonstrate their conformity to the fundamental criterion, reliability
in generating progressive transitions. If the understanding of ethical

progress is disputed, we review the history of our ethical practice, show-ing how the original function has been served and how new functions have emerged.[32] *If* the inferences can be shown to accord with the rules, the rules with the fundamental criterion of promoting progress, and the account of progress to identify genuine functions, ethical disputes can be settled. Assuming pragmatic naturalism can elaborate a substantive ethics furnishing an appropriately constraining method along these lines—*one that will answer worries about the force and relevance of the functions attributed*—the only skeptical option remaining is to question the entire project, and that, as we saw in the previous section, is to take the leap favored by free spirit in his most ambitious and uninhibited guise, a leap into a completely unknown, and doubtfully human, form of life.

32. There is plenty of room for reform here—for challengers might uncover points at which regressive precepts or institutions have been introduced. Pragmatic naturalism is by no means committed to the vindicating ("English") genealogies, of which Nietzsche com-plained. See *On the Genealogy of Morality*.

PART THREE

A Normative Stance

CHAPTER 8

Progress, Equality, and the Good

§44. Two Visions of Normative Ethics

In the classical conception of normative ethics, espoused by religious traditions and philosophers alike, normative ethics aims to offer a set of resources to help people live as they should. Prominent among these resources are sets of prescriptions for guiding action, but there are often also vivid stories depicting what is admirable and what is not. Religious thinkers usually suppose all the fundamental resources have been provided, in an act of revelation, so the task remaining for the normative ethicist is to articulate the principles clearly and precisely, showing how they bear on the circumstances of contemporary life. Philosophers, by contrast, think the principles of normative ethics must first be generated and defended before the work of articulation and application can begin. Rather than invoking divine revelation, they supply reasons for thinking their favored system is correct.

Despite their differences, almost all approaches to normative ethics share a static vision. Correct principles and precepts await discovery, and once apprehended they can be graven in stone. Pragmatic naturalism sees things differently. The ethical project evolves indefinitely. Progress is made not by discovering something independent of us and our societies, but by fulfilling the functions of ethics as they have so far emerged. The project is something people work out with one another. There are no experts here.

From this perspective, normative ethics requires continuing efforts to decide how to live together in a common world. Each generation renews the project, going on from the point reached by its predecessors. The tasks facing normative ethics are those of deciding what should be retained, what modified and how—also to resolve how those decisions are made. If we are like sailors, repairing our ship at sea, we must determine which planks to leave in place and which to move—and our efforts have to be coordinated. The normative ethicist's role is not to offer the grand plan but to help the coordination.

Pragmatic naturalism assigns philosophers the task of facilitating discussion of how we should continue the project of living together. *Philosophy makes proposals.* (That is itself a proposal.) Given the approach to progress offered in Chapter 6, one type of proposal should identify the problems, unsolved and partially solved, to which ethical practice has responded: call this the *diagnostic* proposal. Another type of proposal, the *methodological* proposal, should offer suggestions about how proposals are to be adjudicated, about the rules of the continuing ethical conversation. (That does not close off the possibility of reverting to the classical vision of normative ethics, since one possible conversation is a monologue: we could decide to listen to the advice of a sage.)

Philosophical proposals can be more or less informed, more or less articulated and supported. In light of Parts I and II, we can identify the functions of ethics and try to use factual knowledge to find ways in which those functions could be more effectively discharged. If there were only a single function, or if there were no danger that fulfilling the distinct functions that have emerged would involve choices among them, the importance of deliberation would recede, or even vanish. Ethics could become a fully empirical discipline, whose task was to discover the ways of satisfying a single function or a harmonious set of functions. Because the pursuit of local progress gives rise to functional conflict (§§37–38), the ethical project cannot be turned over to a group of specialists who can work on finding ways to fulfill a function (or compatible functions). A first proposal: we have little alternative to recapitulating the original predicament of the pioneering ethicists. There are no external constraints to which we can appeal, no modes of knowledge to spark ethical innovation. Because the function of ethics is not to articulate the will of a transcendent being, the religious authorities who claim to have special access to the will in question no longer have the last word, or even any word more convincing than that of their fellows. By the same token, secular philosophy cannot lay down fundamental principles a priori, or even through generalization from allegedly basic ethical insights. What resources are left except to renew discussion with one another, to reject the distortions introduced by advertisements of ethical expertise, and to continue the conversation our remote ancestors began?[1]

This chapter and its successors are ventures in facilitation. The present chapter will offer some diagnostic proposals; methodological proposals are postponed to Chapter 9. Any package of proposals should be internally coherent—the diagnostic suggestions and the methodological proposals ought to be mutually supportive. As we shall see (§56), a requirement of coherence plays an important role in addressing residual skeptical concerns.

Those concerns arose (§39) in the form of asking why the "original function" of ethics should concern us. They were generalized (§43) by noting that people committed to continuing the ethical project could give priority to different functions. Pragmatic naturalism's vision of normative ethics identifies resolving controversies of this sort as *the* normative question. Skeptics are not pale figures of philosophical imagination, but alive and among us, offering different ideas about functional priority in going on from where we are. If our efforts are to be coordinated, if the ship is to stay afloat, these differences need resolution. Pragmatic naturalism needs to propose a normative stance, and, if it can do so convincingly, the metaethical challenges will be addressed en passant.

The strategy is to proceed in stages: first some diagnostic proposals, then some methodological proposals, and finally an argument for taking

---

1. Here I allude to a famous remark of Richard Rorty (see the closing passages in *Philosophy and the Mirror of Nature* [Princeton, NJ: Princeton University Press, 1979]). I suspect I assign a larger role to philosophical midwifery than Rorty would allow.

the package to have a coherence its (skeptical) rival lacks. I begin by presenting a useful framework for diagnosis.

§45. Dynamic Consequentialism

The ethical project can evolve indefinitely. Its progress is not measured by the decreasing distance to some fixed goal—the accumulation of correct ethical principles, central to the classical vision of normative ethics. There can be progress *from* as well as progress *to*. Once again, the analogy with technology helps. Inaccessible destinations posed problems to be solved by developing methods of transportation, and technological progress in this domain consists in overcoming those problems (and the derivative problems to which they give rise). No ideal technological system beckons us on, and it would be folly to assess our progress in terms of our proximity to any such thing.

Our predecessors made ethical progress by responding to parts of the problem background, introducing precepts and other ethical resources. They did so by framing (partial) solutions, and their recognition of their innovations *as* potential solutions depended on seeing the state in which the novelties were introduced as better than the status quo. They operated with an implicit notion of the good, fixed and determined through their agreements. The good is local, linked to circumstances and problems; it is constructed through group attempts to solve problems; and it evolves.

Seeing socially embedded normative guidance as operating with a tacit notion of the good permits the introduction of a useful normative framework. The ethical project can be understood as a series of ventures in *dynamic consequentialism*. Participants in it respond to their problems by trying to produce a better world. They have an implicit conception of the good and take the rightness of actions to depend on their promotion of the good as they envisage it. Does viewing them in this way foreclose possibilities? No. Consequentialism is a far more flexible view than its critics usually assume. Not only are there many different ways in which consequentialists can characterize the good, but a consequentialist ethical theory can explicitly acknowledge it has no complete specification of the good, seeing its judgments as incomplete and provisional. *Dynamic* consequentialism makes exactly that admission, supposing that conceptions of the good evolve, that some of the transitions among those conceptions are progressive (in the sense of Chapter 6), and that later conceptions of the good are (sometimes) superior to their predecessors, even though none can claim to be the last word.

According to Mill, the central idea of consequentialism—that the rightness of actions depends on their consequences—is the "doctrine of rational persons of all schools."[2] To see the problem with ethical systems that focus on obedience to prior rules and ignore consequences, consider self-sacrifice. Someone who sacrifices himself typically does so for "the sake of something which he prizes more than his individual happiness," and, if this is not the case, it is hard to see why the alleged hero is "more deserving of admiration than the ascetic mounted on his pillar."[3] There is no ethical justification for following a rule unless one has grounds for viewing that rule as authoritative, and those grounds can come not from labeling the source—either as a divine lawgiver or as its detheologized counterpart "the moral law within"—but only from recognizing the rule as well adapted to producing good outcomes. Following rules *not* well adapted to producing good outcomes is a capricious and irresponsible thing to do: that is why consequentialism is "the doctrine of rational persons of all schools." Ungrounded deontology is dangerous.[4]

There are alternative "schools" because there are very different ways of measuring the value of consequences. When someone acts, he or she changes the way the world would have been in the absence of the action, or would have been if he or she had done something different. At its most inclusive, consequentialism can compare the different total world histories, the courses of the world that run from the beginning to the end, and can, in principle, focus on any or all of the features of those histories, *including those features inspiring deontologists to lambast the simplest versions of consequentialism*. Typically, consequentialism focuses more narrowly—

2. Mill, *Works*, J.S. Mill, *Collected Works* (Toronto: University of Toronto Press, 1970), 10:111, from the essay "Bentham."

3. Mill, *Works*, 10:217.

4. The most spectacular examples of ungrounded deontology occur in invocations of the divine will to prohibit (or command) human actions (§§26, 35).

for utilitarians, the relevant points of comparison are the subjective experiences of sentient beings, specifically their pleasures and pains. In moving from the general idea of consequentialism to utilitarianism, various important assumptions play a role:

1. The only aspects of the world needing to be considered in evaluating the consequences of an action are those subsequent to the action.
2. These pertinent aspects are properties of the lives of existing beings, all of whom belong to a particular class.
3. The existing beings in question are sentient beings.
4. The relevant facts about their lives can be assessed by considering the stream of their subjective experiences.
5. The subjective experiences can be conceived as momentary (or fleeting) states, to which value can be assigned.
6. The relevant states are experiences of pleasure and pain.
7. The value of a pleasure (or pain) is measured by its intensity.
8. The intensities are summed across the interval through which they persist.
9. The value of an individual's life course is calculated by summing the values of pleasures and subtracting the values of pains across all moments of the person's life.
10. The value of a world history is calculated by aggregating the values of the lives of all the individuals from the point of the action on.
11. In practice, comparisons can typically be made by using small segments of the lives of a few people as proxies for the values of world histories.
12. Strictly speaking, an action is right if the value of the world history it generates is at least as high as the values that would have resulted from any of the available alternatives (but using the proxy calculations identified in assumption 11 will almost always provide an accurate test).

Each of these assumptions deserves scrutiny. You might think both assumptions 1 and 2 extremely plausible. After all, an agent can do nothing to change the past, so assessment can concentrate on what happens after the agent's action. That does not follow, however. For *evaluating* the consequences, it should not be assumed that the value of an event or state of affairs is always independent of its causal history. Consequentialists can hold that an effect has greater or lesser value in virtue of its relation to past actions: even if your spending time with the sick would promote greater happiness than expressing your gratitude to someone who has helped you, your overt gratitude and the past aid may confer sufficient value to outweigh the benefits provided by hospital visiting.

Imagine two world histories involving exactly the same distribution of pleasures and pains, satisfactions of desires, and anything else about individual people you might take to be relevant to their individual good, but diverging in the causal relations. Suppose, for simplicity, all that matters is pleasure and the absence of pain. In one world history, many of the pleasures and pains come about because of systematic relations among people—there are relationships of mutual helping; punishment is given for harms caused—much as things happen in the world we know. In the alternative, the causes of the pleasures and pains are quite random: you do something to please a friend, and, instead of the friend, a complete stranger acts to give you pleasure; you cause harm to someone, and, out of the blue, a harm equivalent to the punishment is inflicted on you. If you think something valuable has been lost in this abnormal world, with its unsystematic ways of generating pains and pleasures, you think the causal relations affect the goodness or badness of a world history.[5]

Assumption 2 is similarly vulnerable. Consequentialism is not committed to the view that you can consider lives in isolation from one another: it is quite possible that relations among lives, and relations of living beings to other constituents of reality, are sources of value (consider the value that accrues from higher-order altruism; §21). It is even possible that nonliving things, and the relations among them, could be sources of value (§47).

---

5. Perhaps one might claim that the difference in value is explicable because of the attitudes people would have: their pleasures would be less in the causally unsystematic world. Then grant them the illusion of thinking the causal relations are as they normally are. The difference in value persists.

There is a very general reason for worrying about almost all the assumptions generating utilitarianism from consequentialism. Utilitarianism derives from consequentialism by a series of *reductionist* moves. We aim to compute the values of the worlds that would flow from our envisaged actions. We reduce the problem to one of summing the values of the lives of a class of individuals; we reduce it further by considering only sentient individuals and further still by ignoring most of these and concentrating on those we suppose immediately affected by the actions under scrutiny; we reduce the problem of measuring the values assigned to the individual lives by decomposing those lives into a sequence of momentary states; now we assign values to those states by reducing the aspects we consider to the intensities of pains and pleasures; having reduced the problem in this way, we can start summing, and arrive at the measure utilitarians commend. Any or all of the reductions could be questioned. For there is no reason to think the value of a world will always consist in the sum of the value of the lives of the individuals we consider one by one—distribution might be crucial. Further, we have no reason to suppose that the value of an individual life can be generated by summing the values of momentary states (or even longer experiences that occur in people) taken in isolation from one another.

Consider an alternative view of the value of individual lives. It starts by taking a valuable human life to be one directed by the free choices of the person whose life it is, as something given coherence by an individual conception of the good.[6] However pleasurable a sequence of disjointed experiences (even a repetitive sequence) might be, it would fall short unless it had certain global properties. Valuable lives exhibit a plan.

Similar holistic considerations apply to social states. Rather than thinking the valuable social state is one in which the total sum of pleasure over pain in the population is high, we might conceive a group of people pursuing a wide diversity of projects for their lives, who are nonetheless bound together in relations of dialogue, joint action, and mutual

---

6. As Mill claims throughout *On Liberty and other Essays* [Oxford: Oxford University Press (The World's Classics), 1998]. This essay shows his enormous debt to the Greek thinkers he studied from childhood on.

sympathy.[7] Both the heterogeneity of the whole and the interconnections among the individuals are sources of value. Approximate equality of material resources might obtain value in terms of its contribution to the possibilities of all members of society having fair chances to fashion and to pursue their life plans.

Consider poetry. Discovering a great poet—Wordsworth, say—is not a pleasurable experience like that of a good meal, or even a soother of pain, like an analgesic. If a young man, from whom poetry is withheld, lapses into a state of dejection, from which reading Wordsworth helps him to recover, it is odd to suppose he has the medicine his condition required. An encounter with Wordsworth is not significant for its transient effects, but something that can reshape and redirect an entire life. A calculus of assigning values to passing experiences, based on intensity and duration, is inadequate to measuring the value of Wordsworth, because the Wordsworth encounter resonates in subsequent life.[8]

In the house of consequentialism, there are many mansions. My excavation of twelve different places at which you could make different decisions about the good, and derivatively about the right, reveals abstract possibilities. The central point, however, is that the flexibility of consequentialism allows us to understand ethical progress as the fashioning of conceptions of the good that better discharge the functions of socially embedded normative guidance and the framing of rules as consequentialist projects directed at promoting the good (according to the prevalent conception). Conceptions of the good are *duals* of the functions assembled in the growth of the ethical project.

7. These holistic considerations are evident in Mill's own writings, particularly in *On Liberty, Considerations on Representative Government* in *On Liberty and other Essays* [Oxford: Oxford University Press (The World's Classics)], 1998, and *Principles of Political Economy,* vols. 2 and 3 of J.S. Mill, *Collected Works* (Toronto: University of Toronto Press, 1970).

8. My discussion plainly draws upon Mill's own experiences and his sense of the importance of reading Wordsworth and of meeting Harriet Taylor. His writings make it evident that he understood the holistic significance of these events: see, in particular, the essays "Bentham" and "Coleridge" Mill *Works* vol. 10, the dedication of *On Liberty,* the *Autobiography* (New York: Columbia University Press, 1966), and the closing pages of *On the Subjection of Women* (in *On Liberty and other Essays*).

A natural response: consequentialism can accommodate so much because the alleged "flexibility" makes its claims vacuous.[9] That is already indicated by my concession that many of the features deontologists view as lacking in simple forms of consequentialism can be introduced within a consequentialist framework: one can, for example, insist that the goodness of a state of affairs depends on the character of the psychological processes (including the intentions) that brought it about (this is a special instance of the general point that causal relations matter). The danger of ungrounded deontology shows, however, that the consequentialist framework rules out some things, namely prescriptions of what is right that do not advance ethical functioning (and, in the religious case, often interfere with it). The insights of traditional deontological thinking can be viewed as contributions to a progressive understanding of the good, whereas the deficiencies of other deontological possibilities (those resisting such assimilation) can be exposed.[10]

§46. Failures and Successes

At the early stages of the ethical project, our ancestors focused on the basic desires of their fellows. They recognized some wishes as endorsable and uncontaminated, taking it to be good to divide resources according to some type of equality under conditions of scarcity and to aim to increase the supply so that all might have enough. Part I saw the emphasis on supply as fostering social arrangements found in later societies: division of labor and the development of roles and institutions (§§19–20).

For present purposes of stocktaking and diagnosis, considering how features of our current situation *might have emerged* is not pertinent: what is crucial is *that they have emerged*. Contemporary societies have either abandoned the original emphasis on equality or spectacularly failed

9. Bernard Williams makes some perceptive remarks that anticipate this line of response in his contribution to J. J. C. Smart and Bernard Williams, *Utilitarianism: For and Against* (Cambridge, UK: Cambridge University Press, 1973), sec. 2.

10. So one might say that the idea of the unseen enforcer is unproblematic for ethical progress when the deliverances of the commander are grounded in the prior understanding of the good. Things go astray only when the practice of commanding takes on a life of its own.

to achieve it. Conceive societies as demarcated geographically: start with a set of people who live in the same place, and include all who are in social interaction with any member. Even if you begin with the well-to-do, the society identified not only will be large but will also contain individuals whose endorsable desires are only incompletely satisfied.[11] (This is evident for poor societies, and for affluent societies with pronounced inequality; even the most egalitarian social democracies contain pockets of poverty and want.) If we think of the most elementary form of altruism failure as one in which individuals do not respond to the basic (endorsable, uncontaminated) desires of others, when it is possible to satisfy those desires completely while simultaneously doing the same for everyone in the society, it is evident that *in the most elementary case of the original function of ethics,* contemporary societies do not realize the available ways of discharging that function. On this score, many experiments in the history of ethical practice have done better.

Why is this? Most plausibly, the structure of roles and institutions generated in the evolution of the ethical project has displaced the original conception of the good. New ethical functions have emerged, and, while the roles and institutions we have may be well adapted to discharging these functions, they forfeit the ability to fulfill the original function. Our contemporary codes permit maintenance of roles and institutions, even though those interfere with remedying a very basic type of altruism failure.

The contemporary conception of the good is surely progressive in some respects, for it has absorbed ideas of higher forms of altruism, contributions to joint projects, and richer possibilities of human life (§§36–37). If the original conception of the good were reapplied in the contemporary context, it would have to attend to a broader class of endorsable desires. In terms of offering equal satisfaction of these "higher" desires, the failures of contemporary societies are even more glaring. Inability to satisfy basic desires contributes directly to dramatically lowered chances

11. Here I am imagining societies as defined recursively. One might try to block the conclusions by demanding that any two members must engage in social interaction, or by raising the standards for social interaction. I shall discuss the issue of how to demarcate societies more extensively below. For the moment, the recursive approach will help to fix ideas.

of achieving the "higher" goods. When the most valued social positions go to those who have performed outstandingly on the tasks assigned by an educational system, people often fail because of their inability to satisfy basic needs (hungry, ill-sheltered children do not always concentrate, children who go to decrepit, dangerous schools do less well on the exams devised to test them).

Functional conflict underlies social tensions but is not itself perceived. In practice, the conflict is settled by overriding the original function of the ethical project, and that suppression is thoroughly institutionalized. Equipped with an account of the ethical project and its evolution, we can be more reflective. Dynamic consequentialism sees our normative task as beginning from respecifying the good. That specification cannot be achieved without reflecting on roles and institutions—ethics is social.[12]

The development of a richer repertoire of emotions for responding to the actions of others, especially the negative sentiments of indignation and revulsion provoked by injustice and cruelty, underlies current ways of securing compliance to our codes. People living in a state of blatant inequality, people whose basic needs are unmet, are often not persuaded that there has been no real injustice or cruelty. Indignation and revulsion easily reinforce the anger and frustration that would have been felt by their ancestors under similar circumstances. If the amplified emotional reaction is held in check, it is only because the injunction against initiating violence is powerfully enforced: asymmetries in wealth and power are used to threaten those inclined to protest with even direr consequences for themselves and their loved ones. (This kind of coercion figured in the scenarios about dictatorial rule—§34; impossible though it may have been among the small bands of the Paleolithic, it is readily contrived within contemporary societies.)

Were we to consider the entire human population as a single society, the extent and range of inequality would be enormous. Each day, thousands of people die or become disabled through lack of the resources

12. The phrase is Dewey's, chosen as title for the last chapter of *Human Nature and Conduct* (Amherst NY: Prometheus Books, 2002). The point encapsulated in the phrase recapitulates themes in Foucault and Marx, themes often beyond the horizons of contemporary ethical discussions.

needed to satisfy a basic desire: to preserve the body in a healthy state.[13] The distribution of opportunities for satisfying less fundamental desires, say, for work of a rewarding kind or for cultural development, is even more skewed. Concentrating on smaller human units, the less populous nations or communities within them, makes the inequalities less extreme; but affluent democracies and poor countries alike contain sizable groups of people who can see how the basic and nonbasic desires of others are often well, even fully, satisfied, but whose elementary wants are unmet. Only a few societies, committed to public support and to egalitarian arrangements, reduce the scale of the inequality. Even in them, pockets of deprivation remain.

If the difficulties of profound inequality are not as evident to us as the smaller problems our hominid forebears encountered, it is because those from whom we might expect hostility are often distant or invisible, and because the concentrations of resources have generated forms of power enabling people who enjoy large benefits to threaten the less fortunate with severe penalties for subversive behavior. From the perspective of pragmatic naturalism, that cannot be a solution to our magnified analog of the original problem, for the fragility and tensions of our societies are addressed only in a way that does not identify their root cause, to wit the wide-ranging altruism failures. As with the original powerful dictator of §34, the reduction of social conflict does not work through remedying altruism failure and therefore fails—spectacularly fails—to discharge the original function of ethics.

Not all the conflicts and tensions of contemporary societies (whether the human population is considered as a whole or whether it is split up into smaller groups) are based on the inequalities toward which I have gestured. Some versions of the unseen enforcer interpret the will of the invented being in evangelical terms, so the rules they claim to be expressions of that will are supposed to apply, not only to the society originating the mythology, but also to the entire human species. Societies gripped by this conception will certainly be at odds with other groups whom

13. The annual reports of the World Health Organization provide (sobering) data. See James Flory and Philip Kitcher, "Global Health and the Scientific Research Agenda," *Philosophy and Public Affairs* 32 (2004): 36–65, for a small selection from the statistics.

298 ∽ THE ETHICAL PROJECT

they take to be creating a world inhospitable to their favored ideals of living and who may even be disposed to spread those ideals by any effective means (including force). Some social conflicts and tensions arising within contemporary societies—most evident when our species is viewed as a single society—result from the dangerous ungrounded deontology that articulates the initially useful idea of the unseen enforcer. Codes subordinate concern with the desires and aspirations of individual people to the prejudices of those who advertise themselves as knowing the enforcer's will.

Part of the emancipation of ethics from the concept of the enforcer consists in the recognition of previously forbidden desires as endorsable, the withering of vice (§26). Renewed discussion of the good must be alert to residues of pressures that have made some desires invisible. Yet comprehensive diagnosis should recognize what imaginary transcendent policemen have contributed to the ethical project. Most obviously, they have helped secure greater compliance (§17). Beyond this, they have sometimes played important roles in relieving or in exposing the central feature of our contemporary—and recent—ethical predicament, to wit the presence of vast inequalities, resulting from abundant altruism failures and producing widespread social tension. Religions have sometimes laid great emphasis on recognizing the needs of the poor, the oppressed, and the unlucky: they have, in effect, continued to recognize the desires of those whose aspirations are frustrated, even when these people are excluded from ethical deliberation. Moreover, religious community has sometimes fostered the expression of protest against inequality.

A concomitant of the growth of societies, itself a progressive transition, made possible through including outsiders, is that the original ethical venture, in which an entire social group tried to find common ground, has become practically impossible. Even without deference to people supposedly having special access to the divine will, it would be necessary to assign the task of reviewing and possibly revising a code to a (small) minority. That forfeits an important feature of the extended conception of the good life (§20), the ability to perceive oneself as playing a role in joint projects, conceived by each participant as directed toward a common good. The shared ethical life begins to disappear.

Diagnosis has so far focused on problems and conflicts, places where we might consider moving some planks on the ship we sail together. Yet we should not overlook the successes of the ethical project in achieving directives for individual behavior. There are core ethical truths (with associated rules), vague statements attained by all extant ethical traditions that we reasonably expect will be preserved in an indefinite sequence of progressive future transitions (§38). Whether or not we can make them more exact, or find new concepts for their sharper expression, conduct in accordance with, and out of respect for, these themes is constitutive of any plausible conception of the good.

Consider honesty. Because there is a very large class of actual or potential altruism failures, in which one individual deceives or manipulates another, regulation of speech acts will figure in progressive ethical codes at an early stage. Exchange of information is crucial to human life and has been crucial since we learned to talk. If the informant is to respond to the wishes of the inquirer, it will be important to tell the truth (or to attempt to convey the truth, to be sincere), and discovery that truth has been withheld will generate trouble. "Always tell the truth" looks like a rule to help fulfill the original function of ethics; truth telling is an activity classified as good by a progressive conception.

For a very large segment of human history, the simple formulation "Always tell the truth," was probably apt—just as it is an apt rule for children today and a serviceable one for many people throughout the course of their lives. Nevertheless, truthful responses to questions are not always and everywhere good. A stereotypical scenario involves the demands of the Gestapo officer at the door, when the householder has hidden Jews in the attic. The officer asks, "Are there any Jews in the house?" For the story to work, we must suppose he trusts the householder; given a negative response, he will leave and the hidden refugees will be safe. Here, the consequences of honesty are so dire that many people—perhaps everyone not in the grip of a philosophical theory—thinks lying is good.

Few people encounter situations of this type, but many more face moments when telling the truth has adverse consequences. One of those you love most is seriously ill, and the medical report is grave. You know her determination and resolve will be sapped if she learns the details. Or perhaps there is really no hope, and you know her final days will be miserable if she feels there is no chance of recovery (you know her well; she is sensitive and liable to be crushed by dreadful news). She asks you point-blank what the doctors have said. You are well prepared and know you can convince her. Lying would avoid the bad consequences of telling the truth.[14]

These examples have to be carefully constructed if they are to call into question the universal goodness of truth telling. The potential deceiver must be clearheaded, must have thought through the options, and must have understood that there is no honest alternative to the deception. The officer at the door cannot be fobbed off with an evasive answer, nor will your loved one be satisfied with even the most adroit attempt to change the subject. This means the lie will probably have to be premeditated and its expression carefully prepared—in general, conditions we take to intensify the wrongness of a misdeed.

14. Thinkers inspired by Kant may claim that allowing any exceptions undermines the practice of answering questions, or perhaps that it violates the integrity of the speaker, or perhaps that the good requires that people always act in nonmanipulative ways. When the odds are set very high, as they are in the case of the officer at the door, none of these replies succeeds. The lie to the Gestapo officer is not going to undermine any important human practice, even if all householders in similar situations behave in the same way. If the lies are discovered, the investigating officials are likely to be more skeptical of what they are told and hence may investigate more thoroughly—and that will have to be countered in future practices of asylum granting—but these refugees will have been saved. The example of the seriously sick illustrates the point very well. Those who are extremely ill know that people who care about them will be inclined to spare them bad news. Knowledge of that sort does not undermine any practice of questioning and answering but simply means they are more suspicious than they would otherwise have been about what their loved ones tell them.

Many deontological traditions, religious as well as philosophical, have taken the badness of deception to consist in the corruption of the liar—in effect, they offer an account of the good that gives priority to purity of heart and suppose telling a lie, even to promote otherwise good ends, compromises this purity. Religious teachers and leaders have been advised to respond to questions by discovering ambiguous or evasive or misleading (but not outright false) answers or to take oaths with a private "*reservatio*." Faced with the officer at the door, householders who spend time trying to discover a tricky way out are not so much worthily protecting themselves from corruption as irresponsibly putting others at risk. Even if the chances the officer will penetrate the cleverly evasive reply are slim, the damage done by arousing his suspicions is too great to justify the maneuvers. Further, why should we suppose that those who concentrate their effort on shielding vulnerable victims they have taken in are somehow less pure than people whose first thoughts are divided between the threats to the refugees and their own rectitude?

In both examples, the desires of the questioner are ignored or overridden. The desire to find the Jews is ignored because it is grounded in a spectacular altruism failure; even a rudimentary account of contaminated desires (§34) differentiates the officer's wish to root out and kill from the desires of the refugees for escape. The patient wants the truth, but she also wants the truth to be benign—and because the deceiver understands the strength of that latter wish, as well as the consequences of knowledge that it cannot be fulfilled for other desires the patient has (the desire not to live bereft of hope), she answers untruthfully. Here deception can be seen as altruistic, seeking the most important desires of the sick one and responding to those. (Plainly, this is an instance of paternalistic altruism—§21.)

The simple prescription of truth telling originates in contexts where any foreseeable deception involves only altruism failure. Lies advance the selfish ends of the liar. The desires of the recipient of the false information are ignored. In simpler societies there will typically be other ways of heading off the threat posed by a questioner who wants to inflict harm on a third party—one may summon other group members so complaints can be heard. Similarly, only when special kinds of knowledge are in play does the problem of protecting others from disturbing truths arise.[15] With the evolution of culture, however, altruistic lying becomes possible. Altruistic lying can resemble other forms of altruism we encourage (we admire those who risk their own safety to protect the persecuted). Insisting on the simple rule overlooks the fundamental role reinforcing altruism plays in progressive ethical change.

Yet we do not know precisely how to amend the conception of the good so it is no longer committed to the universal badness of lies. To appeal to obvious aspects of the conditions under which reinforcing altruism is good (support the refugees, not the wishes of would-be murderers), and to declare that lying is permissible when it satisfies these conditions, is not adequate. The Kantian thought that universal permission to lie (in

---

15. It is interesting to speculate about the length of time for which altruistic lying has been possible—perhaps there are occasional analogs of the examples I have used that could have arisen for the small groups of the Paleolithic—but it seems evident that concentrations of power and advances in certain kinds of knowledge (medical knowledge, for instance) generate clear cases with increased frequency.

some specified range of circumstances) would undermine the practice of exchanging information will not support a blanket prohibition of lying, but weakening the rule has consequences. A consequentialist analog of the deontological point: one must attend to the consequences of publicly permitting exceptions.[16] Sometimes we have good grounds for thinking the social consequences slight compared with the benefits of permitting dishonesty—as, for example, with carefully circumscribed versions of the two scenarios considered. That is compatible with ignorance about the costs and benefits in other instances.

This section has offered a preliminary survey on which a respecification of the good might be founded. The next task is to take up the (so far postponed) question of group size. How large should be the scope of our ethical concerns?

§47. From the Local Community to the Human Population

In the beginning, it was simple. The ethical project was undertaken by small groups of deliberators, whose agreed-on rules and whose conception of the good focused on themselves. If we consider renewing the project, how should we divide the human population (the totality of members of our species alive at a particular time) into groups within which versions of the ethical project, "experiments of living," are tried out? A simple suggestion appeals to geography: communities are demarcated by recognizing territories within which people live and out of which they rarely, if ever, venture. Or you can emphasize causal interaction: two

---

16. Mill's treatment of the permissibility of lying appreciates these points. Convinced that "all moralists" would allow for breaches of the prohibition against lying, he points to just the cases I have used: "when the withholding of some fact (as of information from a malefactor, or of bad news from a person dangerously ill) would preserve someone (especially a person other than oneself) from great and unmerited evil, and when the withholding can only be effected by denial" (*Works*, 10:223). He cautions, however, that not every good consequence is sufficient to allow for lying, since we also have to take into account the costs that permitted deception will entail: "inasmuch as any, even unintentional, deviation from truth, does that much towards weakening the trustworthiness of human assertion, which is not only the principal support of all present social well-being, but the insufficiency of which does more than any one thing that can be named to keep back civilization" (*Works*, 10:223).

individuals belong to the same community if they interact with each other at greater than some threshold frequency, and the community is the recursive closure of such interactions.[17] Alternatively, appealing to the altruistic tendencies that make life together possible, communities are groups of people with dispositions to respond altruistically to one another, people who lack such dispositions toward outsiders (or who have them only in reduced form). Or, focusing on the limits of altruism, communities are groups of people for whom altruism failures give rise to social difficulties.

For almost all of human history it would not matter much which of these criteria you selected, for all would agree. Most people who have ever lived have spent most of their lives in the same place, interacting with the same people, responding altruistically to those around them, and living by a shared set of rules that attempt to remedy their potential altruism failures. Nomadic tribes tend to stay together, even in their wanderings, so that, at any given time, the geographical criterion will align with the others, even though, at different times, the places occupied are different. Individuals sometimes leave their native group to join another—up to a certain point in their lives, they belong, according to all the criteria, in one community; afterward they belong, again by all the criteria, to a distinct community.

In the contemporary world the criteria no longer coincide. People who live in one place have some of their most important interactions with others who live at a great distance from them. If one appeals to a criterion of causal interaction, connections will be made across geographical boundaries, and any insistence on transitivity will bind the human population into one vast community: if I interact frequently with someone in Australia, and you interact frequently with someone in central Asia, and you and I also interact frequently with each other, then the four of us are linked by a chain of interaction; if it is true that sharing a

17. That is, if A and B meet the interaction criterion, and if B and C do so as well, A and C belong to the same community, irrespective of whether they also interact above the threshold; the community consists of all the people who can be related to any of its members through a chain whose adjacent members interact at a frequency above the threshold. The minimal notion of interaction is one in which the behavior of one affects the choices available and the prospects of success of the other.

community is transitive, we shall all be assigned to a single community, even though I have no contact with central Asia, and you have none with Australia, and our partners in Australia and central Asia have none with each other.

There is a different way to think about the situation.[18] One can abandon the idea that sharing a community is a transitive relation. Each of us belongs to many overlapping communities. In terms of the illustrative example, there is a community to which my Australian friend and I belong, another community to which you and your central Asian associate belong, and a third that embraces you and me, but none that contains all four of us (or even any three of us). It is possible to introduce community structure at many different levels, using different criteria singly or in combination. A particularly important way to do so is to appeal to history. Those who live in a place in which social life has been regulated by a particular tradition, and who identify themselves as part of that tradition, might plausibly be counted as an ethical community, for whom the continuation of the ethical project consists in amending that tradition.

So there are many possibilities for specifying group(s) of people to work together on respecifying the good. On what basis should we decide among them? The conceptions introduced should be apt for the problems faced. Wherever there is a failure to respond to the desires of another person, with respect to whom there is the potential for interaction, *we have a contemporary analog of the problem that underlies the original function of ethics.* Moreover, as at the beginning of the ethical project, these altruism failures, in a world of far more extensive causal relations among human lives, provoke social tensions and conflicts. These emerge today as counterparts of the fragility of the hominid societies that were transformed by ethics. A proposal: *continuation of the ethical project should include an attempt to frame a conception of the common good responsive to the desires of the entire human population.*

Almost all human beings want a future in which the younger members of their societies, and the future offspring of those younger mem-

---

18. In framing the ideas that follow I have been helped by some acute observations and questions from Erika Milam. There are also links to the framework proposed by Dewey in *The Public and Its Problems* (Athens OH: Swallow Press, 1985).

bers, can grow healthily. All extant ethical traditions count this as an endorsable desire. Today, the actions of people in some areas of the world interfere with the realization of such desires. Practices of commerce, agriculture, industry, and even medical research in affluent countries decrease the probability that people in poorer regions will be able to nurture healthy children.[19] All members of our species face the common problem of avoiding (further) environmental changes that would dramatically disrupt human lives: global warming will, almost certainly, make many heavily settled areas uninhabitable and will leave many others vulnerable to extremities of weather that will challenge available technologies of shelter. We urgently need a conception of the good that considers the desires of all people and that will guide attempts to treat the problems engendered by a thoughtless industrial past.

Causal interaction binds the entire human population together. The altruism failures in that large community are dangerously magnified versions of those that prompted the first ventures in ethics. Pragmatic naturalism's proposal about the good gives priority to the continually more extensive network of causal relations linking us all. It does not follow that there is no place for more fine-grained partitions. There are issues on which a local community could progressively elaborate a conception of the common good by concentrating only on the desires of its members and on remedying the altruism failures arising within it. The areas in which this should occur, however, must be distinct from those covered by more wide-ranging conceptions (ultimately by that conception that takes the entire human population as its province): where there are serious consequences for distant others, there must be an attempt to respond to them.[20] Moreover, the local community's explorations in this regard are properly constrained by the vague central themes adduced as the

19. Commercial practices interfere with the supply of basic goods to the poorest areas of the world; agriculture subordinates the task of feeding the hungry to considerations of profit; industry squanders resources and neglects long-term energy needs; medical research has, until quite recently, conspicuously neglected the problems of the poor.

20. Plainly, this proposal has links to Mill's famous "harm principle" (*On Liberty*); even more directly, it connects to Dewey's reformulation in the opening chapter of *The Public and Its Problems*.

best candidates for ethical truth. Nevertheless, the pluralism already acknowledged (§38) is recapitulated in the idea of domains in which progressive elaboration of the conception of the good can proceed through responses to the desires present in a local community (one that is a proper subset of the human population).

An important objection: focusing on the global *human* population draws the boundary too narrowly. Other things—nonhuman animals, or parts of the biological or inorganic environment, or human artifacts— have ethical standing. I shall take the case of nonhuman animals as central to the objection.

Ethical traditions disagree about the extent to which our conduct toward nonhuman animals should be regulated. Almost everyone agrees, however, that people should not arbitrarily inflict pain on mammals and birds (igniting cats is usually viewed as cruel) and that pet owners should attend to the needs of the animals they have acquired. Some hold that farmers should not breed mammals and poultry under conditions of close confinement that distort the animals' normal patterns of development and normal metabolic functioning; these critics often hold that consumers should not buy and eat animals bred in these cruel ways. Yet others suppose the entire practice of eating nonhuman animals—whether birds and mammals, or all vertebrates, or all invertebrates into the bargain— is wrong. A similar range of positions applies to using nonhuman animals in medical experimentation.

The aim here is not to consider which position, if any, would be adopted in a progressive transition from the current state of ethical practice, but to address the concern that *no* view of any of these types can be reconciled with the framework of pragmatic naturalism. It suffices to focus on the most minimal rules governing conduct toward nonhuman animals, like the proscription of torturing mammals. What ethical function would be discharged by introducing any such prohibition into ethical practice?

It is worth recalling part of the history of ethical experimentation. For a long period in the ethical project, human groups were very small; outsiders were not covered by any of the protections afforded by the rules of the band. The scope of the rules was extended when peaceful exchanges with neighbors effectively created something akin to a broader society.

"The circle expanded."[21] What occurs here is first that a set of interactions takes place with people who have previously usually been avoided, and second that failures to respond to some of their desires count as altruism failures. My "how possibly explanation" (§19) showed *a* way of making that extension. Interactions are set up in an attempt to discharge an ethical function (satisfying the endorsed desires of members of the local group through trade), and the expansion of the rules fulfills another function (the original function of remedying altruism failures; some of the desires of the erstwhile outsiders correspond to endorsed desires of group members, and not responding to these comes to count as an altruism failure).

Is anything similar available in the case of nonhuman animals? Apparently so. Just as the late Paleolithic witnessed first occasions of transient association among neighboring groups and later increases in band size, so, at the very end of this period, people began to set up more regular patterns of association with some kinds of nonhuman animals. The practice of domestication creates something like a society, one including some nonhuman members. That practice refines the ethical function of satisfying the endorsed desires of all. Yet, because the animals newly included differ in some important properties from the people across the river, obvious questions arise. Does this expansion really create anything like a society? If so, to what extent can the rules adopted within the local group be carried over to the new members?

In a minimal sense, a broader society is created simply in virtue of recurrent patterns of interaction. Human beings feed their domestic animals, breed them, work them, and consume some of their biological products. These interactions need to proceed in ways permitting a long series of repetitions. Yet, because of asymmetries in power, that might simply be done by brute force. If the "society" is simply constituted by regular patterns of interaction, its stability requires no more than human skill in confinement and handling.

21. As §33 observed, this is a prominent mode of ethical progress. Peter Singer's title "The Expanding Circle" expresses his conviction that this is the dominant mode of ethical advance, and he argues that this expansion should bring nonhuman animals within the scope of our ethical precepts.

Members of the group across the river may ultimately participate in the ethical project as we practice it. Some generations hence, their descendants and ours may sit at a common campfire and deliberate about how to go on from where they are. That can never occur with our nonhuman animals.[22] Yet there are also human beings who do not participate in ethical deliberation with us and still belong to the scope of our precepts—and this holds even if "we" are members of one of the early bands that initiated the ethical project. Those sufficiently young lack a voice, as do others, perhaps, whose capacities have not grown in normal human ways; additionally, there are descendants, children and grandchildren yet unborn, some of whose desires we endorse, with the consequence that we work to ensure that necessary conditions for the satisfaction of those desires will obtain. The extension of the agreed-upon rules to them does not result from the fact that they have been party to the agreement—for they have not—but rather from a recognition of features of their lives, including the occurrence in them of desires we endorse for others. Despite the differences caused by current incapacity, permanent incapacity, or position in time, there is a common basis on which the notion of an altruism failure can be elaborated.

So too with nonhuman domestic animals. We can recognize their desires for things human beings need (food, rest) and appreciate their wish to avoid particular situations (close confinement, pressure from sharp objects). Once human emotions have been shaped to resent behavior causing suffering to other people, there is sufficient community with nonhuman animals to expect that, among the band of early pastoralists, deliberations in the cool hour will reprove actions making the livestock whimper or squeal.

How far will the progressive elaboration of precepts governing actions toward nonhuman animals go? There are two potential axes of extension: one leading from domestic animals to their wild counterparts, and one encompassing a broader range of taxonomic categories. So long as wild animals are seen as threatening, they will figure in the same way threatening human beings from other tribes did, before the

---

22. Hume saw this point clearly and drew a harsh conclusion. See *Enquiry Concerning the Principles of Morals* (Indianapolis, IN: Hackett, 1983), 25–26.

institution of rules for interacting with outsiders brought them within the circle. Just as participants in the ethical project have learned, over tens of thousands of years, to deal first with nongroup members in a limited range of contexts, to form larger societies governed by a common framework of rules, and ultimately to apply an ethical code to people with whom interactions are negligible or even nonexistent, so too we can understand a progressive sequence of changes identifying altruism failures in inflicting, or even tolerating, pain in animals whose anatomy, physiology, and behavioral reactions resemble those already embraced within the ethical framework. Ethical rules could come to cover animals relatively like those we routinely domesticate—foxes as well as dogs.[23] More difficult is the task of deciding just where the taxonomic boundaries are to be drawn. Which kinds of animals have sensations sufficiently like those of our nonhuman exemplars to warrant counting their pains as states we should try to avoid or relieve? Reptiles? Fish? Nautiloids? Molluscs? Worms?

The idea that it is good to relieve pain, wherever it occurs, is a natural extension of ideas about the good, progressively elaborated during the evolution of the ethical project. Accompanying that idea is a related thought about human beings: it is not good for people to be insensitive to pain, whether it occurs in other people or in nonhuman animals. Inflicting pain—or even permitting it—produces human beings who are debased, whose characters and lives are less good than they might be. As the ethical project evolves, views of the good human life become richer, and engaging in conduct that causes unnecessary pain to others, including nonhuman animals, comes to appear detrimental to living well.

Shared capacities for pain allow the extension of ethical precepts beyond the human sphere. How could pragmatic naturalism go further? Trees and statues do not feel pain. Nevertheless, the same strategy can be deployed. Concern for individuals who do not participate in the ethical

---

23. The justification of blood sports would thus have to show that the hunted animals are threatening in some way (foxes attack livestock) or that the activity of hunting serves some important human purpose (plays an indispensable role in the maturing of young men). I am extremely dubious that any such attempt at justification accords with reflection on the conceptions of the good that have come down to us.

deliberation—juveniles, descendants as yet unborn—has been part of the ethical project from the beginning. Even when the desires endorsed are quite basic, not maintaining the conditions that permit those desires to be satisfied by those who will come later is an altruism failure. For the small bands of the Paleolithic, introducing precepts requiring conservation of parts of the habitat would have been a progressive step, at least once population density was sufficiently high to make a policy of unrestricted exploitation and moving on impracticable. With the more powerful means of environmental devastation available to us today, it would be similarly progressive for us to commit ourselves to preserving the one planet we know we can inhabit.

As more refined conceptions of the good human life emerge, a wider range of desires is endorsed. Communion with unspoiled nature is seen as contributing to the quality of a human life; encounters with human artifacts are regarded as similarly enriching. Preserving the possibilities of experiencing a range of natural environments and of being moved by great human achievements is not as crucial to those who will come later in the history of our species as ensuring that they have an atmosphere fit to breathe, but it is still important. The basic, long-standing precepts enjoining us to care for the world inhabited by our descendants underlie a range of principles of conservation.

It is easy to feel that the approach I have outlined is crassly anthropocentric. Precepts cover behavior toward rain forests and great buildings because future people are expected to gain from experiencing these things. The entities to be conserved have no intrinsic value; their special status is conferred by us, sometimes because of specific ways in which our societies, and the ethical project, have evolved. Can that really be an adequate account of the worth of the environment and of our obligations to conserve it?

Pragmatic naturalism is committed to an even stronger, perhaps shocking, view of the human dependence of value. Anthropocentrism is at the core of the ethical project—even to the characterization of ethics as a *human* project. That project confers *all* values. Parts of the environment are to be conserved because of decisions reached in progressive transitions—but similar decisions lie behind the injunctions that nonhuman animals are to be shielded from pain, that people who belong to different societies are to be accorded the protections available within the local group, and even that altruism failures within a small local band are to be remedied. Valuation is something humans do, and we have been doing it ever since normative guidance became socially embedded. There is nothing else that can ground value—no external source, no divine being. If it seems arrogant to view our species as the source of ethical precepts and values, we should recall that, at least in our corner of the universe, there is no other entity to do the work for us.

§48. Equality and the Good Life

Pragmatic naturalism views us as facing a scaled-up version of the predicament of the original ethicists. The primary challenge stems from the need to address the sources of conflict: the pronounced inequalities of the contemporary world and the clash of different systems, whose religiously based imperatives are both mutually hostile and override desires that might otherwise be endorsed (§46). If these problems are not readily perceived, that is because, in the manner of the envisaged tyrant of §34, a small privileged subset of the human population can insulate itself without attending to underlying causes. It is overwhelmingly improbable that the insulation can be maintained for long, given the technological possibilities for violent retaliation now increasingly available to the poor and oppressed (or to those who claim to represent them). Even if the comfortable few assume their security can be preserved in the long term, the thought of an imaginary conversation, in which they must discuss respecifying the good, on equal terms with the many who live in want, should concentrate their attention.

Dynamic consequentialism begins with the conception of the good world (or the better world) we have and considers how it might be modified in progressive ways. Contemporary ethical and social theorizing offers possibilities. Utilitarianism, generated from consequentialism through a sequence of reductive assumptions, identifies a supposedly objective measure, the degree of pleasure and absence of pain, and aggregates over sentient beings (§45). Inspired by concerns about how differences in ascribed utility could be determined, social theorists either have focused on a supposed proxy for utility (money, subject to various potential discounting factors for risk and future possession) or have attempted to analyze the common good without presupposing the comparability of subjective experiences.[24] One social situation is said to be superior to another just in case there is no individual whose preference satisfaction is diminished and at least one whose preference satisfaction is enhanced. This perspective has been the starting point for some of the most brilliant and illuminating work in theoretical social science and in social and political philosophy of the past century.[25]

Existing consequentialist theoretical traditions offer conceptions of the good in three distinct forms:

A. The goodness of the world is determined by aggregation of some objective measure for the states of sentient (or human) beings (aggregate pleasure minus pain, aggregate desire satisfaction).
B. The goodness of the world is determined by aggregating a publicly observable proxy (e.g., money or surviving people).
C. One state of the world is better than another when no desire satisfied in the latter is unsatisfied in the former, and at least one desire that goes unsatisfied in the latter is satisfied in the former.

Versions of A are notoriously difficult to elaborate. Those elaborations of B focusing on money as the pertinent proxy, while often deployed as the basis for social policy, are of dubious ethical significance; counting heads (or survivors) is only slightly less crude.[26] Conception C, on the

---

24. Lurking behind all this are methodological ideas that have sometimes played a useful role in inquiry (through connecting hypotheses with potential tests and observations) but that have often hardened into restrictive dogma (most evident, perhaps, in some kinds of behaviorist psychology). Significant parts of economics and social theory remain in the grip of sclerotic operationalism, even though its philosophical credentials have long been subject to important critiques.

25. Kenneth Arrow, *Social Choice and Individual Values* (Hoboken, NJ: Wiley, 1951); Amartya Sen, *Collective Choice and Social Welfare* (San Francisco: Holden-Day, 1970); John Rawls, *A Theory of Justice* (Cambridge, MA: Harvard University Press, 1971).

26. It is worth reflecting on the extraordinary combination of a very crude basic measure of goodness (number of survivors) with the excessively refined concentration on wildly contrary-to-fact hypothetical examples (trolley problems) so evident in some philosophical fashions.

other hand, seems to deprive itself of important resources—for even if we cannot make *all* comparisons among the states of different individuals, we can surely make *some* (there is little reason to deny, for example, that the agony you would suffer from a debilitating illness is worse than the discomfort I feel from a mild itch—and thus no reason to think a state, in which I lost by having to endure the itch and you gained by avoiding the agony, would be "noncomparable" with the status quo.)[27]

All three versions are open to charges of hedonism, since all start with a focus on individual states or on variables (like money), viewed as basic because they are proxies for such states. They are opposed by a far more pervasive conception, or set of related conceptions, that start with a concern for the quality of individual lives (not simply for survivorship), understand this in terms of fulfilling the will of a transcendent being, and suppose worlds are good insofar as they contain wider multitudes of people who carry out the divine plan.[28] Conceptions of this sort typically define themselves in contrast to the crudest versions of conception B, juxtaposing the "materialism" of seeking to maximize money (or material goods) with the "spiritual" emphasis on fulfillment. So we have:

D. The world is better insofar as it contains an expanding set of spiritually fulfilled lives.[29]

27. Sclerotic operationalism makes itself evident in the judgments of noncomparability that are questioned here. I should note that Rawls attempts to reintroduce a finer discrimination among states by using the device of the original position to distinguish Pareto-incomparable options; effectively, this is to combine the two approaches I have been separating, taking imaginative deliberation to serve as a further constraint on the elaboration of the good. Since I believe that the operational imperative is overworked in this context, I suppose that the starting point is unnecessarily restricted and that the attempt to articulate dynamic consequentialism has more resources than Rawls's procedure acknowledges.

28. Some secular ethical traditions have a similar structure, taking the world to be good insofar as its inhabitants conform to the moral law. Kantian conceptions obviously fit here.

29. I have formulated the conception in a way that mimics the emphasis on Pareto-comparability found in (C). Proponents of (D) probably rarely consider whether salvation for a greater number at the cost of forfeiting salvation for a lesser would be an advance—and do so because the world modifications that concern them involve what they take to be expansion of the community of the faithful without loss. (It is unclear how they would react if it were shown that sending young people out on missionary work is likely both to increase the number of converts and to involve some lapses on the part of the missionaries.)

Different religious traditions diverge about what it takes for a life to be spiritually fulfilled.

The inadequacies of conceptions A–C were prefigured in §45 and are expressible in terms of an insight of conception D. Instead of breaking individual lives into small intervals in which pleasures are felt, pains avoided, or desires satisfied, conception D views lives as wholes, evaluable as good or bad. Conception D goes astray in grounding the goodness of lives in utterly false doctrines. To develop its insight, we need a surrogate for the dangerous deontology it offers.

Section 45 pointed in a promising direction. Reductionist emphasis on pleasure and pain, or on satisfied desire, should give way to recognition of holistic contributions to the value of lives.[30] Replacing a game of pushpin with a reading of *The Prelude* or a conversation with a soul mate does not boost the value assigned to momentary experiences, increasing aggregate utility across a life, but a life enriched by poetry and intimate conversation *can* have more value *as a whole* than one lacking anything similar.

It is important to tread cautiously. Holism is welcome in accounts of the good, but not at the cost of insinuating elitism: "Only the privileged—those capable of running the polis or of appreciating Wordsworth—can lead truly worthwhile lives."[31] Elitism can be resisted by proposing that almost *all* human beings can advance from a state in which the good life

30. In my interpretation, Mill is innocent of the reductionist errors. Although *Utilitarianism* makes it appear that he is making a minor adjustment in Bentham's position, introducing a third, and problematic, evaluation of pleasures and pains (the higher/lower distinction), he is best read differently. His essay/monograph is an attempt to defend utilitarianism against popular criticisms, and it is readily understandable that Mill would proceed from the version of the position that was most familiar to his readers. Mill's consequentialism is expressed in his entire corpus, and *Utilitarianism* is a response to well-known objections, rather than a definitive exposition of what Mill believes. (See my essay "Mill's Consequentialism," in *The Routledge Companion to Nineteenth Century Thought*, ed. Dean Moyar [New York: Routledge, 2010], for more discussion.)

31. The focus of ancient authors on the lives of the socially and economically privileged is echoed in contemporary discussions and is evident in that liberal democrat Mill. More than his major ethical predecessors and contemporaries, Mill is influenced by Greek thought, so that the question of the good life is central for him, and his formulations often absorb the elitist perspectives of the ancients (witness the language of "higher" and "lower" pleasures).

is understood in terms of the satisfaction of basic needs (nourishment, shelter, protection, and so forth) to a condition in which a richer conception of the good life makes sense for them.[32] This richer conception recognizes lives as structured wholes, with direction and point. Emancipating the conception of the good life from unfortunate elitist formulations, we can suppose that the crucial factors in evaluating lives are, first, whether they reach a conception of the person that can bestow direction and point, and second, whether the aspirations flowing from that conception are sufficiently realized.

Educated people may find the shape of their own lives (and of their selves) in a mix of intellectual work, political activity, appreciation of nature and of art, and (perhaps above all) communion with others. A major contribution to thought about the good life opposes the thesis that this pattern—or *any* single pattern—is appropriate for everyone.[33] Valuable lives can be structured quite differently. For some people, physical activity is more important than it is for many intellectuals—for some, replacing a game of pushpin with reading Wordsworth might be a distortion and diminution of a life.

A significant advance over Greek thought suggests that the pattern one adopts for one's life should be one's own.[34] In opposition to the parent or the community that specifies the model to which a nascent life should conform, we can envisage enlightened choice among many options. Recording and publicizing a diverse collection of "experiments of living" can acquaint those who come later with an ever-wider repertoire of possibilities from which they can make their own free choice. Ideally, once people have the opportunity to advance beyond the stage at which

32. Much of the time, Mill endorses this proposal, thinking of human progress as making the good life ever more widely available. This is evident in the famous phrase from *On Liberty*, "the permanent interests of man as a progressive being," as well as in the closing pages of the *System of Logic,* John Stuart Mill, *System of Logic* (London: Longmans Green, 1959). It is crucial to the egalitarian concerns of *Principles of Political Economy.*

33. Thus, Mill's emphasis on choosing one's own plan of life takes precedence over the very specific ways in which, as an extraordinarily educated man, he conceived of his own life as obtaining point.

34. Mill's writings make this advance in many places. The most well-known occur in *On Liberty*, particularly in chaps. 1 and 3.

human beings struggle to satisfy their basic desires,[35] the young will be invited to draw on the ways of living of many different predecessors to formulate a conception of what matters for them.

This embryonic account of the good life will be developed a little further (§50). The immediate task is to see how an account of this sort could remedy the inadequacies of the conceptions of the good so far considered. Taking the notion of the good life as central, understood in the ways just outlined, preserves the insight offered by conception D, while rejecting the massively false presuppositions of the various religious articulations of it. The holistic assessment of lives is primary, and positive evaluation depends on there being a structure, freely adopted by the individual and, subsequent to that, on the kinds of satisfactions of desire figuring in conception C, and tacitly in conceptions A and B as well.

This possibility arises only at a particular stage in human development—it requires progress beyond the conditions dominant in the early phases of the ethical project (§20). There are *preconditions*, some material, some social, for any state in which holistic evaluation can come into play. When life and health are constantly threatened, it is difficult, if not impossible, to explore possibilities to fashion one's own life pattern. People must be able to interact peacefully with one another (and with outsiders). Furthermore, if their life projects are to have serious chances of success, they will need not only the toleration of those around them but also active cooperation with their fellows.

Consider an imaginary social state: Utopia. In this state, each member of the human population has a serious chance of living a good life, a life in which the person can recognize a number of different possibilities for living, can make a free choice of a project informed by that recognition, and realize a significant number of the plans, intentions, and desires central to that project; moreover, the chances of living such lives are equal across the population. Where the conditions of human life allow no serious chance of bringing about Utopia, it cannot figure in the theory of the good. The good *evolves*, and this is an ideal available to us, but not to our ancestors.

35. A condition Mill characterizes as the "puerile" condition of mankind; closing paragraphs of *System of Logic*.

Pragmatic naturalism's proposal: Once a particular stage of technological development has been reached, a world counts as good to the extent actualizing it would lead us toward Utopia. More exactly, faced with a number of possible outcomes, we should rank them in accordance with the expected sizes of the steps they would take in the direction of Utopia (subject to the proviso that they do not block further progress). The reference to "expected" size accommodates the fact that the results of action may be associated with probabilities, and the proviso takes into account the possibility that a route leading to an unclimbable cliff will not take us to the summit.

The preconditions of Utopia include two sorts of equality. First, all individuals begin their lives in a state of material equality, a state in which they are assured of being able to meet those basic needs that must be satisfied before free choice of a life project can be a serious possibility. Second, enough equality must be maintained so individuals are not coerced into pursuing a particular life project out of material need. Plainly, the present enormously skewed distribution of wealth across our species is glaringly inconsistent with these fundamental forms of equality, and transactions enabling the poor to obtain reliable access to food and water, protection from the environment and from disease, education for the young, and so on would be steps in the direction of Utopia (subject, of course, to the proviso that they did not introduce complications making further advances impossible).

The proposed conception of the good sees progress in the transitions that generated our more complex conception of possibilities for human lives, but it also aims to restore the emphasis on equality abandoned in the most prominent ethical lineages.[36] It thus tries to take advantage of the enhanced options the inegalitarian developments of our past have given us, while simultaneously removing their distortions of the ethical project.

---

36. Some small societies in the contemporary world have preserved the emphasis on equality, but they lack any serious possibilities for achieving the richer versions of the good life available to *some* members of other societies.

§49. Population Size

My references to "Utopia" and to realizing an "ideal" generate an obvious question. Is the proposed conception of the good a feasible goal? A
skeptic might insist that the recommended combination is impossible:
an emphasis on equality must inevitably diminish the possibilities for a
good life; the proposal seeks joint attainment of conflicting values, where
we can settle only for pluralism.

Ideals that guide us should not be beyond the horizon of realizability.
Even if one had no basis for thinking an ideal unattainable, there might
be no clear available method of attempting to achieve it. Under those
circumstances, it could not play any role in shaping action. In the case at
hand, however, our knowledge of how people in affluent societies have
sometimes been able to choose profitable projects for their lives provides
a basis for spreading this possibility more widely. If skepticism (that
genuine equality of opportunity for the good life cannot be achieved for
all) is rebutted, it will prove possible to identify steps toward attaining
that goal.

Skepticism rests on a fundamentally economic thought: the resources
human beings collectively can acquire are too limited to make possible
for all the opportunities for a good life (in the rich sense) currently available to the few. Meeting the basic needs of the entire human population, or
meeting those needs as well as supplying the background circumstances
for good lives (educational and health services among them), would be
possible only if the affluent world impoverished itself. Recent studies of
global poverty cast doubt on this pessimistic assessment.[37] Yet we should
pose a different question: Given pragmatic naturalism's conception of the
good, how many people should there be?

Population size lies at the heart of issues about the good, for two basic
reasons. First, we know that indefinitely continued expansion of the human population would eventually reduce all human lives to a condition
of wretchedness. Second, we know it is possible for local populations to
resist expansion, and even to diminish their size, without giving up opportunities for their members to enjoy good lives. Indeed, statistics show

---

37. See, for example, Jeffrey Sachs, *The End of Poverty* (New York: Penguin, 2005).

that significant advances in the quality of individual lives—for example, advances in educational opportunities, especially for women[38]—produce a decline in birth rate. Even if the skeptic were right about the insufficiency of the world's resources to deliver the preconditions of serious opportunity for the good life with respect to the *existing* human population, the ideal might be attainable for an envisaged future population with diminished size.

Ecologists use the notion of the *carrying capacity* of an environment, with respect to a group of organisms. The carrying capacity is the number of members that group would have at equilibrium: if the actual population size is lower, the population expands; if the actual population size exceeds carrying capacity, the population declines because of the insufficiency of resources. We can introduce an analog of this idea. Say the *proper bound* on the human population is that number at or below which it would be possible, given the material resources of our planet and available technology, to deliver to a population of that size serious and approximately equal chances for a good life, through an indefinite sequence of generations. Pragmatic naturalism proposes that the human population should be at or below its proper bound.

The proper bound is determined by issues not yet precisely resolved. The substantive notion of the various forms of the good life might affect its value. If it is important for people to have the chance to retreat from the presence of their fellows and to enjoy solitary contemplation of the natural world, that might entail programs of preservation (or restoration) limiting the budget of resources. Articulating what counts as a "serious chance" for a good life will affect the amount of resources individuals and groups require and thus constrain the proper bound. Suppose these choices have been made, generating a value for the proper bound. The skeptic is answered. The proposal offers an account of the good for a population at or below the proper bound; the feasibility of the ideal is built in.

Is this cheating? The original worry claimed that meeting the preconditions of equal opportunity for the good life for all human beings

38. See Amartya Sen, *Development as Freedom* (Oxford, UK: Oxford University Press, 1999), esp. 217ff.

would impoverish the lucky few who actually happen to enjoy richer prospects for their lives, in a quixotic venture of trying to satisfy everyone's most basic needs. That worry has been allayed by limiting the size of the human population. Can Malthusian restraint be achieved without threatening the quality of the lives of individual people and families? Quite reasonably, the skeptic insists that begetting and rearing children is central to many people's conceptions of what matters most to them: consequently, confining the human population to the proper bound directly and dramatically reduces the chances for good lives. There are two ways to articulate the objection. One contends that, in a population already at or below the proper bound, maintaining that population size would significantly limit the quality of human lives. A more modest version argues that any possible route to a population below the proper bound requires diminishing the quality of life for the generations traversing that route.

Although family life is central to many people's conceptions of their lives, it does not follow that limiting family size must interfere with their prospects for a good life. As an elementary consequence of the biology of reproduction, in a world where no couple produced more than two children, the human population could not increase (and would almost certainly decline). Thus, the stronger objection can be sustained only if an ethical restriction of family size to two children excludes the opportunity for a good life. Plainly, smaller families are vulnerable to contingent events that might diminish the quality of the lives of their members—because of the impact of the death or disability of children. Nonetheless, larger family size offers no guarantee against such contingencies, nor can it necessarily relieve the pain accompanying the death of a child: children are not intersubstitutable.[39] The most skeptics can rightly conclude is that an ethical constraint on family size might lower some people's chances for a good life, through increased susceptibility to hostile fortune. In worlds where the human population is maintained at or below the proper bound, however, an equitable distribution of

---

39. It should be recognized, however, that the loss of *all* one's children, as in cases where all are killed in war, does strike people as especially tragic. That reinforces the thought that smaller families are more vulnerable to the whims of fortune.

resources lowers the probability of the major kinds of disruptive events—for example, through preventative measures against diseases killing millions of children annually. The stronger version of the skeptical objection fails.

Consider the more modest version of the objection, which focuses on the trajectory the human population would take to diminish the size of the population until it is at or below the proper bound. Suppose the route to the proper bound requires a specific claim about the good: in worlds at larger sizes, it is good for couples to limit their offspring to a single child. Assuming it is a greater tragedy to lose all one's children than to lose all but one, the imagined ethical constraint increases the probability that some lives will be devastated. Pragmatic naturalism must claim that reflective people, envisaging the human future, would be prepared to endorse the ethical constraint, even while recognizing the potential losses. Their endorsement might be based on appreciating that not imposing the constraint would continue a situation in which many people lack the opportunity for good lives, as well as on seeing how a more egalitarian distribution of resources would provide some protection against hostile contingencies. Yet, even if one rejected the ethical constraint and the route associated with it, the ideal would be realizable through a slower process. An ethical constraint restricting family size to at most two, coupled with encouragement of voluntary policies of having only a single child, would steadily decrease the human population. A world in which the more liberal constraint was honored would be a world in which the ideal of a sustainable provision of serious and equal chances for a good life for all members of the population, at or below the proper bound, eventually became feasible.

Does the current size of the human population exceed the proper bound? That is an empirical question, and one for which an extended analysis would be required. Present uncertainties about global warming, and its potential consequences for the provision of food and shelter, would have to be resolved to arrive at any convincing judgment. The appropriate response to the claim that the present human population exceeds the proper bound is the Scottish verdict: not proven.

One of the most important discussions in ethical theory in past decades has been the study of consequentialism when population size varies.[40] Begin from a framework analogous to that of orthodox utilitarianism: the goodness of a world is measured by aggregating the values of the qualities of the lives of the human beings who live within it. Accordingly, a world in which a large number of people who live lives as good as any we can envisage would be inferior to one in which a vaster number live lives that go slightly less well: if N (population size) and U (the value attaching to each individual life) are both large, if U^* is less than U by some tiny amount, and if N^* is vastly greater than N, then $N^*U^* > NU$. Iterating the process of comparison, a world with a truly gargantuan number of people whose lives were just above the level at which they are barely worth living would be superior to the original state. This is the Repugnant Conclusion, and it seems any viable version of consequentialism ought to reject it.

The problem was generated by using a measure of the goodness of worlds that, like classical utilitarianism, adds up the values in the individual elements. Another approach would be to measure by averaging. One world counts as superior to another just in case the average value of the lives lived in the first is greater than the average for the second. Although this revision avoids the Repugnant Conclusion, it entails other unpleasant judgments. Imagine Eden, a very small world, containing just two people who live lives of the very highest quality. Compare that with another world with vastly more people, in which everybody lives a life just infinitesimally less good than that enjoyed by Eden's lucky pair. According to the averaging approach, this latter world is inferior to Eden, despite the fact that the aggregate value of the lives within it is enormously higher. That, too, seems wrong—for, even though aggregation does not count completely (witness the Repugnant Conclusion), it cannot be completely dismissed. Adam and Eve would surely be doing good if they aimed to replace their Paradise with the more abundantly stocked world, rich in lives of extraordinarily high quality (albeit just slightly inferior to their own).

These two thought experiments pose a challenge: find another measure of the overall quality of worlds, avoiding these counterintuitive

---

40. The seminal work here is part 4 of Derek Parfit, *Reasons and Persons* (Oxford, UK: Oxford University Press, 1984). Parfit's brilliant posing of the problem has inspired many efforts to solve it.

implications and giving both aggregation and averaging their due. A formal treatment shows there is no solution to the problem.[41] The impossibility proof locates the fundamental difficulty. So long as it is supposed that the quality of a human life can be represented by a single number, the Archimedean property of the numbers will hold sway: however tiny r may be, and however large s may be, there will be some integer N such that $Nr > s$. This is the mathematical truth underlying the Repugnant Conclusion, and posing difficulties for aggregation, and there is no way to limit the trouble this mathematical feature of aggregation poses.

A *formal* way to respond would introduce a discrete measure of the quality of human lives. At its simplest, this would propose that lives go well or they do not, and the thought of tiny frustrations slightly decreasing the value assigned to a successful life is misguided: if someone's life under particular conditions is assessed as one that goes well, and if we now consider slightly less favorable conditions (an extra headache, a tiny loss, or whatever), the same discrete measure is assigned; small blemishes do not turn a valuable life into a poor one. That, however, is to overlook the need for *some* comparison. Even with respect to two lives that go well, one can make discriminations—but the discriminations are secondary. I propose a two-dimensional account of the value of lives. The dominant dimension accords with a discrete measure: discreteness is the better part of value. The secondary dimension has a continuous measure, one recording the minute satisfactions and discomforts of human existence. Lives counting as successful on the dominant (discrete) dimension submit to secondary comparison on the subordinate (continuous) dimension.

The approach of the earlier parts of this section gives substance to the formal treatment. At early stages of human life, the achievements pertinent to the dominant dimension have not been made. Early in the ethical project, the lives of people are measured only by the variables to which utilitarians point. The two-dimensional account comes into play once it is possible to provide the material basis for the good life for all people, and, at that point, the satisfactions measured on the

41. See my article "Parfit's Puzzle," *Noûs* 34 (2000): 550–77.

subordinate dimension are relevant only insofar as they affect the possibility of doing well on the primary dimension. The formal approach fits the progressivist and egalitarian aspects of pragmatic naturalism's conception of the good. The initial conceptions of the good are directed at obtaining high values on the continuous dimension; later, high-enough values can be achieved for all to make richer considerations of living well come into play; thereafter, the good is achieved by realizing the opportunity for all human beings to have a chance at a good life, and, beyond that, by trying to increase the number of those who succeed in living well.

§50. Aspects of the Good Life

To articulate the proposal further requires that more be said about its egalitarian commitments and about the factors affecting the goodness of lives. I close this chapter by saying a little about both issues.

Aiming at exact equality with respect to material resources or for chances of living well is plainly quixotic. Even in Utopia there would be differences in people's early lives affecting their probabilities of formulating a satisfactory course for themselves and of successfully pursuing it. What kinds of differences matter? I propose that differences do not matter when well-informed parties do not see them as mattering. Two people can know that one has had more chances than the other and still see this as no cause for regret or resentment. Education plainly affects opportunities for choices about one's life. Even though you and I have been to different schools, and even though your school clearly provides a richer range of options for you than mine does for me, I may see the difference as a tolerable accident, not affecting the fairness of distribution. Both schools do well; while your options are richer than mine, mine are also rich. The differences are small in comparison with other features that affect our lives. With similar amounts of effort, you may have slightly more chance of living well than I do, but a bit more exertion on my part would compensate. Furthermore, life has vicissitudes that could swamp the small increment in your chances for success. Struggling to render our opportunities exactly equal seems pointless: my slightly inferior chances of living well are very good, and potential factors thwarting my success would be unaffected by attaining exact equality.

The injunction to promote equality of opportunity recognizes a range of tolerable difference. Utopia requires that the differences lie in the tolerable range—and also requires eliminating them, where that can easily be done. More intricate issues arise with respect to the second zone of vagueness in my account, the idea of the good life.

The possibility of formulating one's own conception of what matters most is one component of a good life—subject, of course, to the proviso that one's choice would not interfere with the like choices of others.[42] It follows that our ability to find our own life project ("our own good in our own way") depends, at least up to a point, on an ability to reflect on the various possibilities, on the characteristics that have proved to be relevant to realizing them, and on our own propensities and talents.[43] Further, it is crucial that the options not be framed in terms of an arbitrary deontological perspective, effectively ruling out the most attractive. It is not simply the law that has often stood in the way of people's choice of the life projects they most wanted, but also a prevailing view that particular ways of living are wicked or sinful, where these evaluations have their basis entirely in the alleged deliverances of the transcendent policeman. An important part of the proposed conception of the good is that the circumstances in choice-of-life conceptions be entirely free from the distortions of religious traditions. Liberation cannot be achieved unless the context in which people choose patterns of life for themselves is thoroughly secular, beyond the myths of the world's religions, beyond their cramping effects on individual choice, and beyond their divisive intrusions among groups of people.

---

42. In shaping the conception, and in the formulation given here, my debts to Mill, and secondarily to Dewey, should be apparent.

43. One might think that our freedom in this regard is always enhanced through the articulation of further possibilities, so that, the more "experiments of living" someone can conceive, the more the choice is autonomous. As a matter of psychological fact, I doubt that this is correct. Too many options can be overwhelming. It remains true that radically new potential ways of living can show us opportunities that had previously been beyond our horizons, so that Mill's plea for further experiments remains cogent.

One particular way in which some religions have distorted our conception of the human good is by allowing for possibilities of valuable lives detached from other people. Solitary communion with transcendent beings (or with the universe), exemplified by hermits who live in remote places or those who pledge themselves to silence, is viewed as one way of living a good life. The source of the value here is surely the attunement of the individual psyche to the transcendent, and when the myths about the transcendental realm are abandoned, the idea that lives can be made significant in this way should go too. It is equally important to repudiate secular ways in which this distorting idea continues to manifest itself. Some of the most militant opponents of the world's religions commend participation in the project of understanding the natural world as an especially valuable way to live.[44] Although there is an important insight here, it needs to be carefully understood. The great discoverers achieve two things: they enjoy private states of recognizing hitherto uncomprehended aspects of nature, and they facilitate the understanding of others. The states of understanding, while superior to the momentary satisfactions of the hermits (because based on genuine achievement rather than on illusion), are not the primary determinant of the value of the discoverer's life. Instead, value accrues through the contribution to understanding on the part of others.

To take this perspective is to emphasize the centrality to the good life of our relationships with other people and of our contributions to their lives. Refined theoretical contemplation has its place among the catalog of factors that promote the good *life* precisely because of its potential to promote the value of good *lives*. Consequently, the life of the priest or scientist, the doctor or nurse, the teacher or social organizer, the tireless participant in the maintenance of community and family, become valuable in similar ways, through the various human relationships the person's actions sustain. The emphasis on individual freedom, on the abil-

---

44. This is prominent in the writings of Richard Dawkins, *Unweaving the Rainbow* (Boston: Houghton Mifflin, 1998). But the idea is very old—Aristotle's final chapter of the *Nicomachean Ethics,* (Terence Irwin trans. Indianapolis: Hackett, 1999) sounds a similar theme.

ity of each of us to choose our own conception of what matters, needs to be accompanied with recognition that any choice that does not incorporate interactions with others and see their good as involved with one's own is inadequate.[45]

This judgment recapitulates the historical story of Part I. Ethics began with our joint reflection on how to live together. From the experiments of living of the distant past emerged the forms of higher-order altruism (§21)—and the thought of joint action as proceeding from attention to the aspirations and interests of one another as especially valuable. The first ethicists were probably concerned with the altruism failures that arose with respect to basic needs, but their efforts led to an evolved conception of the good life, one in which our interactions and relationships with others are fundamental. That aspect of the evolutionary process should be seen as thoroughly progressive.

To recognize progress here is not to suppose that the evolving forms of human life have made it easier to achieve the valuable relationships, marked by higher-order altruism and participation in a rich variety of joint projects, that are principal constituents of the goodness of lives. Indeed, as human societies grow larger, as they emphasize the division of labor, even as they proliferate possibilities for living well, they may make it increasingly difficult to actualize those possibilities and to live lives with real value. Religious myths may distort the conception of the good life—through proposals to the effect that what really matters is our relationship to a nonexisting being—but they are not the only culprits. As we become coordinated parts of larger social machines, it is easy for us to think that what we achieve is determined by the specific contributions we make, the deals we bring off, the things we discover or invent or compose, the tasks of whatever kind we complete, without any reflection on the impact on other people. Our self-conceptions are further debased when we measure our worth in terms of the proxies for success in any of these particular directions, the cash rewards we receive for doing them and the trophies we thereby amass. We miss the fact that all of this effort

---

45. Thus Dewey advances on the Millian formula from *On Liberty*, when he opens *The Public and Its Problems* by posing the issue in terms of the freedom from coercion of *joint* projects.

obtains its significance from effects on, or more exactly contributions to, the lives of others. We miss also the important point that, independent of any large-scale public success, lives may be interlocked in mutual dependence and mutual contribution and thus be genuinely and completely worthwhile.

From the moment at which the issue of how to live well surfaced among the ancients, it has been vulnerable to three types of major error. The hedonist mistake is to decompose our lives into sequences of momentary experiences and measure value by the balance of pleasures and pains. The individualist mistake, prominent in some religious traditions but also retained in some versions of secularism, proposes that some particular nonsocial condition of the individual—the receipt of divine grace, the making of great discoveries, the amassing of wealth—is the major source of value. The elitist mistake, already evident in the restriction of the question to the male aristocrats of the polis, is to suppose that something very large and uncommon is a precondition of a life's going well. By contrast, in the approach I have taken, good lives are in principle available to almost all members of our species.[46] Philosophers tend to talk grandly of "life projects," as if the good life required both a type of intellectual reflection and an exalted focus that can be managed by only a select few. Instead, I offer a schematic account of the good life that celebrates the ordinary. Although in almost all places at almost all times, people have been coerced or led into lives that should not be counted as worthwhile, what they have lacked are certain basic forms of freedom, everyday awareness of possibilities, not exceptional resources or unusual talents. Moreover, in many times and places, ordinary people whose lives are permeated by actions with and for others have sometimes, if not often, lived well.

I have pointed only to *aspects* of the good life: freedom of choice, lack of deontological encumbrance (most evidently the distortions of religions), joint activity and reciprocal relationships with others. To indi-

---

46. The exceptions are those whose cognitive and emotional possibilities cut them off from fully developed relationships with others. This is, I believe, why we find prenatal genetic testing for some sorts of traits a merciful way of proceeding. I have discussed these questions in more detail in the later chapters of *The Lives to Come* (New York: Simon and Schuster, 1996).

cate these features is not to provide an articulated theory, but merely to avoid some dangers that have beset discussions of what makes lives worthwhile. That is, I hope, enough to develop the proposal respecifying the conception of the good, with which this chapter has been principally concerned.

CHAPTER 9

Method in Ethics

§51. Varieties of Ethical Change

Although each of us acquires, early in youth, an ethical code from older members of our society, our ethical convictions and attitudes do not remain constant throughout our lives. There have been societies (most probably in the distant past) in which members were forbidden to add or subtract from what they had been taught, and others, more common, in which only additions, articulations of the group lore, were allowed. That is not the way we live now, nor is it the way in which our species has lived throughout most of recorded history. In recent millennia, societies have equipped their young members with starting points for individual ethical development and exploration, permitting them to add, revise, refine, and subtract. Few people who grew up in the English-speaking world in the 1950s have retained exactly the constellation of attitudes toward sexual behavior originally passed on to us by parents and other ethical teachers. Only the most confident and unimaginative believe they acquired, early on, the complete, correct ethical system.

In ethics, as in biology, ontogeny needs to be distinguished from phylogeny. Ontogenetic changes are those occurring in an individual life, from the beginning of ethical consciousness to its termination. Ontogenetic changes may produce phylogenetic change, either when the explicit discussions of the next generation about what should be passed on involve discussants who have made parallel shifts during their ethical lives, or, without public discussion, when similar modifications occur on a broad scale, so the amended code passed on by parents is reinforced by other members of the community. Plenty of ontogenetic changes have no phylogenetic impact, most notably when people's articulations of an initially shared ethical code diverge from one another. You and I may begin with a vague maxim, one needing more precision in situations we both encounter, but your precise version may be at odds with mine; indeed, society-wide emendations may be so various that the vagueness persists as part of ethical counsel. How exactly the distribution of ontogenetic changes bears on phylogenetic change varies from society to society, depending on the role of prevalent religions or the presence of officially designated ethical teachers. When particular people, or institutions, dominate the training of the young, the ontogetic changes taking place in large segments of the population may be irrelevant.

Much of the time, people are not moved to change. Ethical discussions often fail to mention the obvious fact that large parts of our lives are based on routines and habits. You have acquired an ethical code that set up patterns of behavior, associating them with roles, which, if you reflected, you would understand yourself as playing. Your lack of thought need not be problematic; indeed, to step back and pose questions about habits or roles might be at odds with the ethical code you endorse. Sometimes, however, you do deliberate. The current occasion provokes thought, perhaps because a maxim you accept seems to dictate an action contrary to the one produced by following routine: you are on your way to carrying out your role (as friend, or spouse, or worker), and you encounter someone who plainly needs help.[1] The role-sanctioned activity

---

1. As psychological experiments have shown, commitment to the role may be so strong that people do not deliberate, even though their background ethical code calls for deliberation in the circumstances. See John Sabini and Maury Silver, *Moralities of Everyday Life* (New York: Oxford University Press, 1982).

is interrupted, and you must decide if the maxim has sufficient force to suspend it.[2] Deliberation need not lead to any modification—for the code, as it stands, may give a clear directive; it may, for example, pronounce on the importance of giving aid to people in the predicament you see before you, assigning that priority over the role-associated duties. Yet though episodes of this sort need not lead to ontogenetic change, they can do so.

Imagine your habitual performance is interrupted, and you are brought to think about how to go on. You canvass the code you presently endorse, but there is no unambiguous result. Perhaps there are conflicting vague maxims with no clear priority. Reflecting on other facets of the code, and emotionally reacting to the situation, you introduce a priority, perhaps making one or both of the maxims more precise than they previously were. The judgment about what to do remains stable in the aftermath of your action. As you explain what you have done to others, particularly to those your action affected, their reactions do not prompt a change of mind. From this point on, the claim about relative priority and the somewhat sharpened maxim(s) belong to your code.

Other types of ontogenetic change do not begin with thoughtful deliberation. Your ethical code contains no prohibition against speaking in a particular way to members of a particular class. You are accustomed to talk that way, until one day your remarks are met with protest. Somebody affected by your words shows you, convincingly, how painful it is for him, and, in appealing to the authority of the code, you feel your response is feeble. Suddenly and unexpectedly, you are moved to sympathy with a person whom you have previously seen under the shadow of a label, and your regret about what you said translates into a resolution not to repeat such phrases. You have modified your ethical code by inserting a prohibition where there was none before. The interactions that fuel changes like this are dim echoes of the original ethical discussions.

For most people all the ontogenetic changes made in ethical codes, and for all people most of the ontogenetic changes made in ethical codes,

2. See Dewey, *Human Nature and Conduct* (Amherst, NY: Prometheus Books, 2002), 52ff., 103-5.

are *normal* changes.[3] Given the ethical resources she has, an agent faces
a puzzle, and the ontogenetic changes she makes are the products of her
efforts to solve the puzzle. One cannot, of course, speak of a "solution"
to a puzzle, without some standard of correctness: what makes life hard
for the agent is the sense that there is some way of going on that is right
and others that are not. People do not seek "mere change." Although
ethical lives are largely matters of following the precepts and patterns of
our ethical codes, they are also—and importantly—ventures in trying to
improve those codes, by recognizing and solving puzzles that arise for
us. As individuals, we solve puzzles by making progressive shifts in the
codes to which we subscribe. Normal ethical change consists in attempts
to find a way to amend or extend the code, one refining the functions at
which it is directed.[4] Puzzles arise within an ethical practice in which
a particular set of functions is to be discharged; a solution to a puzzle
modifies the code so that some of these functions are better served and
none is worse served. Puzzle solving occurs only when there is no functional conflict (§36). *Revolutionary* change, by contrast, is marked by the
presence of functional conflict.

How can normal *phylogenetic* change occur? Perhaps as the product
of uncoordinated ontogenetic changes (although this would be highly
unlikely at early stages of the ethical project). Imagine most people encounter a type of situation where the ethical code they endorse fails to provide
clear counsel; through these encounters, they amend their codes in similar ways; in the next generation, the lore passed down from parents to
children absorbs the modifications, and, because the amendments are so
similar, the parental advice is reinforced in wider socialization. Yet if a
type of situation arises frequently, challenging individuals to articulate
their ethical code through puzzle solving, a public response can occur.
Especially when people respond differently, ethicists—including social
critics, religious teachers, and individuals with intimate knowledge of

3. My terminology of "normal" and "revolutionary" change plainly derives from Kuhn,
The Structure of Scientific Revolutions (Chicago: University of Chicago Press, 1962 and
1970).

4. Ethical agents do not conceive things in this way. They want the "right" solution.
According to pragmatic naturalism, the notion of rightness must stem from the concept of
progress (Chapter 6).

the pertinent type of situation, as well as philosophers—attempt to identify the proper solution to the puzzle. If the ensuing conversation
achieves consensus on a way of amending the core ethical code the participants share, the result will be a phylogenetic change.

Chapter 6 identifies a difference between reaching agreement on an
amendment and solving the puzzle that sparked discussion. A *solution*
refines the functions of the ethical practice in which the conversationalists are engaged. There is a set of functions to be addressed, and some
are discharged better by the amendment and none is discharged worse.
Consensus is by no means guaranteed to meet this criterion. Public conversation goes best if the conclusions reached are based on processes
likely to yield functional refinement: to discharge some existing functions better and none worse.[5]

At this point, we can appraise a method underlying some of the best
philosophical discussions of ethical puzzles. Guided by an analogy between ethical and scientific argumentation, philosophers seek *reflective
equilibrium* between general ethical principles and so-called intuitive
judgments.[6] Sophisticated versions of the analogy do not suppose intuitive judgments correspond to the statements provoked by observation,
thereby providing access to sources of external constraint on ethical inquiry. They regard "intuitions" as judgments people socialized within
a particular code would be inclined to make about individual cases.
Effectively, this is to embed the method of reflective equilibrium within
a dynamic approach to ethical practice, one supposing users of the
method already have an ethical code. The task is not to justify the code
from scratch but to identify successful ways of modifying it, to increase
its internal coherence. Puzzles are generated when parts of an ethical
code jar with one another, and the search for equilibrium, when done
thoroughly, places the local conflict within a broader perspective of
principles and responses to particular situations, some real and some

5. Appeal to religious texts is highly unlikely to satisfy this constraint on public ethical
discussion, since there is no positive correlation between the deliverances of those texts
and the refinement of major ethical functions. In fact there is probably a negative
correlation.

6. A classical discussion of this method is given by John Rawls, *A Theory of Justice*
(Cambridge, MA: Harvard University Press, 1999), 18–19, 42–45.

imagined, responses expressing the characteristic habits, emotions, and dispositions the code supplies.[7] Philosophical analysis seeks the best way of achieving overall harmony, not treating the "intuitive judgments" as apprehensions of external constraints, but approaching the problem as one of internal tension (in the extreme case: inconsistency) within ethical practice.

Is this method likely to solve ethical puzzles? Agreement on a modification is distinct from refining the functions of ethical practice, and thus a consensus proposal is not necessarily a solution. By the same token, an optimal smoothing out of tensions within an ethical code, the most harmonious amendment of it, might not improve its ability to discharge the underlying functions (indeed, the modification might make matters worse). In both cases, divergence stems from global trouble: if consensus fails to solve the puzzle, there must be some recurrent blindness among the participants in the conversation; if the broad search for coherence fails to deliver an amendment enhancing the code's functioning, some pervasive feature of the code must be at odds with the underlying functions.[8] If discussants are thinking clearly, we expect their combined efforts to correct for individual biases, errors, and prejudices—we think of many heads as being better than one. If the ethical code is relatively well attuned to its functions, we should anticipate that a global exploration would yield improved functioning—and hence view the method of reflective equilibrium as reliable, unless our background situation is seriously problematic.

So: a qualified endorsement of a prominent approach to contemporary ethical discussion.[9] The method of reflective equilibrium might be further explained and refined through more precise specification of

7. Many of the "thought experiments" beloved of analytic ethics provoke serious doubt whether responses express any socially inculcated skill. The situations are too remote from the mundane situations in which ethical judgment is exercised. I am grateful to Gerd Gigerenzer for discussion of this point.

8. This was probably so for some codes considered in Chapter 6: the Greek honor code and the Puritan code defending slavery seem systematically blind to the bungling of the original function of ethics.

9. One carried out with great skill in the writings of some philosophers, for example, John Rawls, Thomas Nagel, and Peter Singer.

the virtues of coherence to be attained, and the task of improving our understanding and practice of the method might be advanced by using the criterion for puzzle solution to show why particular types of coherence are pertinent. Yet that criterion suggests a different method for normal phylogenetic change, one to which the search for reflective equilibrium can be viewed as a preliminary approximation. Since the fundamental condition of puzzle solution is refinement of functioning, we can envisage a more direct strategy:

1. Identify the functions the ethical code is to serve.
2. Show how an amended code directs action in the situation that gives rise to the puzzle.
3. Show how the amended code improves discharge of some functions without compromising any others.

The indicated method preserves all the virtues of seeking reflective equilibrium, while also forestalling the possibility that some pervasive factor introduces a discrepancy between the functions and the elements of the code. Of course, when there is no discrepancy, global attention to principles and intuitive judgments (formed through exercise of dispositions the code inculcates) will represent the ways in which the underlying functions are served—which is why the method of reflective equilibrium is typically a good approximation.

The strategy just outlined is sometimes available: in contexts of normal change. For revolutionary change it breaks down. The reason is evident: its success presupposes the existence of a way of amending the code that will refine some functions while leaving others no worse than they were before. When functions conflict, the presupposition fails: *there is no modification of the ethical code that will (a) give direction in the situation confronted, (b) refine some function(s), and (c) leave other functions uncompromised.* Revolutionary changes are those transitions in which ethical codes are modified in ways involving both gains and losses with respect to functions the codes endorse.

Ethical codes may tacitly treat their underlying functions as elements of an unordered list: all these functions should be satisfied as completely as possible, and no function is given higher or lower priority. Alternatively, they may already embody judgments about relative importance: these functions are to be satisfied to this degree provided others are satisfied to different (lesser) degrees. In the former case, revolutionary change will introduce an assignment of priorities where there was none before; in the latter, it will be generated by the impossibility of honoring the present ordering of priorities, and it will introduce a revised ordering. Neither type of transition can be defended by appealing to the method for normal change, since there are alternative ways of introducing priorities or of revising the priority ordering, all of which, from the perspective of seeking refinement of some functions without compromise of any, are equally justified (or unjustified) relative to the original form of the code.

Section 40 offered a standard for assessing inferences: good ethical inferences are those likely to generate ethical progress. If all ethical change were normal change, invoking that standard would suffice. The multiple functions of ethical practice, and consequent possibilities of functional conflict and ethical pluralism, opened the door to further worries about pragmatic naturalism (§43). When functional conflict arises and revolutionary changes lead to new forms of ethical practice, can the resultant codes be seen as progressive with respect to the earlier ones, and can their adoption be defended through nonfallacious modes of inference? Consider, again, a challenger who asks why he should be bound by the ethical project. If the challenger questions a normal change, the strategy outlined above will yield nonfallacious modes of inference to answer him. If he is concerned with the ethical enterprise as a whole, it will be legitimate to demand that he present some alternative (other than the hominid state from which the ethical project liberated us). These are the ways in which the challenge was previously turned back (§§40, 42).

Trouble arises from an intermediate case. When functional conflict erupts, the challenger questions any proposed resolution. In doing so, he is neither casting doubt on modifications of the code defensible by inferences meeting the straightforward standard (suited to normal changes), nor rejecting the ethical project in toto. His point is that *within* that project, there are other ways of going on, different ways to assign priorities among the competing functions, and those can serve him as alternatives, from whose perspective he can query the choices actually made. He raises the specter of rampant pluralism (§43).

We need a method for justifying revolutionary change. The following sections develop one.

§52. Method and the Good

Start with an apparently dangerous proposal. Section 45 suggested that the unfolding of ethical practice could be understood in terms of the evolution of the good. The idea that proper ethical inferences are those likely to generate ethical progress founders on instances of revolutionary change, because of functional conflict. Substantive normative conceptions of the good offer proposals to resolve functional conflict. Very well. Use a proposed conception of the good to assess the progressiveness of transitions, and apply the old idea of proper inference as promoting progress, given the determination of progress so achieved. *A proposal about the good can be the starting point for articulating a standard for inferential justification, and methods shown to accord with that standard can then be used to support the proposal.*

The danger is obvious. Using the account of the good to frame the methods to be employed in ethical discussion, and then defending the account by appeal to the methods so generated and approved, looks blatantly and viciously circular. Isn't this simply a matter of justifying a particular conception of the ethical project by taking it for granted at the beginning? Or, to take up skeptical concerns explicitly, isn't this a game challengers can play equally well? Despite the importance of these questions, an attempt to address them will be postponed. The idea will first be developed on the basis of the conception of the good proposed in the last chapter—a concrete instance of the apparently circular strategy will show it to have justificatory force. The conception of the good offered in Chapter 8 will generate ideas about ethical method, and then we shall consider the possibilities for defending the conception by using the method that emerges. Until we return, at that point, to the skeptical challenge, it is a (small?) consolation that, as with the treatment of normal change, there is an apparent analogy with the search for reflective equilibrium.

Consider some fundamental features of the conception of the good offered in Chapter 8. Today we face a scaled-up version of the circumstances that originally provoked the ethical project. Instead of conceiving altruism failures as arising within a small band, and calling for remedy, we should think of the current membership of our species as a society in which altruism failures occur on a spectacular scale—and equally demand attention. *The original function of ethics, that of remedying altruism failures, is thus given priority.* To propose Utopia as an ideal (§48) takes equality of opportunity for a worthwhile life as a central constituent of the good. The human population should be limited so as to render this form of equal opportunity possible. Joint projects and higher-order altruism are important features of worthwhile lives (§50). The world at which we aim would thus be one in which all individuals acknowledge, and try to promote, the existence for each other person of a surrounding set of people with whom the person engages in cooperative activity and with whom that person enjoys long-term relationships.

These features translate the original ethical predicament into the contemporary situation. The first ethicists focused on the altruism failures within a small group, treating the members of that group as equal with respect to the simplest preconditions of the good life (satisfaction of basic desires), and seeking the cooperation of all with all. When the pertinent population (the set of people whose actions impinge on one another) expands dramatically—to include all of us—and when the evolution of ethics has bequeathed to us a richer conception of the good life, the goal is modified: equality with respect to basic needs gives way to equality of opportunity for a worthwhile life; the cooperation of all with all gives way to the fostering of cooperation within subsets of the human population (because, when the "group" is numbered in the billions, the cooperation of all with all on a wide range of ventures is manifestly impossible). Essentially, then, the conception of the good proposed results from an attempt to renew the original project of ethics, while retaining some of the functions since introduced.

Consider the method employed by the early ethicists. Normative guidance was socially embedded: group members discussed potential ways of organizing their lives together. They deliberated under conditions where all were present and all were given equal voice, attempting to arrive at rules all could accept. It will be useful to have a term for these conditions: call them the circumstances of *mutual engagement.* Mutual engagement was well-suited to finding a solution to the social problems generating the ethical project. Acquainting group members with the needs and difficulties of their fellows, giving them equal status in the conversation, and aiming at solutions everyone can endorse constitutes a good strategy for coping with the altruism failures manifested in social tension.

Chapter 8 proposed to renew the original function of ethics and scale up the initial conception of the good. Likewise, our ethical method, for use in cases of revolutionary change, can scale up the original method of the first ethicists. *We should seek a notion of mutual engagement as well suited to the renewed ethical project as the original version of mutual engagement—the deliberations among band members—was to the original venture.* We need an analog of those constructive conversations out of which the earliest rules for conduct emerged.

On the face of it, this suggestion must appear ludicrous, for any *actual* conversation among all affected individuals—that is, among all human beings—is impossible. Public ethical deliberation, however, can proceed by attempting to *simulate* a conversation of the pertinent kind. Faced with functional conflict, so revolutionary change is in order, public contributors to ethical discussion are judged by their ability to ground their proposals in mutual engagement: that is, to introduce the considerations and lines of reasoning that would be brought forward to achieve consensus were the entire human population to participate, *under conditions of mutual engagement,* in a conversation about the regulation of conduct. The italicized phrase is crucial here. An ethical discussion seeking to replicate the conversation that would occur if the entire human population were simply brought together in some vast arena would be a useless exercise in cacophony. Because of their existing dispositions to psychological altruism, limited though these were, because of the pressures on the group and the perceived need for joint action, the original ethicists were forced into mutual engagement with people who lived beside them every day. For us, however, mutual engagement is not automatic (except with respect to small subsets of the human population). If the proposal about ethical method is to have substance, a clear sense must be given to a notion of mutual engagement, suitable for us.

The conditions of mutual engagement will be specified by incorporating insights from the historical understanding of the ethical project, by combining them with a proposed conception of the good, and, on this basis, devising rules to govern the imagined conversation. Worries about circularity emerge precisely here, and, even before we have introduced rules, and so given substance to the concept of mutual engagement, the danger can be made more concrete. One obvious conversational rule, implicit in the idea of scaling up, would require the conversationalists to participate on terms of equality: all are to have a voice, and all to be taken as authoritative with respect to their own wants and needs.[10] The conception of the good drawn from the last chapter, already used to shape the proposed method, also makes equality central. Hence, it might be alleged, equality is built into the method of justification of a proposal to aim at equal treatment for all, and this is simply to beg the question.

Articulating the charge of circularity in this way enables us to appreciate both its force and its limitations. The emphases on equality in the conception of the good and in the putative method of justification are different: the conception of the good aims at equality of opportunity for a worthwhile life; the conversational rule requires equality of participation in the ideal simulated conversation. The difference in the types of equality is important, for an imaginary conversation undertaken on the condition of equal participation is not guaranteed to issue in the endorsement of equal opportunities for worthwhile lives. Substantive work is required to show how the rules governing that conversation would generate consensus on this conception of the good. If a demonstration of this kind were at hand, it would reveal a particular sort of coherence in the two proposals offered: the proposal that ethical discussions justify their conclusions if they rehearse the considerations that would bring ideal conversationalists to consensus, where part of the ideal consists in the equal participation of all, would be seen as providing support for the proposal that the good requires equal opportunity for a worthwhile life,

---

10. As we shall see in the next section, this is too crude as it stands. But the crude version will help to bring out the potential problem of circularity.

through exposing the ways in which discussants under the conditions of mutual engagement would endorse the pertinent conception of the good. Because that sort of coherence is not guaranteed, exhibiting it reveals something important about the package of proposals. It achieves something analogous to reflective equilibrium.

Once this point is recognized, we can glimpse a possible response to the concerns that surfaced at the end of the last section. Suppose my pair of proposals (one concerning the good, one concerning ethical method) is coherent in the suggested fashion. That might be enough to distinguish it from alleged alternatives, recommendations about the good and about ethical method offered by putative challengers. There would be a basis for turning back the challenge and completing the line of argument begun in §§42–43.

Whether this possible response succeeds depends on articulating the account of method. Turn, then, to the concept of mutual engagement.

§53. Mutual Engagement

The search for a method for ethical discussion stems from the wish to make our modifications of ethical practice more reliable than they have typically been. Any defense of the possibility of ethical progress should concede that there may not be very much of it and that what there is may be achieved blindly (§§28, 30). Pragmatic naturalists hope an explicit account of ethical practice might generate a higher frequency of progressive transitions, through the articulation of criteria for the discussion and resolution of ethical questions.[11] In the case of normal change, it is relatively clear how to proceed, and my attempt to refine the method of reflective equilibrium (§51) offers a putative basis. The thought of a conversation under conditions of mutual engagement has been introduced in an attempt to do something similar for revolutionary change.

The search for method can be guided here by comparison with the sciences. According to a popular story, scientific research became more

---

11. This is a major theme in Dewey's writings; see, for example, chaps. 9 and 10 of *The Quest for Certainty John Dewey: The Later Works*, vol. 4 (Carbondale, IL: Southern Illinois University Press, 1984).

successful as it was more self-consciously directed by ideas about method, this joint process occurring in the seventeenth century.[12] It is worth reflecting on this supposed analogy, because attention to the scientific case can temper expectations about method in ethics. Despite the frequency with which pioneers of early modern science linked their hypotheses and discoveries to claims about method, one does not have to read very far in their writings to understand that, first, their methodological conceptions, while related, have important differences, and, second, that their methodological counsel is often imprecise. If we now have a more detailed account of methods of scientific justification, that is because the initial vague thoughts about method have inspired scientific research whose successes could then be used to refine and revise the preliminary ideas about method. To take the comparison between ethics and science seriously should accustom us to the possibility that an initially imprecise account of method might spark ethical deliberations, whose results lead to further precision about method. There are no final definitions, but an evolutionary process.

According to my account of method in revolutionary ethical change, a discussion of an ethical problem (generated by functional conflict) is assessed against a standard of replicating the course of an ideal deliberation under conditions of mutual engagement in which all members of the human population participate. The first condition on replication requires the conclusion to be the consensus the ideal conversation would reach:

(RC) An ethically adequate discussion concludes that p only if an ideal deliberation under conditions of mutual engagement of the issue whether p would reach a consensus on p.[13]

The point of ethical discussion is not, however, simply to *state* a conclusion, but to *show* others why they should adopt it. Hence there is a

---

12. Dewey emphasizes the comparison. See, for example, *Quest for Certainty*.

13. For the moment I focus on what may appear to be the implausible thought of perfect consensus. Later, this requirement will be relaxed.

second condition on replication, directed at bringing to the fore the considerations moving the ideal conversationalists:

(RJ) An ethically adequate discussion discloses those features of an ideal deliberation under conditions of mutual engagement that would prompt each participant to reach consensus.

I postpone issues about how one might apply any such standard, through testing (confirming, refuting) claims that replication of an ideal deliberation has been achieved.

Now to the concept of mutual engagement that lies at the heart of these standards. The conditions of mutual engagement are partly cognitive and partly affective (and in this they resemble the dimensions of psychological altruism, §5). The first cognitive condition eliminates any erroneous factual beliefs from the ideal conversation:

(KE) In their deliberations, the participants do not rely on any false beliefs about the natural world.

This condition seems innocuous and well motivated, for it appears that consensus achieved on the basis of error would be problematic—a group of intensely altruistic people would arrive at quite peculiar conclusions if they assumed that extremely severe pain has all sorts of wonderful consequences for sufferers. (KE) entails that the aspirations and wishes the participants form, and the premises from which they reason, must be thoroughly secular. They cannot announce that certain actions are required or forbidden, or certain elements characteristic of the highest good, on the grounds that there is a transcendent being who commands us in the relevant ways or who offers us an infinitely valuable immortality. Positive beliefs in transcendent beings, and in the arrangements such beings make, are errors that cannot be permitted to distort the ideal conversation.[14]

14. Unlike some contractarian writers (Rawls, Scanlon), I exclude appeals to literalist readings of religious texts not because they introduce reasons not all participants share, but because they are simply false. (KE) embodies the idea that false beliefs will distort the

Two further cognitive conditions run parallel to the cognitive dimensions of altruism (§5). First, in an ideal conversation, the participants know the consequences for all of the types of actions they consider requiring, allowing, or forbidding. Hence:

(KC) Each conversationalist knows the consequences for each other conversationalist of the actions and institutional arrangements under discussion.

Moreover, the empathetic understanding of these deliberators is not faulty; they recognize the wishes of those with whom they converse. As we shall see shortly, those wishes are themselves modified through the interaction of the participants, and ideal conversationalists are able to keep track of these modifications, much in the way that people do under successful conditions of higher-order altruism (§21).

(KW) Each participant has complete knowledge of the wishes of others, and of the ways in which these wishes are modified through the course of their interactions with one another.

Plainly, these cognitive conditions are extremely strong, and unlikely to be realized in human exchanges; because ethical discussions meeting the replication conditions must be sensitive to the desires of all members of the human population (and of the ways in which those desires would be altered through attention to others), the authors of such discussions must apparently be attuned to an impossibly broad range of psychological facts. Once again, I emphasize that I am postponing the issue of how the standards of ethical discussion might be applied to particular cases, and how the credentials of such discussions might be tested.

---

conversation, an idea easily supported by considering a vast number of examples of deliberation based on error (just one of which—the mistake about the consequences of pain—is invoked in the text). Falsehood is to be avoided, and there is no reason to think religious falsehoods should be treated differently from others.

The heart of my account of mutual engagement consists in the affective conditions. Start from the thought that genuine engagement with others begins from an expansion of one's sympathies, in which the perceived desires of those with whom one deliberates are given equal weight with one's own. Because of conflicts, that cannot be carried out consistently across the board. As repeatedly noted, if two people have incompatible desires, it is impossible for a third party to behave as a golden-rule altruist (§5) with respect to both of them. Consequently, expanded sympathy cannot simply be understood in terms of responses to the desires of others that give equal weight to the wishes of each. Nor will it do to seek identification with the wishes of some harmonious majority, for the majority may be blinded by failures of sympathy. How, then, is expansion of sympathy to be conceived?

I tackle this question by introducing the notion of *mirroring* others.[15] The simplest sort of mirroring is that just considered (and found problematic as a general account of mutual engagement). For A to engage in *primitive mirroring* of B is for A's desire to give equal weight to the solitary desire of B, and to A's own solitary desire.[16] Now, in an ideal conversation aimed at addressing functional conflict in an ethical code, the solitary desires from which mirroring begins are not those individuals actually adopt, but rather those in harmony with the functions to which the ethical code responds and sustainable if the cognitive conditions on ideal conversationalists were satisfied. Hence:

15. Here I draw on ideas present in the eighteenth-century sentimentalist tradition, notably in Smith's theory of moral sentiments. I have explored the mirroring metaphor, as he develops it, in "The Hall of Mirrors," *Proceedings and Addresses of the American Philosophical Association* 79, no. 2 (2005): 67–84. As with the original discussion of psychological altruism in §§3–5, my formulations of conditions here incline to a position that attributes conscious recognition of the states of others. It is possible, however, that human beings sometimes engage with the feelings of those around them in more automatic ways (perhaps through the activity of mirror neurons). The account of mutual engagement is easily amended to permit any mechanisms of this sort to play a role.

16. The notion of a solitary desire was introduced in §3. Notice that I am not distinguishing A's perception of B's solitary desire from B's solitary desire itself, since, under the cognitive conditions of the ideal conversation, A has an accurate understanding of B's desires.

(DS) The solitary desires of an ideal participant include only such desires as are compatible with precepts of the participant's ethical code (precepts contributing to the functions the code is to discharge) and also retained when the participant accords with (KE) and (KC).[17]

Effectively, this filters the original solitary desires, eliminating those ruled out by the conversationalist's current ethical code and those present solely because of some form of ignorance.

Primitive mirroring cannot provide a general account of the affective part of mutual engagement, because A can encounter situations in which two others, B and B^*, have incompatible desires, so A cannot accommodate both of them. Given the filtering required by (DS) however, some of these problems can disappear, and primitive mirroring of others may become possible. When that occurs, an ideal participant adopts the pertinent desire (one that gives equal weight to the solitary desires resulting from the filtering). This requires the filtered solitary desires of B and B^* to be the same, so filtering has already done some of the work of primitive mirroring.

The challenge is to understand how to engage with others when differences remain. This is undertaken through *extended mirroring*. In extended mirroring, A attends not only to B's solitary desires, but also to B^*'s assessment of B's desires, B^*'s assessment of A's assessment of B's desires, B^{**}'s assessment of B's desires, B^{**}'s assessment of B^*'s assessment of B's desires, and so forth. Through consideration of a variety of perspectives, a conversationalist seeks the best balance among the ethically permissible and factually well-grounded desires present in the population.[18] So we have:

(DM) An ideal conversationalist forms a desire by extended mirroring of the desires of others, achieving that desire he or she

17. Since we are concerned here only with solitary desires, the relevant constraint from (KC) is that the participant would continue to desire what he or she actually does, if the participant knew all the consequences for him- or herself.

18. This is effectively to construct an analog of the "ideal spectator," but one informed by extensive factual knowledge. It articulates further the synthesis of the methodological ideas of Smith and Dewey I propose in "The Hall of Mirrors."

judges to be the best balance among the varying assessments (indefinitely iterated) made by fellow participants.

If there is complete agreement about how the balancing is to be done, there is no need for further conversation. If there is not, the ideal conversation consists in attempts to support or reject various ways of balancing.

(IC) Ideal conversation consists in attempts to show that proposals that participants desire to implement as ways of responding to functional conflict either accord or fail to accord either with ethical functions all participants recognize or with their shared understanding of the need to respond to the wishes of all.

For the moment, focus on cases in which ideal conversation achieves consensus.[19]

Ethical discussions addressing functional conflict can be genuinely helpful if they can show how the actual considerations adduced in opposition to a potential way of solving the conflict fall foul of one or more of the conditions on ideal conversation. That might be achieved by exposing the fact that particular kinds of solitary desires are at odds with some of the functions endorsed in the current state of ethical practice (one identifies these functions, showing how existing precepts discharging those functions would prohibit the action or state of affairs desired); or it might be done by showing that the desire is undermined by facts about the world (people who express a desire for a particular outcome do not recognize that it has consequences they strongly detest; people have the desire because of some background false belief); or it might be done by showing that the desire persists because of a failure to take into account the wishes of some group of people—most obviously by showing how these people are systematically ignored by those who have the desire, but also by questioning the ways in which the balance among the variety of human desires has been struck). An obvious skeptical com-

19. This apparently optimistic thought about possible consensus will be modified in §55.

plaint about revolutionary ethical change would insist that there is nothing for ethical discussion to do when functional conflict arises, that spoken or written words can be only the expression of an attitude others are free to reject. By contrast, there is plenty to be said, much that can be done to expose factual errors and false presuppositions, disharmonies with background features of the prevalent ethical code, and, most important, shortcomings in accommodating the wishes of classes of other people—failures of mutual engagement.

In the next sections, I attempt to show how this approach to ethical method might sometimes manage to achieve consensus (and also consider appropriate conclusions when consensus proves unreachable). First, however, I acknowledge the considerable idealizations introduced and examine how, given our limited perspectives, actual ethical discussions might be assessed.

§54. Ethical Debate

Typically, we are moved to ethical discussion by disagreement. In recurrent circumstances involving different agents, the actors make incompatible choices and defend what they do with judgments that contradict one another. Some of these episodes are occasions of normal ethical change. The prevalent ethical code needs extension to address the circumstances in question, but there is a way to extend it that refines some underlying function(s) without cost to others. Here the task of public ethical discussion is to apply the method of §51 to demonstrate to all who share the code that this is so.

Awareness of functional conflict typically emerges from efforts to adapt that method to a problem not yet recognized as calling for revolutionary change. As discussants find that those attempts do not succeed in resolving the issue, they come to suspect functional conflict.[20] Their first task is to expose, as clearly as they can, the nature of the conflict. Once that is done, the focus will be on proposals for setting priorities among functions,

---

20. Here I suggest a parallel to Kuhn's claims that normal science may break down as repeated efforts to solve puzzles fail, so what appeared as a puzzle becomes seen as an anomaly.

or for revising attributions of priorities currently in force. The stage is then set for ethical debate, where the aim is to simulate an ideal population-wide conversation.

Those who engage in this debate will rarely, if ever, even approximate the cognitive conditions (KC) and (KW). Although they will argue that the proposal for modification of ethical practice they favor responds to the ideal desires of all (or achieves the best balance among those desires), their presentations will be vulnerable to charges that they have misconceived the consequences for particular classes of people, or that they have failed to identify the solitary desires of some groups of people, or that they have misunderstood the ways in which solitary desires would be transformed under attempts at primitive mirroring, or that they have a misguided view of how to balance in extended mirroring. Even more fundamentally, they may be criticized for basing their claims on factually false premises or for overlooking consequences for parts of ethical practice outside the purview of discussion.

Examples of the last two cognitive failures are easy to find. A defense of some modification of ethical practice invoking the commands of an allegedly transcendent being or introducing suspect psychological entities or processes would rightly be rejected on grounds of violating (KE). Similarly, if someone aims to resolve a conflict between different ethical functions but fails to recognize the ways in which the proposed resolution impinges on functions beyond those explicitly considered, it is important to make these implications manifest. Doing so is a continuation of the search for reflective equilibrium, and it is a familiar feature of ethical debates.

The method of §53 emphasizes other ways in which proposals can be challenged, tested, confirmed and undermined. Often a thesis about the consequences for people will be neither supported nor refuted by the available evidence—or the evidence we have supports only a partial version of the thesis, one restricting attention to the consequences for a small subgroup of the human population. Many debates about the proper distribution of resources involve large claims about the effect of particular socioeconomic arrangements. We are told, for example, that if certain kinds of competition are encouraged, *everybody* will achieve various things they are presumed to want, or that attempts to divide the products of labor along roughly egalitarian lines will interfere with motivation to work, and consequently with production. These claims about the effects of general types of social arrangement—types including a wide variety of possible implementations—are sometimes supported by assuming psychological generalizations, sometimes buttressed by mentioning a handful of specific historical cases.[21] The psychology invoked is simplistic, and the sample of instances ludicrously small. A genuine test of rival ethical proposals, with their variant conceptions of the consequences for people who live in quite different situations, requires either further work in relevant social sciences or a direct experiment.

Some hypotheses about the consequences of social arrangements could be confirmed or disconfirmed through indirect investigations. Without attempting to create conditions under which economic incentives, claimed to be crucial for motivation to work, were absent, one might aim for a general theory of human motivation, testing it in laboratory experiments or in existing social situations, and then deriving from it some conclusion about likely behavior under hypothetical egalitarian conditions. When that strategy is considered more thoroughly, however, it seems unlikely to resolve the question: any combination of laboratory and field studies would be inadequate to the full range of possibilities for realizing an egalitarian ideal. If that is so, the only chance of replacing our—ethically crucial—ignorance about consequences is to bring about the conditions egalitarians envisage and see what happens. Rational ethical debate may require further experiments of living.[22]

21. Adam Smith and John Stuart Mill disagree about the first issue (Smith holding that increased rates of production will translate into increased economic benefits for all, Mill maintaining that distribution is not an immediate effect of increased productivity), but they are united on the second. Smith, *Wealth of Nations* (New York: Modern Library, 2000), bk. 1, chap. 8; Mill, *Works*, 2 [*Principles of Political Economy* (Toronto: University of Toronto Press, 1970), bk. 2, chaps. 1–2]. With respect to this latter question, both appeal to a simple generalization about human motivation. Contemporary critics of egalitarianism are fond of mentioning the prominent instances in which they take egalitarian ideas to have been applied, and they conclude from a few failures that all ways of implementing egalitarian ideals are doomed.

22. Mill envisages this possibility, and it is emphasized even more strongly by Dewey.

The experimental design seems relatively obvious. Those who champion some version of the egalitarian ideal can be encouraged to develop the social conditions they envisage and to discover empirically what it is like to live in this way. In their efforts, they will presumably be mindful of the difficulties previous ventures of this kind have experienced, and perhaps heartened by the well-known difficulty of making even the simplest natural scientific experiments work.[23] Experiments involving human subjects are properly subjected to ethical scrutiny—that is probably a stable part of our current ethical practice—and hence there will be constraints on how such experiments are carried out. One important constraint is already hinted at in my suggestion that the experiment be carried out by those who champion the ideals: commitment to it should be voluntary. The question of appropriate constraints will be considered more systematically in the next section, in exploring situations where the search for consensus breaks down.

The method of §53 also emphasizes the need to engage with the desires of people who live in very different ways. Ethical proposals are subject to the conditions (DS) and (DM), but attributions of desire easily go awry because we misrepresent the circumstances under which other people live and the solitary desires they form.[24] We can make headway, even though ideal identification with all human perspectives remains unattainable, through acquiring detailed information about the lives of others. History and ethnography are pertinent sources, often unduly neglected in the course of ethical debate.

To succeed in replicating the conversations of people who mirror one another, ethical discussants need the ability to mirror others who are often quite remote from them. Knowledge of those others' circumstances and solitary desires is a precondition for doing that, but the bare apprehension

23. With respect to egalitarian experiments, there is a tendency to overlook some of the small societies that achieved partial success (Robert Owen's Lanark colony), or that were set at a disadvantage because of relations to other communities (the Diggers and Levellers in Britain, the Israeli kibbutzim).

24. Hume is rightly criticized for the parochialism of his remarks that we can reconstruct the wishes of temporally and spatially distant people by considering our acquaintances— but it is far from obvious that *anyone* (including Hume's critics) really avoids partiality on this issue.

of particular wants will almost certainly be insufficient. Successful mirroring depends on a capacity for thinking and feeling oneself into another person's life, a form of vivid knowledge through which the imagination is exercised.[25] Good ethnographies and skilled histories can guide us toward that—but they are often usefully supplemented, or even replaced, by imaginative fictions that focus particular aspects of human existence. Great works of fiction and drama aid our ethical debates in two quite distinct ways, first by developing capacities for imaginative identification and for vivid knowledge, and second by distilling aspects of ourselves that enable us to proceed from a bare factual knowledge of another's circumstances to a more vivid awareness of how that person's life is and feels. To cite a simple example: it was surely no news to Victorian readers that impoverished slum dwellers wanted food and shelter, but in his depictions of the squalid conditions of urban life, Dickens made possible a far more vivid apprehension of those wants and of their urgency.[26]

Vivid knowledge is also pertinent to the final (and most intricate) way in which ethical proposals can be tested and confirmed. Extended mirroring involves an attempt to balance the varying assessments people make of one another's desires and aspirations, and (IC) allows for the possibility of achieving this balance in distinct ways. In the envisaged ideal conversation, alternatives are compared, considerations are advanced in support of particular ways of proceeding, and those considerations are debated. That debate focuses on the various ways in which a coherent synthesis of divergent points of view might be achieved, attending to abstract features of the envisaged ways of balancing. In the end, however, the relevance of the abstract features lies in the impact on individuals with their different perspectives. If the wishes of some are overridden, one can seek vivid identification with the people in question, posing the question of whether, from that particular point of view, the approved way of balancing (and the abstract features it embodies) can be understood

25. Here I am indebted to discussions with Moira Gatens and to her illuminating work on imaginative knowledge in philosophy and literature.

26. Perhaps the most striking instance is the description of the life of Jo, the crossing sweeper in *Bleak House*. But there are many others.

and accepted. Suppose the proposal supported by a particular way of balancing is at odds with your wishes, even when those wishes have been refined as the method requires. As you consider the desires of others, you cannot find any obvious flaw—there has been no apparent neglect of others' points of view, you can see their eventual decision as one involving something recognizable as balancing. If so, you should be able to acknowledge the outcome as one you can live with. Convincing ethical discussion, when there are competing modes of balancing, must proceed through vivid identification with those whose wishes would be overridden, revealing that the outcome has been thoroughly considered from their perspective and found to be acceptable.

So abstract a description is vulnerable to the charge that differences in ways of carrying out the process of extended mirroring (different ways of realizing [DM]) are ultimately irreconcilable. Disputes that cannot be resolved at earlier stages—through correction of factual mistakes, through inclusion of all points of view, through vivid identification in mirroring—prove irresoluble. That conclusion is premature—but so equally would be the judgment that more can always be said, that considerations about different ways of balancing can be adduced, that further exercises in identification (particularly with those whose desires are overridden or less fully represented) can settle the debate. The method would be complicated to apply to any serious instance of revolutionary debate, and until one has tried to work through the complexities, it is impossible to decide whether there is hope for resolution or not.

Once again, the analogy with scientific cases may help. Skeptics worry that the large changes labeled "scientific revolutions" prove rationally irresoluble. On my own account of how reasoning works in these scientific episodes, the debates proceed by efforts to show how one position is able to overcome problems, while its rival encounters recurrent difficulties in tackling the problems confronting it.[27] Because both parties can point to successes, and because their problem solving is inevitably incomplete, it is always possible, apparently, to contend that "we" have addressed the crucial issues, whereas "they" have not. Such judgments

27. For details, see chap. 7 of my *Advancement of Science* (New York: Oxford University Press, 1993).

about crucial issues, problems that have highest priority, involve the same sort of "balancing" as that found in the schematic account of extended mirroring. In the scientific cases, we are often confident that some proposals for "balancing," some claims about the truly crucial problems, are defensible and others are not. Our contemporary views about the evolution of life do not resolve every issue in the domain, and yet considerations of the "balance" of successes and failures demonstrate how vastly more successful these views are, in comparison with the positions invoked as challenges to them; it is no less reasonable to judge that Darwinism solves a huge number of significant problems that its rivals do not, than it is to prefer a clean, well-maintained car to a rusty clunker whose only virtue is its fetching hood ornament. "Balance" resists precise definition in the scientific case, and that should caution us against demanding more in the ethical sphere.[28]

I have tried to counter the impression that any search for method to deal with revolutionary change is hopeless. We cannot tell how far we can get until we make a serious attempt at a complicated and taxing venture (one requiring considerable pooling of ideas and recognition of alternative perspectives). Yet it is worth considering, in advance, not only the cases in which consensus can be reached, but also those in which it cannot.

§55. Dissent and the Limits of Tolerance

In characterizing ethical truth, §38 made room for the possibility of pluralism. Different ethical traditions, moved by the felt urgency of discharging different ethical functions, may be drawn to values whose complete reconciliation proves impossible. The members of the traditions, aware of the character of the rival practice, feel themselves drawn to both values. They would like to find ways of satisfying better that function (or those functions) to which, as things stand, they assign lesser priority. Their

---

28. The earliest efforts to specify methods of scientific confirmation were schematic and imprecise. More specific formulations emerged from attempts to put those preliminary conceptions to work. Thoughts about method coevolve with discoveries about the natural world. I envisage a similar process in the ethical sphere: attempts to elaborate the method will yield more precise versions of it.

situation may thus be one of functional conflict, and, for both traditions, the conflict may be resoluble in alternative ways. Because there is no ultimate resolution here, because we can envisage a progressive sequence of changes, proceeding indefinitely without convergence, it would be wrong to require a method for revolutionary change always to achieve consensus. Even an ideal conversation should sometimes leave participants with alternatives.

Furthermore, even along the route to full consensus, even with respect to ethical debates allowing eventual resolution, there are likely to be intermediate stages at which agreement cannot be forced. Consequences hard to predict might be recognized through direct ethical experimentation. To carry out such experiments is, in effect, to fragment a previously united community (although the fragmentation is initially conceived as an intermediate step toward resolution of a debated question). As such experiments are envisaged, how they are to be conducted will itself be a matter for ethical reflection, and the considerations involved here are the same as those figuring in appraisal of the relations among rival ethical traditions. For in both instances, questions arise about the extent to which consensus must be sought or difference tolerated. These questions may be focused by asking when a single community debating a question of revolutionary change can properly allow division into two subgroups that adopt alternative answers.

Imagine a discussion, under the conditions of mutual engagement, that has given rise to two rival parties whose differences cannot currently be resolved. Perhaps this comes about because there are facts about the consequences of the alternative proposals espoused by these parties, facts not determined by the available body of knowledge, and because the only way members of the society can envisage settling these facts is the route of direct experimentation. Or perhaps it is the product of different ways of extended mirroring, the result of alternative conceptions of how to balance the desires of group members (desires conforming to the cognitive and affective conditions of §53). Each party regards the other as espousing an incorrect (or an inferior) modification of their shared current ethical practice. They recognize, however, that no compelling ways of showing the mistake (the inferiority) are at hand, nothing can be done to show the factual error (first case), nothing can be said to bring home the inadequacies of the approach to balancing (second case).

If the debate has reached stalemate in this fashion, then each of the parties must acknowledge its rival to be in a state of *conditional mutual engagement*. Given the perspective adopted by the rival party, it has completely attained mutual engagement with respect to its members. Consider the dispute over empirical facts. Members of each party believe matters are not as their rivals believe (or, perhaps, hope) and, in consequence, the attempt at mutual engagement by the rival party has gone awry (it violates [KE]). But they must also acknowledge that, if the rivals were correct in the disputed matter, they *would have* succeeded in mutual engagement within their party. Similarly, with respect to the difference in ways of extended mirroring, each group thinks the other has done the balancing in an inferior way, but it must also acknowledge that if their extended mirroring *were* adequate, they would have achieved mutual engagement within the party.

When debates reach stalemate, rivals see one another as attaining conditional mutual engagement. Why must this be? The answer is simple. If it were not, the debate would not have reached stalemate. Something more could be said: to wit, that *independently* of the disputed feature of the situation (the disagreement about facts, the alternative approaches to extended mirroring), the rival party fails to achieve an internal condition of mutual engagement. So if the method of §53 is applied to the debate, the point at which differences become irresoluble has to be one at which the contending parties have attained, subject to their conception of the point in dispute, an internal equilibrium that is as good a candidate as any for mutual engagement. In light of this, I propose, each should be prepared to tolerate the other.

This may appear to be a strong, even dangerous, suggestion. There are familiar worries about permitting societies to follow their own customs, and to decide how their members shall be treated. We know that actual societies, inspired by allegedly authoritative religious texts, believe they are entitled to take aggressive actions against other groups and to treat some of their members in ways outsiders view as harsh, restrictive, and cruel. We know also that the recipients of this treatment sometimes endorse what is done to them, genuinely desiring to live in the way assigned them. Is the method then committed to permitting local groups to flout what liberals typically hail as universal values?

No. The conditions of mutual engagement are sufficiently strong to limit tolerance of alternatives so as to exclude these problematic cases. As §§38 and 46 argued, there are imprecise prescriptions, such as those in favor of honesty and opposed to initiating violence, likely to be stable in progressive ethical traditions. These core truths cannot be abandoned if the proposed method is adopted—for dishonesty and unprovoked aggression are, in the overwhelming majority of instances, expressions of failure to engage with the desires of others.[29] This means the core truths will have to be shared among the tolerable rivals, and groups rejecting them will be at odds with the conditions of mutual engagement.

Even more evidently, (KE) rules out appeal to religious texts as the foundation for *any* sort of treatment of group members, and *any* set of attitudes to outsiders. If it is proposed that women should be confined in various ways or that men should proselytize aggressively, and the proposal is grounded in the supposed commands of a deity, a fundamental cognitive condition on mutual engagement is violated. If the ideal conversation is to defend proposals of these kinds, it will have to do so in different ways. It is hard to conceive how any policy of proselytizing could be detached from a foundation in a false belief about transcendent entities, and *aggressive* attempts to spread the word would be debarred by the core truth opposing unprovoked violence. The case of female confinement, however, and analogous proposals with respect to other subgroups, can seem more difficult. For, it may be suggested, the individuals who receive the treatment, unpleasant as it may appear to outsiders, really want to live as they do. Suppose their wish is independent of any faulty religious belief. Is the societal practice then to be tolerated?

Typically not. Although the current expression of the desire is sincere—the woman declares that this is the way she wants her life to be, this is her choice—it is usually infected by a false belief, the belief that

29. The treatment of honesty in §46 argued that there are exceptional instances and that the core truth is thus vague. One can envisage rival ethical traditions proposing different more precise versions of it, but abandonment of it would be ruled out by the method suggested here.

her socialization and education have elicited her wants. If she were to see clearly how her options have been narrowed from the beginning, if others who care about her were to understand how there were alternative possibilities for her, the desires held by her and by them would be quite different. The actual desires owe their presence to a pervasive error, the mistaken belief that the existing processes of socialization are non-coercive. Consequently, there is a failure of mutual engagement: in effect, those who have led her to her present state have engaged not with her but with their preconceived image of her. Only in the case of people who, under alternative schemes of socialization, would remain entirely un-concerned by thinking their choices had been narrowly foreclosed from the beginning can one suppose that the desire is uninfected by error. There may be such people, but we have every reason to think that they are rare.

Discussion of such cases exposes an important facet of the limits of tolerance, one best illustrated by considering the permissibility of ethi-cal experiments. The last section envisaged groups, moved by an egali-tarian ideal, trying out their preferred versions of it. If those who sign on to the experiment are coerced, either through the overriding of their present desires or through prior socialization that has narrowed their choices, the experiment will not accord with the conditions of mutual engagement. At times, however, people voluntarily resolve—by our every-day standards of voluntary choice—to undertake this kind of venture. There is a danger they will coerce others, in particular their children. The forms of coercion making the examples of confining social practices sub-ject to ethical critique may be absent in the first generation, but they can easily emerge thereafter. Do all ethical experiments under mutual engage-ment have to be short-term?

To avoid that conclusion the education and socialization of the young must carefully preserve options. The experiment should allow children to choose from as wide a range of alternatives as those available to their peers, whose parents and teachers are not part of the experiment. The experimenters must recognize and tolerate their children's opting out—and, perhaps, others' opting in. As a result, some possibilities for living (forms of family relationships) may be distorted, a distortion to be recog-nized if the probative value of the experiment is to be properly assessed.

In virtue of their status as minority ventures, ethical experiments are always vulnerable to breakdown. Judgments of failure should take into account the stresses to which they are subject.

To sum up: when conversations under the conditions of mutual engagement break down, competing parties should see one another as in an internal state of conditional mutual engagement. Groups achieving this state can be allowed to pursue their favored ethical practices. Although the suggestion initially seems to allow for overindulgence of alternative ways of living, including those oppressing members of some subgroups, the conditions of mutual engagement are sufficiently strong to debar the cases that provoke condemnation. The method provides a basis for ethical criticism of groups engaging in externally aggressive or internally coercive practices.[30]

§56. The Challenger Revisited

Section 51 closed with a challenge. Functional conflict seemed to make room for ways of rejecting some major functions of ethics in favor of others, and thus for radical differences in continuing the ethical project. To give the challenge clear force, we need only envisage someone who proposes to go in a dramatically different direction from the one pursued in this chapter and its predecessor, someone who has little patience with the original function of ethics. Even if he concedes that the ethical project began with the need to remedy altruism failures,[31] he sees no reason to be governed by that beginning. The subsequent course of ethical evolution proliferated possibilities for human life, and, aware of these possibilities, the challenger emphasizes the freedom of the strong individual, who charts his own way quite independently of the protests (or

30. Whether it will also justify outside pressure to amend the practices, or even the use of force, is a further question, one that would have to be settled by looking carefully at the details in individual instances. To justify ethical criticism is not yet to warrant particular ways of translating the criticism into action.

31. That is, he concedes the story as I have narrated it. The challenger, shortly to be identified as a version of Nietzsche, might prefer an alternative account, one viewing the emphasis on cooperation as a late corruption [see the first essay of Friedrich Nietzsche *On the Genealogy of Morality* (Cambridge: Cambridge University Press, 2007)].

the applause) of others. This challenger can be viewed as a version of Nietzsche, and he sees the principal function of ethics to lie in opening the way for the emergence of higher types of beings, who will carry some of the characteristics of the greatest individuals of the human past to new heights.

What can be said to someone who poses this challenge—and poses it seriously? The method developed in previous sections offers an obvious line of response: the rival possibility spectacularly fails to meet the conditions of mutual engagement.[32] Rejection of the original function of ethical practice goes hand in hand with a repudiation of the method, so the circularity noted in §52 appears to doom the response. Although one could bring the challenger into line, if he agreed to argue according to the conditions proposed, he will claim the conditions take for granted just the emphasis on the original function of ethics he is concerned to reject. In this point he will be entirely correct.

My earlier attempts to confront troublesome characters in §42 generated an understanding of the criteria for success. Turning back challenges requires less than absolute silencing. What must be done is to settle genuine doubts and worries *at least as well as alternative ethical pictures are able to do.* In the present case the comparative approach yields little solace. Some alternative approaches to ethics, Kantianism and other forms of rationalism, seem to have weapons with which to fight the radical Nietzschean challenge. Proponents of such approaches will claim to have a priori methods of establishing fundamental ethical principles, and thus a priori knowledge of the falsity of the challenger's claims. Pragmatic naturalism denies both the supposed a priori methods (or at least the a priori status of the methods) and the "fundamental ethical principles" known on their basis. Hence it may be seen as preparing the way for the challenge—precisely because of the emphasis on the historical evolution of ethical practice, on the emergence of different ethical functions, and on the lack of any a priori insight into ethical truth, it appears to have jettisoned the weapons needed to fight back.

32. The challenger may even flaunt this rejection, comparing the masses whose desires are ignored or crushed to lambs that serve as prey for stronger, "nobler" beasts. See the famous image from *Genealogy of Morality*, essay 1, sec. 13.

Advertising the resources of rationalism has more show than substance. True, rationalists can *produce some sentences* after the challenger
has had his say, but there is no great force in invoking mysterious processes announced as yielding knowledge of an especially robust type.
Swords can be used to threaten people, but the shadow of a sword should
frighten no one. Although rationalists may pride themselves on their
ability to keep talking in a situation where naturalists stand speechless,
the difference is not particularly significant if the phrases they utter are
empty. Nevertheless, I do not want to leave matters here: as I shall now
try to show, pragmatic naturalism has considerably more to say.

Two points from earlier discussions need elaboration. First, even
though the use of a conception of the good to frame an ethical method
threatens to be circular, as we discovered at the end of §52, the method
is not guaranteed to generate the preferred conception of the good. If it
could be shown to do so, the package of proposals would turn out to
be coherent in a way they might not have been. Second, the original
reply to the troublesome characters of §42 demanded the challenger
provide some alternative. The current challenge has begun to do that,
in proposing to dismiss the initial function of ethics in favor of the subsequent function of individual development, but that is only the beginning. This idea needs to be articulated as a full account of the good,
without presupposing ethical functions supposedly dismissed or down-
graded. It must be supplemented with ways (if not a method) of further
ethical development, and the conception of the good must cohere with
the preferred modes of ethical evolution. If pragmatic naturalism can
be shown to be coherent, and if the rival proves to be difficult to articulate, tacitly dependent on ethical ideas it officially rejects, and not easily
combined with any strategy for ethical change, enough will have been
said.

Start with the coherence of pragmatic naturalism. Under conditions
of mutual engagement, deliberators attempt to respond to one another's
desires. Which desires strike them as particularly crucial? They recognize
that we all have common basic needs, and that, as things stand, these are
satisfied for some people but by no means for all. They know, too, that
the more fortunate people have a richer set of aspirations, that from their
own perspectives, some desires are much more crucial than others, and

that the relative centrality of these desires is the result of a conception of what matters in their lives. In light of our shared humanity, they suppose people who struggle to meet their basic needs would also develop richer aspirations of a similar type if they were provided with the chance to do so. From this, they conclude that the appropriate focus of mutual engagement would be the satisfaction of people's central desires, and thus the promotion of worthwhile lives.

This line of reasoning assumes that we try to reach mutual engagement with one another by attending to what each person takes to be most important, and we know enough about human possibilities to move beyond the atomism of hedonism, with its decomposition of lives into sequences of momentary states of pleasure and pain, and to appreciate the *holism* of the desires people form, once they are freed from painful struggles. People who currently are absorbed by the need to survive can appreciate the value of further aspirations, in conditions where their basic requirements are met. Moreover, people who come to see their options as narrowed through social coercion would prefer to form their own assessment of what matters in their lives. If these hypotheses are correct, the deliberators will want primarily (i) the option of forming for themselves a conception of what matters, and of doing so in an uncoerced way; (ii) the satisfaction of the material preconditions of this option; and (iii) the satisfaction of the desires endorsed as central. Mutual engagement takes all three types of desires to matter for all. Does it also require conceiving the good in terms of providing i and ii for all, and offering roughly equal chances with respect to iii?

Apparently. On what basis could the deliberators introduce inequalities here? One possibility would be to maintain that serious chances of satisfying desires of type iii for some people require the failure to satisfy desires of types i and ii for others: one simply cannot give all people serious chances for a worthwhile life, and it is better that some should have such chances, even if it means others live in want. But the conception of the good advanced in Chapter 8 blocked this proposal by taking the question of population size as fundamental (§49). The human population is to be maintained at or below the proper bound, and this means resources are available to satisfy i and ii for all, and to give everybody serious chances with respect to iii. If the deliberators are to arrive at an alternative, they will have to suppose it is better not to impose this condition on population size. That will commit them to holding as good a world in which some people (possibly most people) are doomed not to have the most important desires satisfied. How could this be anything other than a failure of mutual engagement?[33]

Another possibility would be to suppose that a conception of the good can tolerate significant differences in the probabilities of people's satisfying type iii desires. The need for universal satisfaction of i and ii is conceded, but the opportunities for some can be significantly richer than those available to others. Now it is agreed that resources are sufficient to provide serious chances of a worthwhile life for all. Hence there is a distribution of those resources allowing everybody roughly equal, serious chances of satisfying desires of type iii, where the standard of rough equality is set by the recognition of the perturbations uncontrollable factors can introduce (§50). Lowering the chances for some, while raising them for others, would apparently be a failure of mutual engagement.

Perhaps the proponent of inequalities could appeal to an alternative strategy for extended mirroring. From the standpoint of the egalitarian distribution, the inegalitarian proposal divides the population into two classes: the *Augmented*, whose chances are increased, and the *Diminished*, whose chances are lessened. Introducing differences plainly gives less weight to the wishes of the Diminished. Yet it might be viewed as responding to the desires of the Augmented and as supposing that the balancing of conflicting desires is conducted by giving weight to the reaction of the Augmented to the desires of the Diminished. That will happen only if the Augmented judge that the wishes of the Diminished are not to figure in forming their own desires. That could happen if they simply ignored the wishes of the Diminished—a failure of mutual engagement. Or it could happen if they used a strategy for extended mirroring that relied on the judgment of others. But these others would also

---

33. One way in which my conclusion here might be resisted would be to argue that setting bounds on the population is incompatible with major components of worthwhile lives. This line of reasoning was already considered in §49, in confronting a form of skepticism, and I rely on the response developed there.

have to ascribe low weight to the wishes of the Diminished, and exactly the same problem would arise. Thus, the only way for extended mirroring to generate the inegalitarian desires is for a failure of primitive mirroring to occur somewhere among the class of the deliberators—and that means that the discussion cannot satisfy the conditions of mutual engagement.

Application of the method of §53 is likely to yield pragmatic naturalism's conception of the good, as the provision for all of serious, and roughly equal, chances for worthwhile lives. Further, the conception of human relationships as central elements in worthwhile human lives can be viewed as in harmony with the method's insistence on the conditions of mutual engagement. Yet the conclusion should not be overstated. Any rehearsal of an ideal conversation must appeal to psychological assumptions about human reactions, and these can be falsified by actual reactions to the line of reasoning people are supposed to adopt. In advance of those reactions, a tentative conclusion: pragmatic naturalism's proposal about the good coheres with its view of ethical method.

Turn now to the challenger's position. The challenger's conception of the good sees the development of the few as crucial and the lives of the many as unimportant. There are special people—call them "free spirits"— supposed to be capable of a form of human existence superior to any attainable by the rest. A world is good to the extent that it allows for the full development of the free spirits; what happens to the others is irrelevant.

This conception is evidently incompatible with the conception offered in Chapter 8, and the incompatibility descends from a difference about the functions the ethical project is to discharge. The egalitarian conception is embedded in an attempt to take the original function as primary, to see ethics as continuing to be directed at remedying altruism failures (and to view ethical method as an attempt at mutual engagement). Its rival views the original function of ethics as a trigger that eventually led to enhanced possibilities of human living. Whereas the egalitarian conception attempts to integrate the emergent possibilities with the original function (through the focus on equal opportunities for worthwhile lives), the challenger takes the emergence of the richer possibilities to be a turning point, after which the original function can be forgotten, in favor of ever further enhancement of these possibilities, even if the resultant opportunities are available only to the few.

So far, the rival approach has no method for ethical discussion—perhaps because there is to be no further ethical discussion. It is not obvious, however, how we arrive at a point beyond the need for further consideration of what to do. Note first that the enhanced human possibilities realized by free spirits had better be considerably richer than those available to the human population generally, for, if they were not, we would all be approximations to the free spirits, and the challenger's position would collapse into the egalitarian view it is supposed to replace. Now the human possibilities open to ordinary people under favorable circumstances—the kinds of things people take to be important elements of worthwhile lives—are typically dependent on extensive social coordination. Material resources must be provided and social institutions must be in place for us to enjoy those possibilities. Even without any definite understanding of the splendid ways in which free spirits are able to live, there is no basis for thinking their predicament is different. They will depend on a society and rely on coordinated efforts. Questions will arise about how the requisite basis is to be achieved, how the relations between free spirits and the herd are to be governed, how the potentially conflicting projects of free spirits are to be reconciled. Apparently, then, ethics cannot come to an end, and there will have to be some way of continuing ethical discussion (even if it is supposed that the discussion is one in which only free spirits take part).

We do not know what the possibilities are to which the free spirits (and only they) can aspire—hence, the conception of the good is woefully indefinite. We also have no ethical method (despite the need for ethics to continue), and hence no possibility of showing that the method and the conception of the good are mutually coherent. We cannot even decide whether the enhanced possibilities for the free spirits presuppose a society, whose coordinated efforts already rest on ways of discharging the original function of ethics (the function allegedly transcended). These are serious defects, but, because of elementary facts about human biology and psychology, further trouble arises.

Any attempt to promote the splendid flourishing of the free spirits would first have to identify them. Perhaps the challenger hopes they will identify themselves, soaring above the mediocrities around them. Yet a banal truth about us is that our potential is not marked on our foreheads at birth. Human beings develop, and, as they do, their talents and capabilities sometimes become clear. Not always, however, since it is a sad and familiar fact that people are often limited and even crippled by inadequate social environments. No free spirit will be able to soar unless he has previously had a sufficiently supportive environment. One way to allow for that (hypothetical) self-identification would be to provide ample opportunities for all—but that would absorb the central features of the conception of the good currently subjected to challenge. *Promoting the development of free spirits would then presuppose the provision of equal opportunities for worthwhile lives for all.* What other possibilities are there? Could we use the identified free spirits in one generation to pick out the likely free spirits of the next? Even with an understanding of the phenomena of heredity far less articulated than that provided by contemporary genetics, it is easy to see that this is hopeless.[34] Without a background provision of equal opportunities, the supposedly incompatible approach cannot even get off the ground.

Section 42 responded to radical challenges by arguing that they deserted the ethical project without offering any clear alternative to it (except for the fragile hominid social state from which that project liberated us). Section 51 concluded with a way of reviving those challenges, in the exploitation of functional conflict: it appeared to be possible to develop rival ways of continuing the ethical project, ways highlighting or subordinating different functions. One version of the challenge has been considered, and we have discovered (tentatively) that proposals grounded in the original function of ethics cohere with one another, whereas elaboration of the alleged alternative encounters severe difficulties. Not only is it unclear what method of ethical discussion it can offer, or how method and conception of the good can be coherently developed, but there is

34. As Plato already did. The thought of concentrating on biological heredity is central to his way of providing different courses of education for differently talented people [*Republic* (trans Grube) (Indianapolis: Hackett, 1992)]. Yet Plato understands that this will at best be an approximation: golden children can emerge from leaden parents, and golden parents can produce baser offspring.

significant danger that the alternative will presuppose the function it is supposed to transcend.

Consideration of a single radical challenge can easily inspire the objection that the alternative has been carefully chosen: pragmatic naturalism succeeds only because it is juxtaposed with an implausible rival position. On the contrary: this version of the challenge is important *precisely because it demotes the original function of ethics* and thus undercuts the idea of ethical decision through mutual engagement. Less sweeping rival approaches would allow the importance of the original function. Once remedying altruism failures and mutual engagement are taken seriously, the range of alternatives for continuing the ethical project is greatly narrowed. We are left with those functional conflicts that give rise to tolerable pluralism.

Both the conception of the good presented in Chapter 8 and the ethical methods outlined here seek ways of replicating, within the contemporary context, characteristics of the earliest stages of the ethical project. They can be viewed as endeavoring to undo the distortions introduced in later phases. Moreover, the conditions imposed on mutual engagement deliberately reflect the account of psychological altruism (§§3–5, 21). Two motivational corollaries follow. First, the renewal of the ethical project should proceed by demanding of deliberators that they exhibit, in the most complete possible form, the characteristics of psychological altruists, for that is to take the original project, the remedying of altruism failures, as primary. Second, to the extent one is concerned that the concept of an altruism failure is sufficiently open or ambiguous to allow different elaborations by participants in ethical discussion, it is appropriate to demand of those discussants wideranging forms of psychological altruism. The judgments about progressiveness should be vindicated by submitting them to discussion under conditions of mutual engagement. Those conditions can be invoked to fill out the account of progress of Part II: to the extent that there are questions about whether a proposed transition remedies altruism failures, questions turning on what is to count as an altruism failure, those questions are resolved by identifying the altruism failures as those recognized by discussants who conform to the conditions of mutual engagement.

Section 44 denied the existence of ethical expertise. The considerations offered in the past two chapters are hardly knockdown arguments. They are invitations to consider our ethical choices in a particular way—as a large-scale emulation of those who began the ethical project—and they are to be tested by reactions, based on a variety of human perspectives. That is entirely in the spirit of the method proposed, a method emphasizing conversation, simulated or perhaps real.

CHAPTER 10

Renewing the Project

§57. Philosophical Midwifery

How does the normative stance developed in the last two chapters, the
egalitarian conception of the good, and the method for ethical decision
that aims to simulate wide-ranging deliberation under conditions of
mutual engagement, apply to our contemporary predicament? To recapitulate: it is not for any single author to answer. Philosophers can make
proposals, attempting to facilitate the conversation that would deliver
answers. Call the work of facilitation *philosophical midwifery*. The most
obvious forms of philosophical midwifery consist in proposing topics
for consideration (places on our common vessel where planks might deserve attention) and suggestions about those topics (specific ways of
rearranging the timber in those places). Almost all of this chapter will be
devoted to efforts in this vein. First, however, a more basic type of midwifery is needed if the ethical project is to be reborn.

The ancients wondered if virtue could be taught. Pragmatic naturalism takes the catalog of virtues to be something generated from ideal
conversations. A prior issue, then, is how to produce good simulations
of those discussions. To renew the ethical project in the ways suggested,
deliberators capable of approximating the conditions of mutual engagement are needed. Where will they come from?

When the ethical community becomes the entire species, all the involved parties can no longer sit down together and discuss potential modifications to a shared code. Socialization of the young can no longer stop with transmitting the code as it has been passed on, leaving its further evolution to conversations in which they will engage with one another. At least some people in each generation, perhaps the entire cohort, need to understand what the ideal of mutual engagement is and how they might exemplify it. The proposals of the last two chapters call for further proposals about ethical education.

The cognitive conditions on mutual engagement require recognition of the predicaments of others. In many affluent societies, the general idea of learning about people who live under very different circumstances receives official approval, but, for serious ethical discussion, it deserves more systematic emphasis. Young people need to acquire *vivid* knowledge of a range of lives lived by their fellows. They should be conscious also that their power of vivid representation reveals only a sample of the actual forms of contemporary existence. When they reflect on proposals for action, they should be aware of how those proposals would be felt by people whose situations are very different from their own. If they eventually conclude that nothing can be done to improve the predicaments of the less fortunate, they should do so with an intimate understanding of how the neglect would be experienced. The aspirations of others should be present in their own psychological lives.

Education should also foster an ability to satisfy the affective conditions of mutual engagement. Besides a general encouragement of empathy, potential deliberators need an ability to shape altruistic responses along the lines of the inherited ethical code, to recognize endorsable and contaminated desires, insofar as these have been discriminated. They should have practice in mirroring, in contemplating an outcome from various angles. They should become aware of the value of higher-order altruism. All these skills can be developed by simulating the early phases of the ethical project. Children can learn, from early in their lives, how to work out a life with others, to find their way to rules all can accept—and they can experience the joys of fluent cooperative interactions according to those rules, as they attune their own conduct to that of their peers.

Although some issues, most notably the custody of our planet, arise at a specieswide level, others can be negotiated by smaller groups, and these provide practice in reinforcing capacities for ethical conversations at any scale. Sensitivity with respect to the largest questions, where discussion involves a small number of (with luck) representative voices, and in which eventual action depends on voting, can be cultivated if those who frame the options and publicly debate them, as well as those who respond to the conversation, have practice in working on smaller topics, where the coordination of social life can be carried on as it was by the ethical pioneers. Ideally, as people grow, they should gain experience of ethical deliberation at many different scales.

To the extent that this already occurs in contemporary societies, it often does so within religious communities. Joint action to address the problems felt locally emerges from conversations in the synagogue, the mosque, or the church. Previous chapters have characterized the embedding of ethics within religion as a distortion of the ethical project, and to insist on that description seems doubly at odds with the educational proposals of this section. For not only does secularism undermine the most prominent institutions in which local ethical deliberation is found, but the emphasis on mutual engagement appears at odds with dismissal of aspirations and ideals figuring centrally in most human lives.

It is possible to admire the fact that ethical discussion occurs in local religious communities, and even to approve its responsiveness to human problems, while maintaining that the conversation should be freed from the potentially damaging reliance on taking particular texts, interpreted by leaders credited with special insights, as overridingly authoritative. Secularism should aim to replicate the dedicated attempts to work through problems together, without succumbing to the distortions induced by supposing particular commands to represent the divine will. Surrogate institutions are needed to do the work once done in the "cool hour" around the campfire.[1] Nor is the empha-

---

1. The secular humanism I defend cannot be a purely negative position, one that stops with the declaration that religion is bunk. For elaboration of the point, see the closing pages of my *Living with Darwin* (New York: Oxford University Press, 2007), as well as some further essays ["Beyond Disbelief" in Russell Blackford and Udo Schüklenk (eds) *Fifty Voices of Disbelief* (Oxford: Blackwell, 2009), "Challenges for Secularism" in George Levine (ed) *The Joy of Secularism* (Princeton, NJ: Princeton University Press, 2011), and "Militant Modern Atheism" *Journal of Applied Philosophy*, 28, 2011].

sis on secular discussion a failure of mutual engagement. Insisting that appeal to religious ideas not be part of ethical conversation, embodied in the requirement (KE) of §53, is based on three important ideas. First, false belief can misdirect the discussion; second, proselytizing religions systematically fail to respond to the aspirations and desires of large classes of people (those currently outside the faith, as well as insiders whose propensities diverge from those taken to suit the divine taste); third, the reasons advanced in deliberation should be shared among the ideal discussants, and for this a focus on the filtered desires of others is appropriate. Yet it is not enough to leave matters there. Religion *is* central to the lives of many people, and, for them, to remove it from the ethical forum can be felt as disenfranchisement. A mutually engaged secularism should take seriously the psychological and social needs religion, and religious community, satisfies, recognizing and responding to the desires and aspirations out of which religious commitment grows. Especially for the world's poor, for whom basic material needs are not met and for whom it is difficult to think in terms of a freely chosen structure for their lives, religion can provide both consolation and framework. The conception of the good, proposed in Chapter 8, is intended to recognize and respond to the predicaments of the poor, including the desires that have made the world's religions so attractive. This egalitarian conception is a deeper embodiment of mutual engagement.

The proposals sketched here recapitulate themes often found in educational theorizing during the past century. Seen in the context of the ethical project, their systematic importance may be appreciated. Ethical discussion, in undistorted forms, is urgently required of us, and the discussants have to be prepared. In a world dominated by crude consequentialist visions of the good (A and B of §48), it is easy to marginalize educational programs fostering mutual engagement: so much firm technical knowledge needs to be impressed on the future worker! Pragmatic naturalism suggests a different vision, one in which learning has its place as a component of the worthwhile life.

§58. Scarce Resources

The original problems of sharing basic resources, providing food, security, and shelter for all, do not arise for many people, so long as they focus attention on those with whom they have most contact. Around the world, however, these necessities are often lacking, and there are pockets of poverty even in the most affluent societies. Pragmatic naturalism's conception of the good yields an imperative to correct this situation, as immediately as possible. Making food and clean water available to all is almost certainly not beyond contemporary human ability. Problems of shelter are harder to solve, especially in a period when many areas of the world are increasingly vulnerable to natural catastrophes. Most difficult of all is the provision of security for many of the world's poor, especially women and children. The constant threat of violence arises from many causes: lack of reliable supplies of basic necessities, historical tensions among groups, and religious differences. Only when it becomes clear that shelter is available for all, that escalating vendettas are avoidable (and better avoided), and that satisfying lives are possible on a secular basis, are the risks likely to decline to tolerable levels. For that to occur, significant educational resources will have to be provided, enabling rival groups to appreciate the tortured history of their antagonisms and sapping the power of militant religions (§60). Even if the will to tackle the problem of security were present, it would have to be sustained over a long term.

To recognize these issues of global inequality and the need to remedy them is hardly novel. For pragmatic naturalism, however, they are an absolutely necessary first step toward realizing the conception of the good. Until food, water, shelter, and security are provided, talk of "opportunities for a worthwhile life," supposedly available to all, will be idle. Should we then direct all our energy toward satisfying the basic needs of the world's population? If so, would the affluent world have to make severe sacrifices, possibly even undermining our struggle to advance the good?

There is no reason to think an energetic and resolute effort to provide basic resources for the poor would bankrupt the affluent world. The issue is, however, too many-sided to be resolved here. Instead, I want to identify an important constraint on attention to global inequality. The fundamental task is to bring about a state of affairs in which the necessities of life can be *stably maintained* for all. To reach that goal, some institutions and practices in the affluent world must be maintained. Specifically, it will be crucial to support schools, colleges, universities, and research centers, because of their role both in providing sustainable techniques for agriculture, hydrology, and engineering, and in serving as the basis for realizing further aspects of the conception of the good (specifically, enabling young people to grow into sensitive ethical deliberators and to learn to make free choices about worthwhile life projects for themselves). The conception of the good may well demand that affluent people sacrifice certain luxuries, for doing so could ease the burdens of the poor: the reorientation of biomedical research toward tackling infectious diseases, in the environments in which they kill and disable millions, rather than providing diet pills and cosmetic treatments, is one specific example.[2] By contrast, educational systems are likely to serve as sources of new technologies for solving the problems of providing basic necessities for all, and they provide models for the extension of education to give all young people opportunities to develop their life projects.

Is this to desert pragmatic naturalism's conception of the good? Committed egalitarians may suppose too much has been conceded to the notorious trickle-down defenses of capitalism: let competition thrive and entrepreneurs enjoy enormous revenues, for out of this will come benefits for those who are currently poor. My claim seems to run in parallel: let funds be invested in high-quality schools and universities in the affluent world, for out of these may come new ideas and technologies to help transform the conditions of the world's indigent people. The defenses are parallel but by no means identical, for there are two different classes of empirical facts, one about money flows under unfettered capitalism, and one about the genesis of helpful ideas and techniques under high-quality education. We have plenty of evidence that the first class of facts does not favor the trickle-down defense: the condition of the poorest citizens is better in those countries in which inequalities are curbed. By

2. See James Flory and Philip Kitcher, "Global Health and the Scientific Research Agenda," *Philosophy and Public Affairs* 32 (2004): 36–65.

contrast, even though the transfer of technologies to the impoverished areas of the world is imperfect—largely because of the intrusions of unregulated capitalism—it does offer actual successes, and the potential for many more, if the orientation of research toward the needs of all were taken more seriously (for example, by penalizing research and development that caters to trivial luxuries at the cost of delivering necessities).[3]

Ultimately, pragmatic naturalism proposes that further resources, including quality education and medical care, be distributed to all. Focusing on these cases should remind us that serious problems about scarce resources remain, felt even in the affluent world. Only the children of lucky parents receive the education they deserve. Here, then, we find ourselves in a situation analogous to that of the first ethicists. Three different desiderata need to be weighed in a decision about sharing scarce educational resources: first, there is the question of what to do, here and now, with these children, for whom variable educational opportunities are available; second, there is the question of improving the local supply, that is, of trying to ensure that the educational prospects for future groups of children, in this segment of society, are better than those currently existing; third comes the issue of realizing the long-term conception of the good, a world in which all people are provided with the basic necessities and all children have the educational opportunities they deserve.

The first question requires a plan, a scheme for dividing up the educational goods, which all those affected would acquiesce in as the best available, provided they were to deliberate with one another under the conditions of mutual engagement. Yet to concentrate on it alone would be inadequate, for the consensus view obtained under that deliberation might slight the wants and aspirations of another affected group, those who will face the same problem in coming years. They might protest that their own choices were constrained because of the inaction of predecessors who *merely* focused on a scheme for sharing and did not think

3. To cite just one example, techniques for genetically manipulating crops might dramatically increase the sustainable food supply in the parts of the world most subject to famine, if agricultural companies were less concerned to secure an indefinite return of large profits by modifying plants so poor farmers could not harvest their own seeds.

about increasing supply—and suggest that increasing the supply of educational opportunities for later generations should factor into the determination of a policy for sharing educational opportunities now. By the same token, deliberators who embodied the viewpoints of the world's poor, included in a broader discussion under conditions of mutual engagement, might criticize the results of narrower conversations on grounds that they achieved local improvements without doing anything to advance the universalist egalitarian conception of the good. The challenge is to discover integrated solutions acceptable to the broad group of ideal deliberators as the best available—a package of proposals deliberators under conditions of mutual engagement would endorse.

Consider briefly various factors that might play a role in devising a scheme for apportioning children to schools, or young people to universities. One possibility would be to institute a lottery. Another would allow access to those who can pay—in the extreme form, permitting an auction in which parents bid for positions. Yet another would propose that all children be evaluated by administering tests and the distribution be chosen according to what maximizes their potential for development. Still another would look to the effects of distributions on the future of various subcommunities, on considering, for example, whether people of particular backgrounds can emerge to play respected social roles (or perhaps emerge as future teachers who will raise the overall level of education). No deliberation that fails to consider these potential factors (and several others) can be considered ideal, but each of them brings in its train a large set of empirical questions. We still lack firm knowledge about how providing educational opportunities for underprivileged groups enhances motivation for members of those groups—and what other changes in the social environment might be needed to magnify the effects existing under the status quo.[4] In fact, much current discussion

---

4. An obvious biological analogy is frequently unappreciated. When opponents of schemes offering educational opportunities to the underprivileged complain that the motivational effects on young people in these groups are small, they fail to recognize that providing such opportunities may be analogous to watering a plant—without water the plant will surely die, but even with water, it will not thrive unless other nutrients are available in the soil. Debates about educational policies are entangled with other issues about the handicaps and obstacles faced by the poorer members of competitive capitalist societies.

of the proper modes of sharing scarce resources, like access to the best schools and universities, proceeds by guesswork.

That said, the ways of considering the issue adopted by educational institutions and by policy advisers are not entirely bad. Instead of trying to draw conclusions from abstract philosophical principles, they typically proceed by involving a number of people who represent different points of view. Procedures of this sort can be regarded as an approximation to pragmatic naturalism's method—even though they fall far short of the ideal conditions of mutual engagement. There are obvious ways of improving them: the population of those represented in the conversation could be far more varied and far less skewed toward the affluent than is currently the case; the deliberators could make far greater efforts toward understanding rival perspectives; and they could benefit from experiments to resolve questions currently answered by guessing.

Issues about scarce medical resources seem simpler. A society committed to offering health care for all its citizens must consider the ways in which a budget of medical resources is to be divided up. Focusing on giving equal opportunities for a worthwhile life inspires the proposal that preventative procedures promoting health in childhood and maturity are more important than late rescue efforts enabling old people to hang on for a few weeks or months at greatly reduced capacity. Accepting that proposal would reshape the character of medical expenditure in the United States, in which, at present, half of the money spent on a person's health is expended in the last years of life.[5] One might even go further, proposing that, for some elderly patients, the principal task of medical care is to ease the process of dying. A good death is, for many people, a component of a good life.

Public appreciation of that fact would mark a salutary shift in our ethical life. In many hospitals, many doctors quietly appreciate that preserving life is not a fundamental goal. A more basic aim is to help lives realize the projects to which those who live them are committed. Many active, passionate, and thoughtful people reflectively view an envisaged future state of lingering in a state of dementia, or subject to a condition in which

5. See, for example, Berhanu Alemayehu and Kenneth Warner, "The Lifetime Distribution of Healthcare Costs," *Health Services Research* 39 (2004): 627–42.

a failing body is patched together with a tangle of tubes, as a travesty of who they are. For people who conceive their lives in this way, the solace of medicine is not that physicians can maintain an absurd life but that they can offer the gift of a death suited to the life, sometimes by letting the patient die, sometimes by accelerating the process. Moreover, those who view medicine as protection against a bad death will often also regard themselves as remedying an altruism failure: instead of receiving expensive forms of life support, they wish to give to others who might benefit more from the supply of medical resources. Proposals of this sort must be evaluated by people with different points of view, fully informed about the possibilities and committed to mutual engagement.

Consider a medical case where scarcity of resources is likely to endure. However successful a society may be at recruiting people for organ donation, it is hard to have organs with the right properties available at the right times in the right places to be transferred to those whose own organs are failing. Hence, priorities have to be set. Doctors have to decide which patients will be chosen to receive the hearts, livers, and kidneys that can be transplanted. (Their predicament is, in fact, the original "doctor's dilemma."[6]) Many of the criteria to which they appeal look to the future: they assess the probability that the transplant will be successful, evaluate the future life prospects of the patients, consider how the patients' lives impinge upon others, and so forth. Yet one historical consideration comes into play. It is often considered relevant to distinguish cases in which the failure of the organ has been caused by something beyond the patient's control (a genetic defect, or a trauma early in life) from cases in which the patient's own behavior has generated the problem (the liver is failing because of a career of excessive drinking).

Pragmatic naturalism allows us to understand and refine the forward-looking criteria. If our actions should promote equal opportunities for the good life, it is appropriate to favor those whose life projects are more gravely threatened because bad luck has struck early, rather than those

---

6. See George Bernard Shaw, *The Doctor's Dilemma* (New York: Penguin, 1956), in which a doctor has a newly discovered cure in a quantity sufficient to save one patient—and there are two potential recipients.

who have already been able to realize many of their aspirations before the current misfortune. If the lives of many others are bound up with the patient's continued existence (as with parents of young children, or people whose work contributes to the welfare of many others), it is appropriate to give the patient more weight. The ideal doctor—or the ideal committee of medical planners—would thus attend to the various ways in which patients' failures to survive would affect the future quality of lives, both their own and those affected by their presence in the world. Ideally, there would be a process of mirroring, in which the possible futures for an inclusive group of people were surveyed, out of which a decision could come.

So far, however, no attention has been paid to the past. If the alcoholic scores more highly on the future-oriented measures than his or her temperate counterpart, he or she will qualify for the single available liver.[7] Can the recommended form of consequentialism be sensitive to the thought that the carelessness of a patient's earlier life is pertinent, that people who neglect their health, or even actively cause their own medical problems, are less worthy as recipients of corrective treatment? It can. A commitment to equal sharing of scarce resources must consider whether, in the initial distribution, everyone received the same. In the original context of food sharing, for example, if the portion given to one member of the group were to spoil quickly, our first ethical ancestors would likely try to make up the lack, perhaps by having the others contribute from their own allotted shares, viewing this as producing the equal distribution they had aimed at, but failed to achieve, in the first place. Similarly, a person whose liver has failed because of a genetic defect was not granted an equal portion of the resource *healthy liver*, and a commitment to equality (equal opportunity for a worthwhile life) has to account of this fact.

An ideal distribution of the scarce medical resources would attend to the variety of forward- and backward-looking factors, balancing them in

---

7. In practice, of course, the alcoholic's past excesses may diminish his or her future prospects, and thus reduce his or her score on the forward-looking measures. But to exclude the alcoholic in this way would respond to the *future consequences* of his or her behavior, not to the behavior itself. Many people would maintain that, even if this yields an acceptable result, it does so in the wrong way.

ways representatives of different perspectives, committed to terms of mutual engagement, would endorse. In contemporary contexts of medical decision, however, an attempt to work out how the idealized discussion would go is only rarely appropriate. Judgments frequently have to be made quickly, and those affected by them need to be assured that their own cases are not prey to capricious or partial decisions. Hence, knowing there will rarely be time to investigate all the details of individual cases, to understand the complicated ways the continued existence of one person affects other lives, well-informed individuals committed to mutual engagement are likely to prefer rules and guidelines elaborated in advance. These guidelines are to be applied in the urgent cases, subject to the proviso that more refined judgments—judgments closer to the ideal survey of particular details and consequences—can be introduced when time permits.

§59. Habits and Their Limits

Turn next to mundane contexts of ethical judgment, as they arise for people whose lives are comfortable. Although most of what these people do is a matter of routine, the exercise of habits that go unquestioned, each day brings occasions for judgment. There are tasks to be carried out, and agents need to organize their activity so as to achieve those things others rely on them to produce. Despite the fact that many ways of proceeding will typically suffice, some minimal planning is usually required. Besides the discharge of their duties, however, there are also occasions for doing more, for helping those around them or for giving unanticipated pleasures. People whom others regard as particularly thoughtful are more alert to these opportunities and take them into account in planning their daily actions—they make time to telephone a distant friend who has had some setbacks, to prepare a surprise for the children, to ask the elderly neighbor whether he or she needs a ride to the doctor. Those not so sensitive to the possibilities of helping often chide themselves for allowing their lives to be so constantly guided by habit—and make resolutions to do better.

This is the humdrum stuff of ethical life for many well-to-do people. The social environment consists of an interwoven tapestry of routines, which can be followed to promote what the people involved identify as the well-being of all. Institutions (such as marriage and private property), roles (wife, teacher, citizen), and conceptions of good individual lives (economic comfort, family harmony) are taken for granted. Given these institutions, roles, and assumptions about what promotes the good life, the routines work smoothly together. If each of the people involved carries out the duties associated with his or her role, the individual components of welfare will be advanced and good lives for all in the local circle will ensue. If people are sensitive to the lives of others, taking advantage of opportunities to provide extra aid and comfort, there will be additional benefits. Some roles—family member or friend, say—may even require a certain level of sensitivity, expressed in a willingness to deviate from routine, even though *more* than the minimum would often be welcome.

The anatomy of everyday ethical judgment in contexts like these is easy to discern. The agents draw on the ethical resources passed down to them, general injunctions to do certain things and avoid others, prescriptions associated with roles and institutions, often illustrated by stories of exemplary figures, vague ideas about what makes for human welfare. In everyday planning, these ethical resources mark out some goals and rule out particular means—food must be bought so the family can eat, but the money cannot simply be taken from someone else's pocket. Even within the framework supplied by the accepted ethical perspective, however, there is an enormous range of possibilities. The agents know that particular things have to be done, and they can recognize, more or less clearly, some possibilities for going beyond their assigned duties. How should they budget their time, dividing it among various domains of action? Because the options are so numerous, a thorough exploration of what plan of action would be best would preclude any activity whatsoever. Some thought is required, but not too much. As we judge our own days, and those of others, we criticize the thoughtless who run entirely by routine and their polar opposites who are paralyzed by the difficulties of choosing, praising those who give enough consideration to pick a plan that allows discharging responsibilities to be combined with some extra benefits to others but who stop deliberating in time to carry it out. Praise can be combined with the recognition that some alternative might have been a little better, and that many would have been equally good, provided the superior possibilities would have been hard to appreciate in the time available for decision. Reflections on everyday life are pervaded by concessions to our limitations.

The mundane character of ethical deliberation and ethical judgment is easy to overlook and takes for granted so much that might be examined and questioned. Talk of ethics gravitates too quickly to striking cases, often to cases both striking and artificial. The reality of everyday ethical life, however, consists in attempts to use and extend the collection of resources previously acquired, and, in doing this, there is no single answer an ideal system of ethics might in principle supply, nor do we judge ourselves and others from the perspective of any such system. Against the background of what we already have, an ethical code that presupposes institutions, roles, facets of the good life, and ways of allowing for human limitations, each of us considers what to do, and, retrospectively, appraises his or her own conduct and the deeds of others. Judgment turns on a number of obvious questions: Is the conduct in accordance with the general prescriptions of the code? Does it discharge the duties associated with the agent's role(s)? Does it appreciate opportunities for contributing to welfare that go beyond those roles? Does it exhibit too much thought or too little? Are any lapses understandable as consequences of human limitations affecting (almost) all of us? Answers typically allow a significant range of plans of action to count as permissible, even praiseworthy, applications and extensions of the code.

The everyday practice of ethical judgment might itself be judged from a more extensive perspective. Is there occasion for modifying the code itself, the prescriptions it offers, the institutions and roles it presupposes, its conception (tacit or explicit) of the good life, its recommendations about the proper extent and the proper occasion of deliberation, its ways of accommodating human frailty? Stepping back and judging all aspects of our code at all times is plainly impossible. Decisions about what to do next have to be made, and they cannot await protracted scrutiny of everything we have inherited. Equally, failing to raise questions about the various features of our codes, about the ways we answer the questions on which everyday judgment turns, is the unthinking devotion to routine our mundane practice condemns. While most of our ethical

life consists in using the ethical resources we have to deal with novel situations, our current ethical code—with its appreciation of the dangers of
thinking too little as well as those of thinking too much—requires us to
probe its constituent resources. How does the conception of the good (in
terms of equal opportunity for the good life) bear on institutions and
roles we normally take for granted? How does the method of deliberation (in terms of ideal mutual engagement) affect the principles we try to
apply? Are we too slack in allowing people to steer by their routines, and
not to engage, more frequently, in attempts to mirror the attitudes of
others? Is the practice of everyday judgment itself too comfortable?[8]

Almost certainly, contemporary members of affluent societies (philosophers included) are often caught up in the unexamined life. Operating by
habit, we narrow the horizons of ethical concern, so the plight of many
people becomes invisible to us—hence the tolerance for worldwide inequalities, even those as glaring as those just considered (§58). Routine
inclines us to consider a restricted range of options, among which we seek
one as good as any rival. Often, everyday ethical judgment involves problems to which there are many adequate solutions. The challenging examples, those able to jar us into considering reform, are occasions on which,
given the resources of the ethical code, there appears to be *no* available
solution. Ethical dilemmas are not the principal part, or even a very large
part, of ethical life—at least, not for those who live comfortably.[9] Their
value lies in causing us to pay attention.

Dilemmas can be more or less isolated, or conversely more or less systematic. Systematic dilemmas are those that can erupt for different people

8. Although we have seen (§27) why ethics cannot be founded on the will of the deity,
there is an important question about the positive impact of embedding ethics within religion. William James argued that a connection between ethical precepts and ideas about
powers transcending the human makes ethics properly "strenuous" ("The Moral Philosopher and the Moral Life," in *The Will to Believe* [New York: Dover, 1956], 211). For all the
problems of that connection, one might maintain that, as a matter of fact, the discipline of
religious participation provides structures that encourage people to examine their conduct
more closely and thus prevent certain types of slackness. If this is so, secular societies require substitute ways of fostering self-examination. (The example of Socrates, whose references to deities are peripheral to his probing interrogations, shows us that uninstitutional-
ized surrogates are possible.)

9. They are surely more prominent in the lives of those who live in want.

under a range of contexts, and if this is the case, a proper diagnosis may require us to understand accepted institutions or roles as responsible. Suddenly, the background to our habits needs examination. The well-brought-up conversationalists imagined in §57 might be less dependent on these occasional provocations, more constantly attuned to the possibility of habit-induced myopia.

§60. Conflicting Roles

Many people in affluent societies, predominantly but not only women, find themselves torn between the demands of the workplace and those of the home. As §24 reported, late-nineteenth-century opposition to the entry of women into public life centered on claims that working women would inevitably fail to discharge important family duties—and that argument is often heard today. More to the point, almost every mother who works outside the home has felt, at some point in the early life of her children, that she must choose between doing things required by her job (often things that must be done to retain her job) and things good for her children (often things her children need). To a lesser extent, the same dilemmas are felt by fathers too, and one might even assess the degree of equality within a relationship by gauging the frequency of these pressures on both partners. Often the solutions are suboptimal, achieved by cutting corners in the workplace (in ways that curtail opportunities for advancement or even put employment at risk) or by substituting extended periods of inferior care for parental attention. Richer families sometimes buy the services of caregivers (usually poor women) who can offer closer approximations of the focus the parents would like to provide themselves, but, in doing so, they frequently transfer the problem: the caregivers must themselves provide substitute care for their own children.

Oversimplifying, one can distinguish two cases. There are women whose central project for themselves involves raising children in ways equipping them to pursue their own chosen course, one selected from a wide range of possibilities. Because of their economic circumstances, these women are forced to take on a secondary role, that of worker in the public sphere. A job outside the home is simply a necessity if they are to provide for their children and make the envisaged opportunities available. Yet to *keep* their job they are required to frustrate their central purpose. Sometimes, even when the children are sick or distressed, they must be left with overburdened caregivers or sent from school to after-school activities because parents cannot leave the workplace to provide support and comfort. The internal logic of examples like these is that a chosen role (child rearing) requires a secondary role (working) that sometimes can be pursued only at cost to performance in the primary role.

The second class of cases involves women who hope to combine career satisfactions with responsible motherhood. In many of these instances, more options are available: the aspiring lawyer, doctor, businesswoman, or researcher has enough disposable income to purchase better substitutes for her own presence. If, however, the career goals are to be realized, the same squeeze is felt in a different place. Maybe a ten-hour workday can be managed, if sufficiently hefty expenditures on child care secure nurturing and stimulating environments for the growing children. Career advancement, even career maintenance, may demand even more. Ten hours at the desk, in the hospital, on the road, or at the bench is not sufficient. The childless—and some of the fathers— are able to devote more time and energy, and they are likely to receive the most prized opportunities. Eventually, perhaps, the career of the promising young woman stagnates—she has put in "only" her ten-hour days, leaving for a few hours with the children as others remain—and as she recognizes stagnation, even the ten-hour day becomes unbearable. Here, the logic of the situation is that a life centered on filling two roles, both viewed as important and neither prior to the other, is constantly affected by their competing demands.

Given luck in the form of sufficiently favorable circumstances, women (and men) who fall under these types may stagger through, achieving something they retrospectively identify as satisfactory. The parent for whom work is simply a means to support the children sees them happily launched on lives they have chosen; the parent who hopes to combine career and family reaches a similar goal while also managing a satisfying professional life (perhaps the career could have gone farther without the "distractions" of the family, but it has gone far enough). With great good fortune, women and men can feel even that they have discharged both roles to the best of their abilities: their child raising has gone as well as it would have done, even given more "investment" in family life; their career has gained whatever rewards[10] it would have garnered, even if full-time efforts had been devoted to it. Yet the most successful ways of balancing may require squeezing out other roles, limiting the range or depth of friendships, for example.

If we compare the women and men who strive to combine family and professional life, and who constantly feel these dilemmas, with the class of men (and a smaller number of women) who either pursue careers single-mindedly or else rely on another (a spouse or a partner) to take care of the family responsibilities, successes, especially at the highest levels, will be found predominantly in the latter group. These people, after all, do not require great good fortune to achieve what they might have attained with maximum effort—for maximum effort is just what they devote. Even among the most egalitarian communities in the affluent world, the sex ratios in the two groups are different: more women than men find themselves confronting serial dilemmas, generated by clash of roles; more men than women are freed from those serial dilemmas because they can depend on someone else to discharge the potentially clashing role. Hence, an imbalance in performance will persist. The most spectacular successes in public life will typically be achieved by men rather than by women, even though opportunities are, supposedly, now equal.[11]

The recurrently arising dilemmas stemming from clashes among domestic and public roles give rise to two distinct ethical problems. The more obvious of these is that which confronts the individual agents as they must make their everyday decisions. So long as the demands of the conflicting roles are not too severe, these people can resolve the sequential predicaments by judicious accounting. Those who stagger through to success typically do so by keeping a running tabulation of their

10. These need not be matters of salary, or public honors, but the sense of having made a contribution to an important enterprise.

11. For a brilliant summary of the manifold ways, many of them small, in which equality is undermined, and of the large cumulative effect of apparently minute differences, see Virginia Valian, *Why So Slow?* (Cambridge, MA: The MIT Press, 1998).

shortfalls in each department of their lives: the urgent need to work late this evening requires the children to be left longer at the day-care center or after-school program, but that can be made up to them on an upcoming state holiday; staying with the sick child this morning can be compensated by working late at night, once the other parent is home. Ethical decision making under these relatively benign conditions is straightforwardly consequentialist: one has a pair of goals to be attained, and by borrowing time here and returning it there it is possible to achieve tolerable levels with respect to each. But things easily go wrong, as severe demands from both roles make any such flexible accounting impossible. An accident of some kind requires the parental presence for an extended period; during that same period, significant time must be put in if the work of years is not to be entirely wasted. Dilemmas of this intensity are felt by a not inconsiderable number of parents, and they are as difficult as any of the famous classical examples.

The second ethical problem arises against the background of a commitment to pragmatic naturalism's conception of the good. To adopt the ideal of equal opportunity for a worthwhile life is to insist on genuine freedom of choice in life projects for all. Even in the most affluent societies, opportunities have traditionally been limited to one sex, and the mere declaration of permitted access to all modes of public life is only a formal advance, so long as the most severe clashes of work and family roles remain unresolved. Free formation of aspirations depends on expectation that the goal is attainable, and, in a situation in which attempts to combine career and family are always and evidently under threat from dilemmas of varying severity, clearheaded young women will feel that they must make an early choice. As they contemplate the imbalance in achievement at the very highest levels—the predominance of men in the most rewarding positions in many areas of public life— they believe that certain career choices simply cannot be successfully combined with the parental life they want. Formally available they may be, but women simply do not have equal opportunities for successful pursuit of them. Even the luckiest women, whose efforts open up a wide spectrum of possibilities for their daughters, cannot free those daughters from the shadow cast by less fortunate parents, whose patent difficulties offer cautionary lessons to young women who are deciding on the direction of their lives.

When dilemmas arise so systematically for a particular class of people, the principal ethical challenge is to consider ways in which the underlying conditions might be changed so future generations are not similarly afflicted. It is worth being completely explicit about the structure of the situation. For some people in affluent societies who meet the formal conditions for choosing goals others are able realistically to pursue (women, for example, who are formally able to choose, like their male counterparts, to combine family and career), the consequences of pursuing those goals will be to face a recurrent series of dilemmas. At best, that series of dilemmas will be navigable by keeping a firm eye on both aspects of the combination and maintaining a balance in both accounts. Quite frequently, however, that strategy will not be available, and the dilemmas arising demand a selection of one goal at the expense of the other. The effect of these unresolvable dilemmas erodes further the condition of genuine equality of opportunity for later members of the group. Thus any acceptance of the egalitarian conception of the good should lead us to seek changes in the underlying social conditions that generate the series of dilemmas. We should scrutinize the pertinent roles and the social institutions in which they are embedded.

In the evolution of the ethical project, roles and institutions have been introduced to solve prior problems.[12] When we discover that the roles and institutions we have generate difficulties in realizing our conception of the good, it is necessary to reexamine those roles and institutions. Can we secure the benefits of having roles of *caregiver* and *worker*, while pursuing sexual equality with respect to opportunities for choice? Under some kinds of conditions of work, that does not seem to be very difficult, for, when the needs to be satisfied are quite limited, the strains on equal participation in both roles are diminished. Agrarian societies in benign environments, for example, can, if they choose, combine high-quality parental care with sexual equality. The challenge arises for contemporary

---

12. The "how possibly" account of §20 proposed particular problems as giving rise to specific roles. Even if that account is historically inaccurate, the more general thesis that roles and institutions respond to problems would not be falsified.

societies in which citizens aspire to far more, societies usually dominated by intense competition for scarce goods and forms of distinction.
Serious thought about the conflict of roles ought to consider the character of these aspirations and the value of the supposed goods for which
people compete. We also need to ponder the value of social distinctions
we routinely make about kinds of work and the ways those distinctions
are reflected in individual choices about what is worthwhile.[13] Empirical
investigation—and possibly experimentation—should settle unresolved
issues about what, if anything, would be lost by relaxing the constraints
on the role *worker*, or by making the care of children a more collective
commitment. Differential distribution of material rewards and of social
approval is often viewed as a necessity for high productivity: if no attention is paid to differences in effort, we are told, workers will be lazy and
the fruits of their labor will be shoddy and scanty. Energetic workers
prepared to work longer hours should be rewarded for their exertions. It
follows, then, that those who give no time to caregiving will, other things
being equal, receive the largest rewards. Arguments of this sort presuppose a number of theses for which there is only impressionistic evidence:
we do not know enough about human psychology to determine whether
differentiation of rewards is necessary for involvement in work; we do
not know the extent to which differentiations have to be made to generate high performance (do we really need highly skewed systems of rewarding people?); we do not know whether the differentiations must
be intrinsically connected with the work done, so that those who do not
receive them effectively lose the opportunity for engaging in particular
types of work.

The last point is especially pertinent to the clash between the roles of
caregiver and *worker*. From the perspective of the aspiring doctor, lawyer, or scientific researcher, the especially high salary, the gongs, and the
plumes may not matter. What she wants is the opportunity to carry on a
particular kind of work at a satisfying level. If an inegalitarian distribution of material rewards were combined with an egalitarian distribution

13. This is a point that has been emphasized in feminist analyses for decades. It is eloquently made, for example, by Virginia Woolf. See *Three Guineas* (San Diego: Harcourt
Brace Jovanovich, 1966).

of opportunities for work, she would be happy to trade the *external* goods for the satisfactions of contributing to family life. Trouble arises because the differentiation extends into the work itself. She does all the routine work at the understaffed hospital with the personally (but not intellectually) demanding patients; she works on mundane legal cases; or she is able to do scientific research only as an assistant on somebody else's project. None of that is required by a commitment to differential rewards. For it would, in principle, be possible to pay more to her male counterparts (who are freed from burdensome family commitments), and to honor them with prizes and titles, while distributing the actual work more evenly. (Of course, whether those prizes and titles are themselves worthy of retention is an issue deserving scrutiny.) Without pursuing questions about the balance of work and leisure, the roles could be revised so that putting in the longest possible workday was no longer a requirement for having the opportunity to do the most satisfying kinds of work.

This possible reshaping of the role *worker* is a very conservative one. From the stance of pragmatic naturalism, it would be unwise to leap at it without both acquiring more empirical information and also considering more extensive revisions. It is, after all, entirely possible that a world in which many people are educated to aspire to lives in which they work at the expense of leisure time with their partners, parents, and children is not quite the best world from the perspective of the quality of the lives achieved.

§61. Ethically Insulated Spheres

As §§38 and 46 emphasized, our ethical codes contain vague maxims, likely to remain as we make ethical progress: "Tell the truth" and "Do not initiate violence" have served as exemplars. Yet our practice divides conduct into two spheres, in one of which the maxims that hold sway in the other are relaxed.[14]

---

14. Here, I consider only the toleration of violence in warfare. Concealment, even dishonesty, is allowed in some commercial and political contexts. For preliminary analysis, see my "Varieties of Altruism," *Economics and Philosophy* 26 (2010): 121–48.

Modern warfare employs techniques often leading to the deaths of numerous civilians, and contemporary ethical codes tolerate a certain amount of "collateral damage." There are limits. Almost nobody thinks relentless policies of raping and murdering people who belong to an enemy nation, but who have played no role in combat, can be condoned. Yet if an overall campaign is viewed as justified, perhaps because it resists those who would subjugate or destroy others, we allow means to pursue it that are recognized, in advance, as likely to cause the deaths of noncombatants—and even of people not associated with the enemy. There is an intricate philosophical defense of such uses of violence, one that turns on claiming that the deaths in question are "foreseen but unintended." Even though their occurrence is regrettable, actions causing them are excused because of a larger purpose.[15]

When the allegedly greater end is dubious, any such defense of violence is undermined: the dropping of bombs on innocent civilians unfortunate enough to belong to a nation falsely accused of threatening another is an act of murder, as culpable as any other (and the criminals who knowingly falsify facts to gain support for their aggressive acts deserve punishment). Even when the end to be achieved has properties justifying initiating aggression, when (for example) the goal is to stop a dictator bent on a systematic policy of genocide, there are serious questions about acceptable means. Is it permissible to take a greater number of civilian lives in order to avoid a smaller number of deaths on the part of combatants? Can one legitimately save "our" troops by dramatic explosions of bombs that kill (immediately or eventually) large numbers of "their" noncombatants (and even many lives of people unaligned with either side)? We think there are relatively clear cases, determined by the balance of the accounting. When the threat posed by the enemy is very large, so the lives of millions of people are already at stake, a policy causing the deaths of thousands of noncombatants might appear acceptable (assuming no alternative would require a lesser loss of life). A decision to fire-bomb a city containing hundreds of thousands of non-

---

15. See the essays collected in P. A. Woodward, ed., *The Doctrine of Double Effect* (Notre Dame, IN: University of Notre Dame Press, 2001), for lively debate about this issue.

combatants, and a similar number of refugees who have suffered at the hands of *both* sides, in order to shorten a war by, at most, a couple of weeks, does not.[16]

You might think there is some intermediate point where the loss of civilian life exactly balances the gain from promoting the just cause of resisting violence and aggression. The egalitarian conception of the good makes that thought problematic. Appreciating the value of human lives does not fit the accountant's perspective. Is there some more precise version of our ethical code that would introduce specific numerical ratios marking the limits of "collateral damage," a function assigning to the number of soldier-lives spared, or the number of war-weeks averted, some definite maximal number of civilian deaths to be allowed? Would that specification be progressive, in revealing where the limits of permissible violence lie? A defense might tout the advantages of knowing just how far military tactics can go. Nevertheless, the defense ignores the fact that the search for a precise limit is part of the callousness about human life to which it seeks to respond. A world in which policies were straightforwardly justified by showing how they fell within the permitted range would be one failing to recognize the real problem, a world with too little regret. Hard choices need to be made, but in making them we should not forget that they are hard.[17]

For the dominant ethical task is not so much to find some way of drawing lines, as to find ways of reconfiguring human life so the lines become unnecessary. From the perspective of pragmatic naturalism, the clashes of nations (and other groups) in the contemporary world are a scaled-up version of the intrasocietal conflicts of our ancestors. At the earliest stages of the ethical project, groups devising the most successful experiments would have had to confront the threat of violence that is constant in chimpanzee social life (and was prevalent in hominid social

16. The bombing of Dresden was probably inspired by the need of the Western Allies to satisfy Stalin's demand for action. The official claim was that this act would break the German morale (in odd neglect of the fact that the Blitz had only stiffened British resistance to Hitler). Those who crafted the policy knew that Dresden had become a sanctuary for many people from Eastern Europe who were fleeing the advancing Russian army.

17. My formulation here is intended to allude to Isaac Levi's pioneering work on this topic. See his *Hard Choices* (Cambridge, UK: Cambridge University Press, 1985).

life).[18] Part of the ethical task was to frame rules for dealing with aggressive interactions once they had broken out. Even more crucial was the formulation of rules for preventing conflicts from erupting in the first place. Egalitarian arrangements were essential at those early stages because the group could not manage without the services of any of its members and because failing to attend to the needs of any member would lead to defection or outright conflict.

Pragmatic naturalism treats the contemporary world as a scaled-up version of this initial stage in our ethical life, one in which our inherited social technology is inadequate to the expanded field on which conflict has emerged. Huge inequalities between groups that have become aware of one another's relative situations inspire the have-nots, when they can, to wrest what they lack by force. When the economic differences become highly skewed, the most disadvantaged often outnumber those who live in comfort, and, when they are able to band together, they have sufficient chances of success to make aggressive action pay. As technology produces weapons that can be used to inflict massive destruction, great numbers of people are no longer needed. Armed with atomic bombs, even in crude and "dirty" versions, small groups of disadvantaged people can threaten to take hundreds of thousands of lives. Our predicament exacerbates that faced by our human ancestors—as if, in their small bands, each member, even the weakest, had access to tools that would realistically threaten the life of any four or five others the person selected. The challenge to amend our ethical practice so as to forestall situations provoking such threats is more urgent than the fragile social life that inspired the origin of ethics.

The egalitarian approach to the good addresses that challenge as our ancestors did when they tackled their problems: they introduced rules for sharing, and so should we (§58). But this is only part of the difficulty we face. Not all human conflict arises out of the protests of those hitherto

18. As recognized in §10, the evolution of hominid behavior after the split with the common ancestor of human beings and chimpanzees may have broadened or intensified the original tendencies to psychological altruism. See, for example, the proposals made by Sarah Hrdy about cooperative parenting (*Mothers and Others* [Cambridge, MA: Harvard University Press, 2009]). Despite these steps, there remained plenty of limitations of hominid altruism and resultant tensions within hominid societies.

deprived. Human history and the human present are full of wars sparked by religious differences. Equally, some of the most brutal hostilities result from the ambition of megalomaniacs who succeed in inspiring others. To address all the factors generating conflict in the global society to which we now belong, egalitarianism must be combined with secularism, at least to the extent of undercutting the military enterprises of zealots who would impose their conception of the divine will by force. If we could achieve a world in which there were no economic causes to invade others, and in which any religious exhortation to conflict was viewed as illegitimate, if not absurd, two of the major sources of warfare would be dammed up.[19]

Even though we can hope to reverse the biblical (and Malthusian) dictum that the poor will always be with us, the same is surely not true for megalomaniacs and psychopaths. Their power to generate wars would be radically diminished if they were deprived of their two major means of securing support: through rallying those who suffer from various forms of inequality and through invoking a divine mission. Pragmatic naturalism does not suppose that the pacifist project of working to eliminate war is unrealistic, and that we have to settle for articulating the precise conditions under which various kinds of lives should be traded. It can draw inspiration from the thought that our first ethical ancestors learned how to solve the problem in the small—and also from the more recent fact that, on a continent dominated, through centuries, by fierce wars among neighbors, European states have entered into relationships that are broadly (if insufficiently) egalitarian and secular. To many who look at their accomplishments, the hostilities of the past seem no longer possible. It is not entirely utopian to think their achievement could be more broadly emulated.

A package of proposals is required. In a world constantly scarred by acts of violence, it is necessary both to find ways of removing the causes of conflict as speedily and as thoroughly as possible, and to institute rules for responding to the brutality that continues to occur. Pragmatic

---

19. Achieving that kind of world might also enable us to address a third source of conflict—the shadow of the past, with its recurrent, often escalating, violence between particular groups.

naturalists do not think there are magic wands to be found and waved, but there may be coordinated proposals that would both take steps toward the diminution and eventual eradication of violence and supply directives for coping with the imperfect precursors of the goal state. Its optimism derives from the hope that we can learn about the causes of social conflict and find strategies for enlarging human sympathies.

§62. Maintaining Equality

Pragmatic naturalism commends equality. Yet a familiar concern is so far unaddressed. At least from the Enlightenment on, critics of egalitarianism have argued that, however one might institute equality of wealth, the distribution would be transitory.[20] Differences of talent, inclination, and effort would quickly make some people richer than others, and the resultant differences would be transmitted across the generations, typically in amplified form. Maintaining even approximate equality of wealth would require intrusions that would limit human freedom.

The egalitarian conception of the good does not demand equality of wealth, merely a distribution of resources allowing everyone a serious, and approximately equal, opportunity for a worthwhile life. Nevertheless, since that opportunity will rest on vastly improved systems of education and other forms of social care, it will depend on drawing from people who amass the most. If these people are permitted to pass on large amounts to their children (or to whomsoever they designate as their heirs) there will be a diminution of equality from generation to generation, just as the original argument suggested. If they are not, their freedom to do what they like with their property is limited. So the challenge persists, even when we shift from egalitarianism about wealth to the egalitarian conception of the good.

To address it, we should scrutinize our institution of property. Ordinary thought operates with a very simple distinction: either there is private property, and owners have an absolute right to do what they want

20. Hume makes the point eloquently; *Inquiry Concerning the Principles of Morals* (Indianapolis, IN: Hackett, 1983), 28.

with what they own, or property is held in common, and only the group is permitted to make decisions about how it is used.[21] Nobody should endorse any such dichotomy. Conceptions of property lie on a continuum. At one pole are those conceptions that subordinate all decisions about the use of an item of property to the group (impossible in anything but the very simplest society); at the other are conceptions that subordinate all decisions about the use of an item of property to the individual (subject to legal constraints: owning a gun does not give you the right to shoot somebody). Between these poles lie a host of positions that specify two kinds of contexts, one where the individual owner's will holds sway, another where the community judgment is decisive.

An institution of property consists in a set of rules about how one may acquire it, and a set of rules about what one may then do with it, without intervention from others. In the Anglo-Saxon world, the dominant institution derives from a seminal discussion by Locke.[22] Primarily focused on the example of land, Locke famously proposed we acquire property by "mixing our labor" with it—that is, making use of it in some way— and we are entitled to acquire anything previously unclaimed by others, subject to two provisos: first, we take no more than we can make use of, and second, we leave as much and as good for others. Once the property is acquired, however, we are entitled to do what we want with the entities we own, given that we do not break any law by doing so. In particular, we are allowed to give our property to anyone we want, so there can be unlimited transfer across generations. The stage is set for arguments that egalitarian distributions must be transitory.

This conception is incoherent because of the asymmetry between its account of the acquisition of property and its relaxed attitude toward property transfer. When property is acquired, the provisos constrain. That accords with egalitarianism: initial distributions of property must allow all to have approximately equal shares. Transfer does not

21. This oversimplified conception underlies the polarized rhetoric that emerges in some political discussions: advocates of communal property think of (private) property as theft; contemporary conservatives sometimes view taxation as a form of theft.

22. *Second Treatise on Government* (Indianapolis, IN: Hackett, 1980). For contemporary exposition and defense, see Robert Nozick, *Anarchy, State, and Utopia* (New York: Basic Books, 1974).

have to meet these constraints. How can the asymmetry be justified? If the wants and needs of others are relevant in the original context, why do they become irrelevant later? Why do people alive at the time of acquisition receive the protection of the provisos, while latecomers do not? Can temporal position properly affect your life prospects in this way?[23]

Imagine a world in which generations of property owners are discrete and synchronized. In each generation, the elders own all the property and support their dependents. When they die, property is transferred to one of the dependents, who takes over the role of elder. Initially, the world is divided into equal portions, and the first elders each control one of these. Their successes are unequal, and, in the next generation, the inherited portions diverge from one another. So it goes, until after many generations, some are extremely rich, while others have no ability to meet the most basic needs. In the later intergenerational transfers, property is passed on without leaving as much and as good for others. For those who inherit from impoverished elders, the situation is exactly the same as it would have been for people in the original situation, if there had been no requirement to allow for the needs of everyone. If it was appropriate for that original situation to debar the *acquisition* of property that shortchanged others, why should one now allow for the *transfer* of property that leaves some, perhaps many, in want? Are the sins of the fathers (presumably profligacy and sloth) visited upon the children?

Any such doctrine of original sin is a serious error. Later generations are not accountable for what earlier people have done, and the chances of a worthwhile life should not be held hostage to the vagaries of history. So: restore the symmetry. Intergenerational transfers are permitted only to the extent that they provide for the heirs what they can use and leave as much and as good for others. The most straightforward mechanism for doing this is a rigorous system of taxation setting limits to what can be transferred across generations. Tax revenues support the public structures needed to give all young people serious chances for a worthwhile life.

23. In discussing these points, I have been helped by conversations with Rohan Sud.

More than a century ago, Mill argued for a proposal of this kind.[24] His reasoning was double sided. In addition to the more obvious concern that children born to poor parents should be provided for, he worried about the effects on the recipients of dynastic wealth.[25] Instead of offering enhanced opportunities for choosing and pursuing a person's own life project, acquiring a very large fortune can prove limiting, sapping the desire to exert oneself. Whether that psychological generalization holds, there is an important point behind it. If what is fundamentally valuable is not wealth per se, but the opportunities wealth brings, history reveals many worthwhile human lives that began in far from wealthy circumstances, as well as children of the extremely rich who lost their way. Pending more detailed analysis, it is not obvious that leaving one's children a fortune is more likely to bring them a worthwhile life than providing them with the means (protection, nurture, material necessities, educational opportunities) to discover their own way.

Does the proposal wrongly curtail human freedom? "What I have earned is *mine,* and I have the right to do what I like with it. When that right is abridged, I suffer a loss of freedom." The protest begs a crucial question. For the issue is what it should *mean* to say that something is "mine," and one cannot choose by fiat from a continuum of possible positions. If it would be ethically progressive to introduce a modified conception of property, amending the asymmetry infecting Locke's proposal, the claim of right lapses.

Behind the protest stands one plausible thought. If we should aim for a world in which all people enjoy a serious and approximately equal opportunity for leading worthwhile lives, it is important to consider whether the modification, one that appears to enhance the opportunity for some, does so by interfering with the life projects of others. Is the revised conception of property hostile to certain ideas about the worthwhile life, which it eliminates as possibilities for people?

What might these views about worthwhile lives be? Here are the central possibilities:

24. *Principles of Political Economy* (Toronto: University of Toronto Press, 1970), *Works,* 2:218–27, 3:755–57.

25. *Principles of Political Economy, Works,* 2:221.

1. A life can be worthwhile through its enjoyment of rare and expensive material goods.
2. A life can be worthwhile through the effort to accumulate material goods.

No doubt, enjoyment (use, viewing, consumption, etc.) of material goods can be *part* of a worthwhile life. Nevertheless, we should make a twofold distinction: some material goods are rare and correspondingly expensive, although they have counterparts that are common and relatively cheap; some material goods are enjoyed privately, while others can be shared among a large group of people. In light of this distinction, we can separate scenarios that accord with the first vision of the worthwhile life. All of these share the feature that what makes the life worthwhile is the *consumption* of the goods—the ownership itself—for this is the difference between the two perspectives. (If the enjoyment does not lie in consumption, the good will persist and be available for enjoyment by others.) Imagine, then, a life centered on consumption of rare and expensive material goods that can be enjoyed only privately—caviar and fine champagne, for example. The list of such goods cannot be extended sufficiently far to defend this form of existence as valuable. On the other hand, if one starts with rare and expensive goods whose "consumption" (or, in many instances, contemplation) would more plausibly make for a worthwhile life, there would be no need to make the enjoyment private. Imagine the life of the connoisseur, who lives in an architectural masterpiece, on whose walls hang masterpieces of world art, and whose evenings are devoted to hearing great music played by the most gifted performers. Would a decision to make these goods public seriously detract from the value of his or her life?[26] Passive consumption of even the most beautiful things, whether natural products or human artifacts, is not obviously sufficient for a valuable life, but, even if one concedes this point to the first vision, claiming the enjoyment must be private cannot be defended.

26. It is important to be mindful here of Mill's point about the need for solitude (*Works*, 3:756). Public resources can be shared without introducing such dense crowds that enjoyment of them becomes impossible.

The second vision sees value accruing not from the toys one has, but from the activity of accumulating them (already a positive step). Lives of this sort are not threatened in any way by the envisaged modification of the conception of property. For the duration of a person's existence, the activity of accumulation is allowed to go undiminished (subject to whatever taxes are levied before the intergenerational transfer). If there is to be any frustration of the accumulator's plan, it must result from the centrality of a further desire: to give what has been amassed to a younger heir. Assuming the desire is directed toward the perceived interests of the heir, the thought must be that bequeathing the wealth increases the opportunities for the heir to live a worthwhile life. Yet, under the egalitarian conception, serious chances are already available. Only if the life envisaged for the heir were one of consumption—after the pattern of vision 1—would the inheritance be necessary. It would be strangely schizophrenic for someone to hold the active work of accumulation worthwhile for himself, whereas a life of passive consumption would be suitable for another. If the life envisaged as valuable were one of active accumulation, it would be counterproductive to provide great wealth for the young.

Vision 2 is more promising as an account of a worthwhile life than is vision 1 because it emphasizes the work of achievement rather than the material goods garnered. Material goods are important bases for most activities around which valuable lives are pursued, but those required are typically not the expensive and the rare. A world in which all were given serious (and equal) opportunities for living valuable lives would be one in which anxieties about accumulating wealth for one's heirs would evaporate.

§63. The Challenges of Technology

How might pragmatic naturalism be applied to the novel options technology has made possible? Consider an area of human intervention in which secularism and an emphasis on equality of opportunity for a worthwhile life would make an evident difference. Biomedical technology has offered abilities to carry out in vitro fertilization, to obtain details about the genomes of adults, children, and embryos, and to clone mammals—and these opportunities have sparked debates about legitimate ways of using the new techniques. Contemporary societies face the following questions:

1. Is it permissible to produce early-stage human embryos for purposes of developing treatments that might aid people with degenerative diseases?
2. Is it permissible for prospective parents to obtain genetic data about an embryo or fetus and to terminate a pregnancy based on what they learn?
3. Is it permissible for parents to produce a child by deploying the techniques of mammalian cloning?

A striking feature of the discussions about these questions, particularly in the United States, but to a lesser extent elsewhere, has been the tendency of large groups of people to give blanket answers, based on their readings of religious texts.

The passages of scripture usually cited supply no direct answers to these questions: the issues lie beyond the conceptual horizons of their authors. People inspired by those passages believe a human zygote is inviolable: it is wrong to destroy a zygote or to manipulate it, or even to bring it into being in a deliberate way (as would be done by reproductive cloning). Religious people prefer less abstract, more homely, language. The embryo is, from the moment of conception, a "human being," it has an immortal soul and is sacred; to decide on what features it is to have is to arrogate powers belonging only to God.

Pragmatic naturalism regards claims like these as having no place in ethical discussion. They depend on very particular ways of interpreting the religious texts, and alternative ways of reading are offered by other devout people, for whom the texts are canonical. Even supposing the interpretations are well grounded, the texts have no ethical authority. If there is a basis for crediting the texts with insight, it must lie in the fact that the ideas attributed to the nonexistent transcendent authority contain elements that would be taken up by progressive ethical traditions. In other words, the passages could obtain authority *only on the basis of a prior ethical analysis.*

The appeal to scripture stops the conversation just where it should begin. We do better to start from what we know. So we would arrive at the following:

1. A blastomere is an early-stage embryo produced after several divisions from the original zygote; further cell divisions are required before gastrulation, the stage at which the pre-pattern of the central nervous system is laid down. Is it permissible to produce blastomeres in order to find ways of generating stem cells for research into degenerative diseases?

2. It is possible to obtain DNA samples from a fetus at the fourth month (and refinements of technique may make this possible significantly earlier), and to determine whether or not that fetus possesses genetic markers associated with known probabilities, in the usual human environments, with various phenotypic properties. With respect to what properties, and what probabilities, if any, is it permissible for parents equipped with this knowledge to terminate the pregnancy?

3. Mammals can be cloned by inserting the DNA from a chosen cell into an enucleated egg, stimulating the product to begin cell division, and inserting the early-stage embryo into the uterus of another mammal of the same species. When this is done, there is a known frequency (quite low at present) with which the pregnancy will proceed to term, producing a mammal with the nuclear DNA originally inserted. Under what conditions, if any, is it permissible for human beings to bring about a child in this way?[27]

Given these descriptions, an *initial appeal* to pragmatic naturalism's conception of the good yields preliminary answers.

With respect to question 1, the commitment to providing equal opportunity for worthwhile lives is expressed in helping people whose

---

27. In each of these formulations, I give only the bare bones of the scientific account of the options made available by biotechnology. The intention is to draw attention to important features without drowning readers in technical details.

degenerative diseases, especially if they strike early in their lives, threaten their life projects. The chances of giving aid are obtained through bringing into being clusters of cells that are, in one clear sense, human beings: these clusters of cells are organisms, and the species to which they belong is *Homo sapiens*.[28] The blastomeres are not, however, organisms to which one can ascribe the capacity for having any kind of life project, or even organisms capable of feeling pleasure or pain—they are a major embryonic stage away from having even the *pre*-pattern of a nervous system. There is no violation of the conception of the good in using them for the specified purpose, and the purpose itself accords with that conception of the good. While there is no sense in which the proposed intervention ignores or overrides the desires of the organisms manipulated (they lack any desires), neglecting the aspirations of those who suffer from neural degeneration (for example) appears as a form of altruism failure.

The second question invites a more complex answer. With respect to some knowledge about a fetal genome, termination of the pregnancy seems evidently permissible (perhaps even mandatory); with respect to others, it appears just as evidently impermissible, and a range of instances demand further exploration and discussion. Consider first the most devastating and excruciating developmental abnormalities—Lesch-Nyhan syndrome and Hurler syndrome can serve as examples. In these instances, massive retardation is inevitable, and the child will have a short life, with no capacity for forming a sense of how that life is to be pursued. Furthermore, pain will be a necessary part of it (in Lesch-Nyhan, this will stem from the intense urge to self-mutilate), unless the drugs administered render the child comatose. However the conception of a worthwhile life is articulated, these children lack the opportunity of having such a life. Here, then, pragmatic naturalism recommends sparing them pain by terminating the pregnancy as soon as possible. Contrast that example with that of cases in which the genetic test reveals a condition that does not interfere with the opportunity for a worthwhile life—female

28. Similarly, the mutant embryos used by Christiane Nüsslein-Volhard in her brilliant investigations of the genetics of development in fruit flies belonged to the species *Drosophila melanogaster*. For an account of this work, see Peter Lawrence, *The Making of a Fly* (Oxford: Blackwell, 1992).

sex, say, or relatively short stature.[29] Here, any wish to terminate the pregnancy would have to be defended against charges that it embodies a frivolous view of what matters in human lives, and, in general, we can expect the defense to prove inadequate. Hence, pragmatic naturalism frowns on termination in these instances. Finally, there are genetically identifiable conditions with respect to which the probability of worthwhile life is significantly diminished. Down syndrome (strictly speaking a chromosomal abnormality) covers a wide range of possible courses of life. On the one hand are children (and occasionally adults) whose mental retardation is severe and whose physiological problems cause difficulty and pain; on the other are people who achieve significant things. Although great strides have been made in providing environments allowing people with Down syndrome to develop more fully, we still lack knowledge of the distribution of phenotypes across the kinds of environments we can provide. Prospective parents cannot even calculate the chances that their child with Down syndrome will have the opportunity for a worthwhile life. Under this situation of extreme uncertainty, it is hard to blame them for choosing termination—even though, in the future, we might arrive at more refined understandings that would change our perspective.

With respect to the third question, it is important to ask why reproductive cloning might appeal. We can distinguish two types of cases: there are circumstances in which cloning would provide a unique opportunity for two people to conceive a child with a close biological relationship to both of them;[30] alternatively, someone (or some pair of people) may want a child with the genetic material of an admired individual. With respect to the latter case, the conception of the good tells against the use of reproductive cloning. If children should be provided with opportunities for a worthwhile life, and if having a worthwhile life involves making free decisions about what is most central, there are serious concerns about parental pressures that appear to be implicit in the idea of having a child

29. A caveat is necessary here. There are plainly environments in which characteristics like these do interfere with the possibility of having a worthwhile life. The primary ethical imperative in such instances is to do something about the crippling environments.

30. Consider, for example, two lesbians who have no biological relatives who propose to have a child by inserting the nuclear DNA of one into an enucleated ovum from the other, and reintroducing the embryo into the second woman's uterus.

with *this splendid DNA*.[31] To want a child with the phenotype of someone one admires is already to set one's own ideas about what matters
ahead of those the child might choose for him- or herself—the same sort
of coercion James Mill inflicted on his son. By contrast, when reproductive cloning would be a unique means to biological continuity, matters
cannot be settled so simply. It is worth asking whether people should be
so wedded to biological continuity with those they rear that they resort
to such technologically delicate ways of achieving it. Under conditions
where mammalian cloning requires the use of a very large number of ova
to obtain a single birth, a couple with a desire for children should choose
some more reliable method, even if biological continuity is sacrificed.

These are only *preliminary* answers to issues about the uses of biotechnology. Answers ought to be constrained by the broader commitment
to egalitarianism. Sophisticated forms of reproductive technology are
currently available to a minority of the human population, and we should
ask if the costs of making them available to all would be worth the benefit.
Even if one were to conclude that some instances of reproductive cloning
would be permissible, providing opportunities to all those who would
want them might be incompatible with satisfying more urgent medical
needs: biomedical research might use its resources more profitably. The
nonaffluent world has higher priorities. Conversely, there might be increased pressure to engage in genetic testing and early termination of
pregnancies because providing appropriate environments for children
born with various disabilities could not be carried through on too large a
scale—without using genetic tests to reduce the population of the disabled, maintaining good environments for children with particular genetic conditions would draw resources away from meeting the needs of many
other children. Adding these points about the background commitments
of the view of the good does not end the discussion but merely reveals
reasons for possible modification of the initial answers.

Yet a principal concern about those answers, whether in their original
form or modified, is that something is lost in substituting the bare *bio-*

---

31. Prospective parents who aim to replicate someone they admire also have radically
misguided views about genetic contributions and are victims of a crudely mistaken genetic
determinism.

logical characterizations for those offered by religious opponents of bio-
technological interventions. Although we should not accept the appeal
to scriptural texts, or the tendentious descriptions religious people offer,
it is not obvious that talk of blastomeres and zygotes is ethically adequate.[32] The bare scientific language, one might charge, cultivates an
important kind of ethical blindness.[33] Simple appeals to the scriptures,
conversation stoppers founded in myths, need to go. Yet we also need
concepts enabling us to recognize human worth that do not lead to practices that ignore some groups of people and their desires, treating them
as "commodities" or as "vermin." The deliberation pragmatic naturalism recommends should be alert to this potential problem. In countries
where repulsive eugenic programs have been carried out, many people—
thoroughly secular people—urge caution with respect to the uses of biotechnology.[34] They should be part of the conversation.

How exactly? Clinical language permits interventions—and it is important to ask if people who accepted those permissions would also be
led to conclusions that, from their present perspective, they would completely disavow. Can one find ways of making sharp and clear distinctions between using blastomeres for medical benefit and engaging in the
eugenic experiments of Nazi medicine? I believe one can, but the issue
should be thoroughly and carefully analyzed, attending to possible aspects
of the "worth" or "sacredness" of early human life that too simple and
crudely secular formulations overlook. Empirical issues about human

32. This point is made with great eloquence by Ronald Dworkin in *Life's Dominion*
(New York: Knopf, 1993). Dworkin's discussion of issues about the "ends of life" in that
book is an exemplar of the wide-ranging canvassing of different points of view that I see as
the proper form of ethical conversation. Very few writers on ethical topics, whether religious or philosophical, have offered so sensitive and insightful a treatment of any contemporary ethical question.

33. The point here is akin to one affecting the "two-sphere" approach to trade. Sus-
pending maxims fostering altruism toward others within a particular sphere (between
buyer and seller) might erode responsiveness to others more broadly. Talking clinically
about embryos might similarly promote a more callous attitude toward (for example) children with severe disabilities. Once again, there are psychological facts to be explored.

34. This is particularly true in Germany. The German translation of my book on ethical
consequences of genomic research *(The Lives to Come* [New York: Simon and Schuster,
1996]; *Genetik und Ethik* [München: Luchterhand]*)* aroused far more controversy than its
Anglo-American original.

attitudes under alternative types of socialization lurk here—would people who systematically thought in terms of the biological concepts be inclined to insensitivity toward a broader class of human beings? Behind the literal conclusion of the would-be conversation stoppers, in the emphasis on the "sacred" character of human life—may lie an ethical insight. For, even when that language is thoroughly detached from the idea of a creative deity, it may have normative force, supporting attitudes toward human life progressive ethical traditions want to cherish.

Conclusion

§64. Summing Up

Tens of thousands of years ago, our remote ancestors began the ethical project. They introduced socially embedded normative guidance in response to the tensions and difficulties of life together in small groups. They were equipped with dispositions to psychological altruism that enabled them to live together, but the limits of those dispositions prevented them from living together smoothly and easily. Out of their normative ventures have emerged some precepts we are not likely ever to abandon, so long, at least, as we make ethical progress, the vague generalizations that embody ethical truths. Besides those core themes, we have also inherited a conception of the good that includes conflicting elements, as well as providing for us a far richer conception of human life than any the first ethical pioneers could have apprehended. Our ethical task is to decide how to go on.

The three preceding chapters recommend renewing the project by emulating the early phases and expunging some of the distortions introduced in later transitions. Pragmatic naturalism's normative stance consists in an egalitarian conception of the good, focusing on equal opportunity for a worthwhile life, and a method for ethical discussion in terms of mutual engagement within a comprehensive population; both proposals advocate disentangling our ethical practices from myths about supernatural beings. Neither religion nor philosophy can pronounce with authority. Ethics is something people work out together, and, in the end, the only authority is that of their conversation. The final chapter has offered, *in a highly preliminary way,* some suggestions—in the spirit of philosophical midwifery—about how we might continue that conversation.

The greatest philosophers of the past focused on problems salient in their societies, framing the issues in light of what they took themselves to know. Philosophy today should address, not technical questions spun off from investigations that no longer touch human lives, but the deepest challenges of our contemporary predicament.[1] Because the totality of what we know is so hard to survey, our task is daunting—and may well require cooperative interactions among scholars attuned to different realms of expertise. We might, however, be encouraged to think of such cooperative inquiries as the continuation of an enterprise, the ethical project, that has occupied us for fifty thousand years, and that has been a major part of what makes us human.

1. See Dewey, *Democracy and Education* (New York: Free Press, 1966), 328; *The Quest for Certainty,* vol. 4 of *The Later Works* (Carbondale, IL: University of Southern Illinois Press1984), 204.

Acknowledgments

I entered ethics by the back door. In the early 1980s, as I tried to become clear about the pitfalls of human sociobiology ("pop sociobiology," as I uncharitably called it), it seemed important to assess the famous claim that the time had come for ethics to be removed from the hands of the philosophers and "biologicized." The final chapter of my book *Vaulting Ambition* criticized sociobiological proposals to provide ethics with a biological basis. Yet, as I reflected, I was not entirely satisfied with my own response. The negative arguments seemed sound, but I imagined the complaint of an aggrieved biologist: "You have found fault with efforts to reconcile ethics with a Darwinian picture of life, but do you have any positive view to offer? Is any of the available philosophical conceptions satisfactory?" I had no good answer to this complaint, and, around 1985, I set myself the task of finding one.

The present volume is my attempt to complete the task. It has taken a long time, and a very large number of people have helped me. My views have evolved in ways I would not have anticipated. Along the way I have found myself increasingly at odds both with efforts to carry forward the program Darwin outlined in *The Descent of Man*, and with the standard perspectives from which contemporary philosophers pursue ethical and metaethical questions. I have often worried that the book I have tried so many times to write would please nobody.

My first efforts were thoroughly Darwinian, inspired by the brilliant work on altruism begun by William Hamilton, Robert Trivers, and Robert Axelrod. At early stages, I was impressed with the insightful books of Allan Gibbard and Simon Blackburn, and I envisaged extending the theoretical framework for understanding the evolution of psychological altruism and connecting it with the kinds of metaethical considerations Gibbard and Blackburn advanced. My work was aided by conversations with Brian Skyrms, to whom I am indebted for his interest and encouragement, and it culminated in some articles of the early 1990s. If Skyrms, Gibbard, and Blackburn do not figure largely in this book, it is not because my respect for their ideas has diminished, but rather because my investigations have led me to different questions and different ways of conceptualizing the issues.

By the mid-1990s, my study of primate social life had convinced me that, while we probably shared some ethically pertinent psychological capacities with our evolutionary cousins, an ability to regulate conduct by explicit rules, formulated in discussions among members of a small group, had transformed the possibilities for living together. I came to view ethics as an evolving practice, founded on limited altruistic dispositions that were effectively expanded by activities of rule giving and governance. Since 1996, I have tried out versions of this approach in many forums, giving lectures or series of lectures. I am most grateful to the members of many audiences, whose questions, criticisms, and comments have helped my ideas evolve: prominent among them are Michael Baurmann, Richard Bernstein, Akeel Bilgrami, Martin Carrier, Nancy Cartwright, Kate Elgin, Sam Freeman, Bob Goodin, Paul Guyer, Stuart Hampshire, Dick Jeffrey, Edie Jeffrey, Mark Johnston, Arthur Kuflik, David Lewis, Steffi Lewis, Bill Lycan, Karen Neander, Fred Neuhouser, Richard Rorty, Alex Rosenberg, Carol Rovane, Geoff Sayre-McCord, Jerry Schneewind, Jack Smart, Kim Sterelny, Sharon Street, Thomas Sturm, Pat Suppes, Bas van Fraassen, Achille Varzi, Jeremy Waldron, David Wiggins, Torsten Wilholt, Meredith Williams, Michael Williams, David Wong, and Allen Wood.

During my years at University of California, San Diego, I learned from many colleagues and students. Patricia and Paul Churchland offered good advice about my models of psychological altruism; John Batali was a wonderful interlocutor and collaborator; George Mandler provided pointers to psychological discussions I would otherwise have missed. As the project evolved, I was greatly aided by conversations with Jessica Pfeifer, Gila Sher, and Evan Tiffany, and especially by the expert guidance of Dick Arneson and David Brink.

The move to Columbia brought me a new set of colleagues and students. Sidney Morgenbesser was a principal influence and was responsible for leading me to Dewey. My Columbia friends and colleagues have been constantly helpful and supportive, and I am most grateful to them for creating such a wonderful forum for exchanging ideas.

An important influence on the ideas of this book has come from my involvement in Columbia's famous core course, Contemporary Civilization, in which I have had the opportunity to discuss ethical and political texts with students from a range of disciplines.

A first full draft of the present book was written during a sabbatical year in 2007–8, spent in Berlin. I learned much from conversations at the Wissenschaftskolleg with Catriona MacCallum, Randy Nesse, Bob Perlman, Sascha Somek, and especially Moira Gatens and Candace Vogler.

During that year, I was affiliated with the Max Planck Institut für Wissenschaftsgeschichte, spending the year in Lorraine Daston's department (Abteilung II). I am most grateful to Raine for her warm hospitality, for the extraordinarily cooperative atmosphere she creates, for organizing a workshop on my manuscript in progress, and for innumerable insights and suggestions, offered on many formal and informal occasions. Thanks also to supportive friends and astute advisers: Michael Gordin, Maria Kronfeldner, Erika Milam, Tania Munz, Adrian Piper, and Thomas Sturm.

During that sabbatical year, I received much warm hospitality and valuable feedback. I would like to thank Michael Baurmann, Mario Branhorst, Harvey Brown, Martin Carrier, Nancy Cartwright, Christel Fricke, Ulrich Gähde, Oren Harman, Paul Hoyningen-Huehne, Matthias Kaiser, Evelyn Fox Keller, Anton Leist, Tim Lewens, Chrys Mantzavinos, Kirsten Meyer, Felix Mühlholzer, Susan Neiman, Chris Peacocke, Suzann-Viola Renninger, Peter Schaber, Thomas Schmidt, Bettina Schöne-Seifert, Tatjana Tarkian, and Gereon Wolters. Chrys and Mario also gave me extensive written comments that have prompted significant changes.

During the many years I have been thinking about this project, there have been several people who have helped me on a number of occasions. Conversations with Allan Gibbard have invariably been wonderfully instructive, and I am delighted that he decided to spend a sabbatical leave in New York City. Isaac Levi has shaped my thinking in many ways, most likely more than either of us could enumerate. Interactions with Frans de Waal, both on the occasion of his Tanner Lectures at Princeton and during a visit to Emory have enlightened me greatly about crucial facets of primate behavior. Chris Peacocke has been a splendidly open-minded interlocutor and a constructive critic. Continuing discussions with John Dupre, Michael Rothschild, and Elliott Sober have been illuminating and always enjoyable.

My debts to my students at Columbia, especially those who have taken my seminars on evolution, altruism, and ethics, are extensive. Particular thanks go to current and former graduate students: Dan Cloud, Laura Franklin-Hall, Michael Fuerstein, Jon Lawhead, Katie McIntyre, Heather Ohanesen, Herb Roseman, and Matt Slater; also to some truly brilliant undergraduates: Lauren Biggs, Leora Kelman, Jonathan Manes, Howard Nye, Michael Roberto, Sam Rothschild, and Rohan Sud. Many thanks to Jon Lawhead for preparing the index.

The penultimate draft, prepared in 2008–9, was somewhat longer than its predecessor—and considerably longer than the final version. I am extremely grateful to two perceptive and constructive readers for Harvard University Press who gave me excellent advice—in particular to make it shorter. It helped that their thoughts coincided completely with those of the person who has always been my best reader, Patricia Kitcher, to whom I owe more than I can say.

Lindsay Waters has been a helpful and supportive editor. He has had to wait a long time, and I appreciate his patience.

My thinking about these topics began before our sons were teenagers, and that is reflected in the nickname the book in progress still bears. Charles continues to ask me how "Nice Monkeys" is going—even though I have told him many times that it is no longer much about nonhuman animals, that the relevant nonhuman animals are apes, not monkeys, and that part of the point is that they are not particularly nice. For many years, I have been entertained, enlivened, challenged, and informed through their increasingly sophisticated conversation—and in recent years by that of Sue-Yun Ahn as well. It has been wonderful watching them grow. From Andrew, now a doctor, I have learned much about medical care in New York City—and have been moved by his compassion for his patients, many of them indigent and most baffled by an alien, monolingual English, environment. He embodies many of the qualities I have tried to articulate and defend, and I dedicate this book to him with love and admiration.

Index